Electrochemical Approach to Selected Corrosion and Corrosion Control Studies

Papers from 50th ISE Meeting, Pavia, September 1999

EARLIER VOLUMES IN THIS SERIES

Number 1: **Corrosion in the Nuclear Industry**
Prepared by the Working Party on Nuclear Corrosion

Number 2: **Practical Corrosion Principles: a manual of corrosion experiments**
Prepared by the Working Party on Corrosion Education

Number 3: **General Guidelines for Corrosion Testing of Materials for Marine Applications**
Prepared by the Working Party on Marine Corrosion

Number 4: **Guidelines on Electrochemical Corrosion Measurements**
Prepared by the Working Party on Physico-Chemical Methods of Corrosion Testing

Number 5: **Illustrated Case Histories of Marine Corrosion**
Prepared by the Working Party on Marine Corrosion

Number 6: **Corrosion Education Manual**
Prepared by the Working Party on Corrosion Education

Number 7: **Corrosion Problems Related to Nuclear Waste Disposal**
Prepared by the Working Party on Nuclear Corrosion

Number 8: **Microbial Corrosion**
Prepared by the Working Party on Microbial Corrosion

Number 9: **Microbiological Degradation of Materials — and Methods of Protection**
Prepared by the Working Party on Microbial Corrosion

Number 10: **Marine Corrosion of Stainless Steels: Chlorination and Microbial Effects**
Prepared by the Working Party on Marine Corrosion

Number 11: **Corrosion Inhibitors**
Prepared by the Working Party on Inhibitors

Number 12: **Modifications of Passive Films**
Prepared by the Working Party on Surface Science and Mechanisms of Corrosion and Protection

Number 13: **Predicting CO_2 Corrosion in the Oil and Gas Industry**
Prepared by the Working Party on Corrosion in Oil and Gas Production

Number 14: **Guidelines for Methods of Testing and Research in High Temperature Corrosion**
Prepared by the Working Party on Corrosion by Hot Gases and Combustion Products

Number 15: **Microbial Corrosion (Proceedings of the 3rd International EFC Workshop)**
Prepared by the Working Party on Microbial Corrosion

Number 16: **Guidelines on Materials Requirements for Carbon and Low Alloy Steels for H_2S-Containing Environments in Oil and Gas Production**
Prepared by the Working Party on Corrosion in Oil and Gas Production

Number 17: **Corrosion Resistant Alloys for Oil and Gas Production: Guidance on General Requirements and Test Methods for H_2S Service**
Prepared by the Working Party on Corrosion in Oil and Gas Production

Number 18: **Stainless Steel in Concrete: State of the Art Report**
Prepared by the Working Party on Corrosion of Reinforcement in Concrete

Number 19: **Sea Water Corrosion of Stainless Steels — Mechanisms and Experiences**
Prepared by the Working Parties on Marine Corrosion and Microbial Corrosion

Number 20: **Organic and Inorganic Coatings for Corrosion Prevention — Research and Experiences**
Papers from EUROCORR '96

Number 21: **Corrosion-Deformation Interactions**
CDI '96 in conjunction with EUROCORR '96

Number 22: **Aspects of Microbially Induced Corrosion**
Papers from EUROCORR '96 and the EFC Working Party on Microbial Corrosion

Number 23: **CO_2 Corrosion Control in Oil and Gas Production — Design Considerations**
Prepared by the Working Party on Corrosion in Oil and Gas Production

Number 24: **Electrochemical Rehabilitation Methods for Reinforced Concrete Structures – A State of the Art Report**
Prepared by the Working Party on Corrosion of Reinforcement in Concrete

Number 25: **Corrosion of Reinforcement in Concrete — Monitoring, Prevention and Rehabilitation**
Papers from EUROCORR '97

Number 26: **Advances in Corrosion Control and Materials in Oil and Gas Production**
Papers from EUROCORR '97 and EUROCORR '98

Number 27: **Cyclic Oxidation of High Temperature Materials**
Proceedings of an EFC Workshop, Frankfurt/Main, 1999

European Federation of Corrosion
Publications
NUMBER 28

Electrochemical Approach to Selected Corrosion and Corrosion Control Studies

Papers from 50th ISE Meeting, Pavia, September 1999

Edited by
P. L. BONORA AND F. DEFLORIAN

Published for the European Federation of Corrosion
by IOM Communications

Book Number 733
Published in 2000 by IOM Communications Ltd
1 Carlton House Terrace, London SW1Y 5DB

IOM Communications Ltd
is a wholly-owned subsidiary of
The Institute of Materials

ISBN 1-86125-110-6

Neither the EFC nor The Institute of Materials
is responsible for any views expressed
in this publication

Design and production by
SPIRES Design Partnership

Made and printed in Great Britain

Contents

Part 4 – Passivity 223

Part 5 – Corrosion Testing and Evaluation in Special Environments 283

European Federation of Corrosion Publications

Series Introduction

The EFC, incorporated in Belgium, was founded in 1955 with the purpose of promoting European co-operation in the fields of research into corrosion and corrosion prevention.

Membership is based upon participation by corrosion societies and committees in technical Working Parties. Member societies appoint delegates to Working Parties, whose membership is expanded by personal corresponding membership.

The activities of the Working Parties cover corrosion topics associated with inhibition, education, reinforcement in concrete, microbial effects, hot gases and combustion products, environment sensitive fracture, marine environments, surface science, physico–chemical methods of measurement, the nuclear industry, computer based information systems, the oil and gas industry, the petrochemical industry, coatings, automotive engineering and cathodic protection. Working Parties on other topics are established as required.

The Working Parties function in various ways, e.g. by preparing reports, organising symposia, conducting intensive courses and producing instructional material, including films. The activities of the Working Parties are co-ordinated, through a Science and Technology Advisory Committee, by the Scientific Secretary.

The administration of the EFC is handled by three Secretariats: DECHEMA e. V. in Germany, the Société de Chimie Industrielle in France, and The Institute of Materials in the United Kingdom. These three Secretariats meet at the Board of Administrators of the EFC. There is an annual General Assembly at which delegates from all member societies meet to determine and approve EFC policy. News of EFC activities, forthcoming conferences, courses etc. is published in a range of accredited corrosion and certain other journals throughout Europe. More detailed descriptions of activities are given in a Newsletter prepared by the Scientific Secretary.

The output of the EFC takes various forms. Papers on particular topics, for example, reviews or results of experimental work, may be published in scientific and technical journals in one or more countries in Europe. Conference proceedings are often published by the organisation responsible for the conference.

In 1987 the, then, Institute of Metals was appointed as the official EFC publisher. Although the arrangement is non-exclusive and other routes for publication are still available, it is expected that the Working Parties of the EFC will use The Institute of Materials for publication of reports, proceedings etc. wherever possible.

The name of The Institute of Metals was changed to The Institute of Materials with effect from 1 January 1992.

The EFC Series is now published by the wholly-owned subsidiary of The Institute of Materials, IOM Communications Ltd.

A. D. Mercer
EFC Series Editor,
The Institute of Materials, London, UK

EFC Secretariats are located at:

Dr B A Rickinson
European Federation of Corrosion, The Institute of Materials, 1 Carlton House Terrace, London, SW1Y 5DB, UK

Mr P Berge
Fédération Européene de la Corrosion, Société de Chimie Industrielle, 28 rue Saint-Dominique, F-75007 Paris, FRANCE

Professor Dr G Kreysa
Europäische Föderation Korrosion, DECHEMA e. V., Theodor-Heuss-Allee 25, D-60486, Frankfurt, GERMANY

Preface

The 50th meeting of the International Society of Electrochemistry (ISE) was held in Pavia, Italy, on the occasion of the bicentenary of Alessandro Volta's invention of the 'Electric Pile' (1799–1999) and featured the main theme "200 Years of Electrochemical Energy Conversion". This volume contains selected papers presented at one of the ten monothematic symposia of which the meeting was composed, namely No. 7, which was devoted to "Materials Corrosion and Protection". The symposium was co-sponsored by ISE Division 6 (Corrosion, Electrodeposition and Protection) and by the European Federation of Corrosion (EFC Event No. 232) (Organisers: P. L. Bonora, V. A. Macagno, F. Mansfeld).

The ISE has, as one of its aims, the promotion of international co-operation and communication in all fields of scientific and industrial electrochemistry; the Society has eight scientific divisions each dealing with a special field of electrochemistry. Matters concerning corrosion, electrodeposition and surface treatment are handled by Division 6.

The European Federation of Corrosion (EFC) is an Association whose purpose is to contribute to the general advancement of the science of corrosion and protection of materials by promoting co-operation in Europe between scientific and technical societies and associations devoted significantly to these areas of activity and by collaborating with similar associations throughout the world.

The Symposium, a result of a more general agreement recently signed by the two bodies, is likely to widen the horizons of the twin disciplines of electrochemistry and corrosion. Corrosion is one of the driving forces of the pile (cell) and a knowledge of the mechanisms of corrosion as well of the ways to control both its morphology and rate can enhance the performance of a cell. On the other hand, electrochemical cells, which may be difficult to identify and locate, are hidden inside many corrosion phenomena.

These proceedings contain papers on the industrial application of the electrochemical approach to corrosion research, for example in testing and monitoring, and on corrosion prediction, prevention and control as well as fundamental research. Scientists from both industry and academia can be updated about the availability and reliability of electrochemical approaches to corrosion problems and related industrial applications. The papers, all carefully reviewed, are grouped under the headings of inhibitors, organic coatings, inorganic coatings, passivity and corrosion testing and evaluation in special environments. The contributions of both the Advisory and the Organising Committee Members, of the plenary and key- note lecturers, of session chairmen, authors, reviewers and all attendees are gratefully acknowledged. Professor Sergio Trasatti, Chairman of ISE 99, may here receive our thanks for the warn hospitality given to the co- sponsoring EFC Officials and for the great help he gave to the success of the Symposium.

P.L. Bonora
F. Deflorian
Editors

Foreword

In recent years the number of meetings and symposia dedicated to corrosion or some aspects of corrosion has been steadily increasing, making it more and more difficult for researchers in the field to keep abreast. The International Society of Electrochemistry (ISE) and the European Federation of Corrosion (EFC) took the welcome intitiative to coordinate some of their activities and to organise together a symposium on "Materials Corrosion and Protection" at the 50th ISE meeting in Pavia in September 1999. The present book published by the Institute of Materials is an outcome of this collaboration.

At the joint symposium a total of 45 oral papers were given of which 5 were keynote lectures and one an award lecture. In addition 51 posters were presented. The present volume contains a selection of these papers, namely 16 oral and 10 poster papers. The choice of published papers is somewhat arbitrary, because it was essentially left up to the authors whether to submit a manuscript or not for inclusion in the present volume. In spite of this limitation the papers in this book illustrate well the different topics and approaches presented at the joint symposium and they give the reader a flavour of the types of problems studied by researchers in corrosion using electrochemical methods. All manuscripts have been carefully edited and this has resulted in a certain uniformity of presentation that greatly increases the value of the book.

The papers presented here cover five important aspects of electrochemical corrosion and protection research: inhibitors, organic coatings, inorganic coatings, passivity, corrosion testing and evaluation in special environments. The first part of the book contains four papers on inhibitors. Two main driving forces govern modern inhibitor research; (a) the need for environmentally safe compounds which can replace some of the more harmful traditional inhibitors such as heavy metal oxides or chromates, (b) the availability of ever more sophisticated experimental methods for *in situ* and *ex situ* studies which permit information to be gained on adsorption phenomena and electrochemical mechanisms on an ever finer scale. The first paper from the Academy of Sciences in Budapest presents an interesting new concept of inhibition, namely the use of self-assembling molecules. These are organic chain molecules with a functional group at each end. One of them permits adsorption of the inhibitor on the substrate, while the other faces outwards and thus determines the surface properties of the metal-adsorbate system. Such molecules have been studied previously by surface electrochemists using inert electrodes such as gold, but they have not been extensively investigated for purposes of corrosion inhibition. The authors show that the inhibition efficiency for iron depends markedly on the concentration of the self-assembling molecules (alkane-diphosphonic acids) and on their chain length. It will be interesting to perform further studies with this type of molecules to get more insight into their specific inhibition properties on a molecular level. Yao *et al.* for the first time present Raman spectroscopy data on adsorption of model inhibitors on iron obtained *in situ*. Two other papers deal with electrochemical testing of organic inhibitors for iron and copper in acid media.

Part two of the present volume contains six papers dedicated to organic coatings. Deflorian *et al.* apply the electrochemical impedance method to the study of the loss of integrity of organic coatings due to plastic deformation of the substrate. Impedance spectroscopy is the most widely used experimental method today for the electrochemical study of the protection efficiency and of organic coatings and the underlying reaction mechanisms, although the interpretation of results obtained with this method is often not straightforward. Most papers on organic coatings presented here use impedance in one way or another, but other methods are also applied. Mansfeld compares impedance results of sea water exposure tests with electrochemical noise analysis and Buttery *et al.* present an original method for measuring penetrating porosity by electrochemically generated chemiluminescence. Nazarov and Thierry apply the Kelvin probe to the investigation of loss of adhesion by monitoring time-dependent changes in the Volta potential. The Kelvin probe is a relatively new tool for coating research, which yields correlations between the measured Volta potential and the reaction behaviour at the metal–coating interface.

The third part of the book includes a variety of papers on different aspects of metallic and non-metallic coatings including a study of the corrosion behaviour of tungsten based hard alloy coatings fabricated by sputter deposition and studies of metal matrix composite coatings fabricated by electrodeposition. The mentioned coatings are primarily of interest for their wear resistance, but they must also resist corrosion. While Medeline *et al.* codeposit diamond or amorphous boron with nickel using saccharin as an additive, Benea *et al.* study the effect of co-deposition of zirconium oxide on the corrosion properties of nickel coatings. Anodic oxides are investigated by several authors including stress generation during anodisation of titanium alloys and an ellipsometry study of corrosion of tungsten oxide.

Corrosion resistant engineering alloys rely on passivity for protection. The papers presented in the chapter entitled 'Passivity' do not have the ambition to cover the most significant recent developments in the field of passivity; other publications dedicated specifically to passivity are available for this. Rather, the papers in this book deal with a few specific topics related to anodic films and to passivity. Two contributions from groups in Argentina study the electrochemistry of anodic films on copper and silver, respectively from a fundamental point of view. The corrosion behaviour of selected passive engineering alloys is the subject of three papers. They include a study of the effect of heat treatment on the corrosion resistance of a nickel-chromium-molybdenum high temperature alloy and a study of the effect of structure and composition on the electrochemical behaviour of amorphous or nanocrystalline alloys for magnetic applications.

The fifth and last part of this book entitled "Corrosion testing and evaluation in special environments" contains both experimental and theoretical contributions. Two experimental papers deal with the corrosion of iron in concrete and one with corrosion in polyethylene glycol–water mixtures. At the high pH values typical for concrete iron is passive. Andrade *et al.* find that the passive oxide film has a dominating effect on the electrochemical response of rebar in concrete. Polarisation and impedance measurements are presented, which indicate that under certain conditions redox reactions take place in the oxide film. They may dominate the electrochemical response of the system and can obscure the effect of the actual corrosion reaction. Impedance is a powerful method for monitoring corrosion phenomena and for studying the mechanisms of corrosion reactions in different environments. One has to keep in mind, however, that the interpretation of impedance measurements is always based on some kind of a theoretical model, often represented by an electrical equivalent circuit. To evaluate the experimental data in terms of a theoretical model one applies statistical methods to regress the measured data points. The paper by Orazem *et al.* theoretically explores the manner in which stochastic errors in the time-domain signals propagate through the instrumentation to the frequency-domain. Different experimental arrangements are simulated using LABVIEW™ software. Results indicate that independent of the distribution function in the time-domain, stochastic errors in the frequency-domain have a normal distribution.

The present book published on behalf of the European Federation of Corrosion contains a good collection of original papers dealing with the application of electrochemical methods to the study of corrosion and protection of metals. It will be of interest to researchers who want to get first hand information on current approaches in this field.

D. LANDOLT
Lausanne, March 2000

Part 1

Inhibitors

1
Corrosion Protection Properties of Self-Assembling Molecules

P. PÓCZIK, I. FELHÖSI, J. TELEGDI AND E. KÁLMÁN
Department of Surface Chemistry and Corrosion Research, Chemical Research Center, Hungarian Academy of Sciences, P.O. Box 17, H-1525 Budapest, Hungary

ABSTRACT
Self-assembling molecules have been applied recently as anticorrosive additives. The present research focused on the corrosion inhibition behaviour of α,ω-diphosphonic acids ($H_2O_3P–(CH_2)_n–PO_3H_2$; $n = 5–10$). The inhibition effect of these compounds on the corrosion of Armco iron was studied by mass loss and electrochemical impedance methods. The effect of the chain length as well as the diphosphonate concentration on the corrosion prevention of iron is discussed. The effect of exposure time on the inhibitor layer formation was also studied. It was observed that the adsorption of diphosphonates on iron in neutral aqueous solution is strongly dependent on the solution concentration, thus, inhibition could be attained only above a certain concentration limit. A significant difference was obtained in inhibition effect between compounds having chains with odd and even numbers of carbon atoms, better protection could be achieved by the use of diphosphonates containing odd numbers of methylene groups.

1. Introduction

Because of environmental requirements, the main aim of corrosion inhibition research currently is to develop new, non-toxic inhibitor compounds. Phosphonic acids are chemicals that are extensively applied in cooling water systems as corrosion inhibitors for the protection of iron and low alloy steels. They may be used without risk due to their low toxicity.

Previous investigations and applications have been mainly based on short-chain type phosphonates. The present research is focused on the investigation of the inhibitor properties of alkane-diphosphonates having self-assembling properties [1–6]. The study of the structure, physical and chemical properties of mono- and multilayers formed by self-assembling molecules (SAMs) opens up a new perspective in surface science.

Self-assembled monolayers are molecular assemblies that are formed spontaneously by the immersion of an appropriate substrate into a solution of an active surfactant. Self-assembly results in supramolecular hierarchical organisations of interlocking components. From the energetic point of view, self-assembling molecules consist of two functional groups and a relatively long, mostly alkyl, chain in between. One of the functional groups is responsible for chemisorption on the substrate surface. Van der Waals and electrostatic interactions of alkyl chains lead to the formation of a compact well-ordered structure of monolayers. The other terminal group determines the physical and chemical characteristics of the modified surface and the further applications. The ability to tailor both head and tail groups of the molecules provides the desired control of the structure and chemical properties of surface at the molecular level and thus should be considered as a potential technique for the construction of future organic materials [1]. Although the area is not limited to long-chain compounds, self-assembling molecules of functionalised long-chain hydrocarbons are the most frequently used for building supermolecular structures. Among the various self-assembling systems, oriented, densely packed monolayer films of *n*-alkanethiols on the surface of gold, silver and copper have been those most investigated [7].

Formation of multilayers requires further modification of the monolayer surface. Most common ways to build up multilayers are either to apply terminal groups that are able to interact chemically or to use transition metal ions as a bridge between phosphonate monolayers.

Due to their dense, ordered and stable structure, SAMs have potential applications in corrosion prevention. These chemicals are capable of forming a protective layer with appropriate stability.

Self-assembled monolayers derived from the assembly of n-alkanethiols onto copper have been successfully applied for prevention of corrosion of the underlying copper in various conditions [8–11]. Chemisorption of alkanethiols on a reduced iron surface has been shown to suppress iron corrosion by hydrophobic film formation [12,13]. A new, promising field of application of the self-assembly process is in the replacement of chromating procedures on reactive metals like aluminium. Chemical modification of aluminium [5,6] and zinc surfaces [5] by self-assembled layers of alkane-phosphonic and -phosphoric acids can improve the organic coating adhesion and corrosion protection. Self-assembling molecules with two reactive groups, one attached to the metal surface and the other connected to an organic coating, can inhibit electrochemical reactions at the interface and result in very stable bonds between the organic coating and the metal surface.

Our interest was directed towards the corrosion inhibition properties of α,ω-alkane-diphosphonic acids ($H_2O_3P{-}(CH_2)_n{-}PO_3H_2$; $n = 5{-}10$). Here, we report some preliminary results obtained in neutral aqueous solution. The inhibition effect of these compounds on the corrosion of Armco iron was studied by mass loss and electrochemical impedance spectroscopy methods. The effect of the number of methylene-groups in diphosphonates as well as the concentration of diphosphonates on the corrosion prevention of iron is discussed. The effect of exposure time on the inhibitor layer formation was also studied.

2. Experimental

Diphosphono-alkanes under investigation were prepared from triethyl phosphite and α,ω-dibromo alkanes by the Michaelis–Arbusov method [14].

For gravimetric measurements, precisely weighed coupons (previously acid-cleaned, polished, degreased with acetone and washed with distilled water) were immersed into a test solution (0.2930 g $CaSO_4.2H_2O$, 0.1395 g $MgSO_4.2H_2O$, 0.0675 g $NaHCO_3$ and 0.0758 g $CaCl_2.2H_2O$ dissolved in 3 L distilled water, referred to as the model water and representing a hard water) at room temperature for 24 h. At the end of the experiment the coupons were cleaned and reweighed for the determination of corrosion rate and inhibitor efficiency.

Electrochemical experiments were performed using a standard three-electrode cell of 500 mL volume at room temperature in 0.1 mol L^{-1} $NaClO_4$ solution at pH 7. The working electrode was Armco iron, with a surface area of 0.5 cm^2, embedded into epoxy resin. Pretreatment of the electrode was carried out by abrasion (emery paper of 600 to 1200 grit), rinsing with bi-distilled water, ultrasonic cleaning in bi-distilled water and then immersion into the aerated solution. The counter electrode was a half-cylindrical platinum plate with an area of 90 cm^2. The reference electrode was a saturated calomel electrode (SCE) and the measured potentials are referred to this electrode.

Electrochemical impedance measurements were carried out using a Solartron 1286 Potentiostat and 1250 Frequency Response Analyzer. Impedance spectra were obtained in the frequency range of 10 mHz to 60 kHz with seven points per decade at the corrosion potential. A sine wave with 10 mV amplitude was used to perturb the system. The quantitative analysis of the experimental impedance data was performed by the non-linear least square minimisation method using a theoretical transfer function.

3. Results and Discussion

3.1. Gravimetric Test of Inhibition Efficiency of Diphosphono-Alkanes

Gravimetric corrosion tests were performed to determine corrosion inhibition properties of α,ω-diphosphono-alkanes ($H_2O_3P{-}(CH_2)_n{-}PO_3H_2$; $n = 5$–10) in the model water. Figure 1 shows the concentration dependence of the corrosion inhibition efficiency of different diphosphonates on carbon steel corrosion after 24 h exposure. The results show that the inhibition effect and thus the adsorption of diphosphonates in neutral aqueous solution strongly depend on the solution concentration. A significant inhibition effect could be achieved only above 2×10^{-4} mol L^{-1} concentration. At lower concentrations ($c \leq 10^{-4}$ mol L^{-1}) accelerated corrosion was observed as a result of deficient adsorption, so that these kinds of chemicals behave as dangerous inhibitors. This behaviour is frequently observed among adsorption type inhibitors. Increasing solution concentration resulted in a sharp increase in inhibitor efficiency near a certain concentration limit ($c_{lim} \cong 1.5\text{–}2 \times 10^{-4}$ mol L^{-1}). No major change in the efficiency of compounds could be observed on further increase of concentration.

This behaviour can probably be ascribed to the inhibitor layer formation properties of diphosphonates, the higher concentration being necessary to ensure the formation of a continuous, dense inhibitor layer.

The inhibitor layer is probably relatively thick and contains calcium and magnesium complexes of diphosphonates in different compositions. The following electrochemical measurements were carried out in indifferent $NaClO_4$ solutions to distinguish any effects of complex formation of phosphonates with bivalent cations on corrosion inhibition behaviour [15].

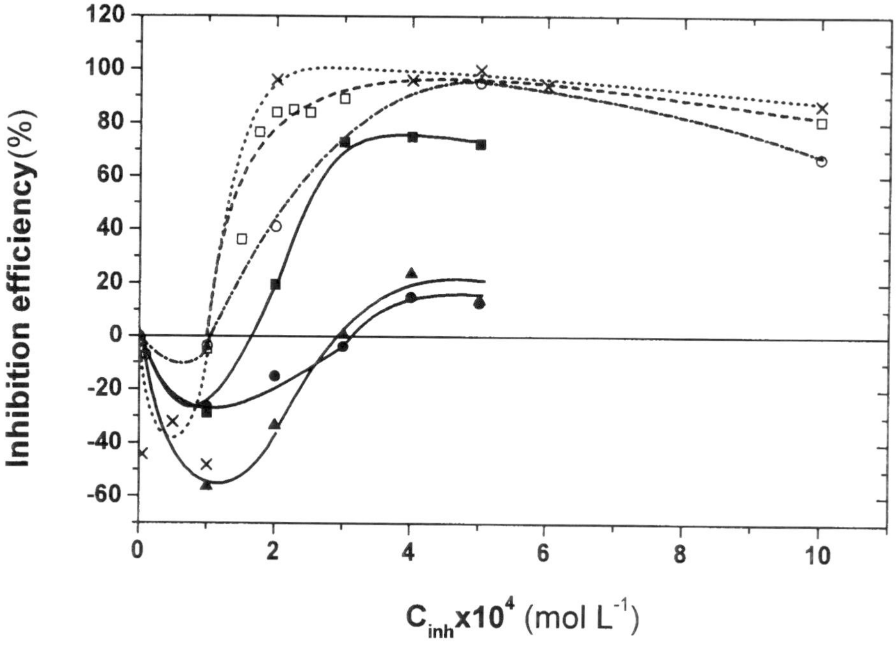

Fig. 1 Concentration dependence of inhibition efficiency of different α,ω-diphosphono-alkanes determined by gravimetric corrosion test. (- -□- -): 1,5-diphosphono-pentane; (—■—): 1,6-diphophono-hexane; (···×···): 1,7-diphosphono-heptane, (—▲—): 1,8-diphosphono-octane; (- · -○- · -): 1,9-diphosphono-nonane; (—●—): 1,10-diphosphono-decane.

3.2. The Effect of Alkyl-Chain Length of Diphosphonates on Corrosion Inhibition Properties

Figure 2 shows polarisation resistance derived from impedance spectra as a function of carbon chain length of diphosphonates. Impedance spectra of Armco iron were measured after different exposure times (2 h, 5 h and 1 day) in neutral 0.1 mol L^{-1} $NaClO_4$ solution in the presence of 5×10^{-4} mol L^{-1} alkane-diphosphonates. Experimental data were fitted using the equivalent circuit shown in Fig. 3. The charge transfer resistance of iron dissolution (R_p) and constant phase element of the double layer (Q_{dl}) characterise the corrosion process taking place at the metal surface. The low frequency end of the impedance spectra is determined by the corrosion process. On the assumption of a porous structure, the properties of the surface layer were taken into consideration in the model by the elements of layer resistance (R_L) and constant phase element of the surface layer (Q_L).

The results obtained here clearly show a significant difference of inhibition effect between compounds having chains of odd and even numbers of carbon atoms, thus, better protection was achieved by use of diphosphonates containing odd numbers of methylene groups. The best corrosion inhibition could be achieved with 1,7-diphosphono-heptane (1,7-DPH). The influence of the number of methylene groups on the inhibition properties of these molecules shows that the inhibitor layer formation

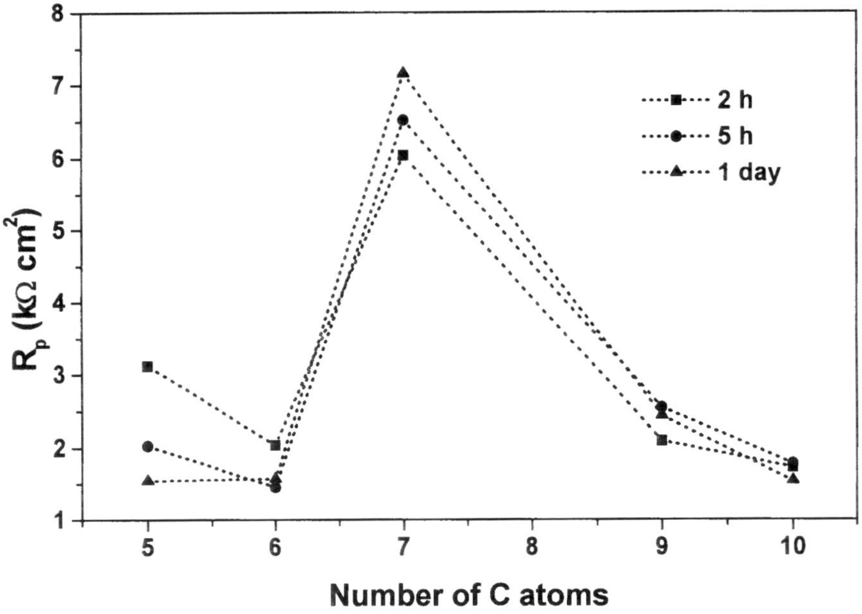

Fig. 2 Polarisation resistance of Armco iron as a function of carbon chain length of α,ω-diphosphono-alkanes determined by impedance spectroscopy in solution of 0.1 mol L^{-1} $NaClO_4$ + 5×10^{-4} mol L^{-1} inhibitor.

is strongly affected by the geometry of the molecules. The odd number of carbon atoms in diphosphonates is presumably preferred for the formation of an ordered structure of the inhibitor layer. Such differences of physical and chemical properties between odd and even carbon-chains of homologous series are well known in different fields of chemical science.

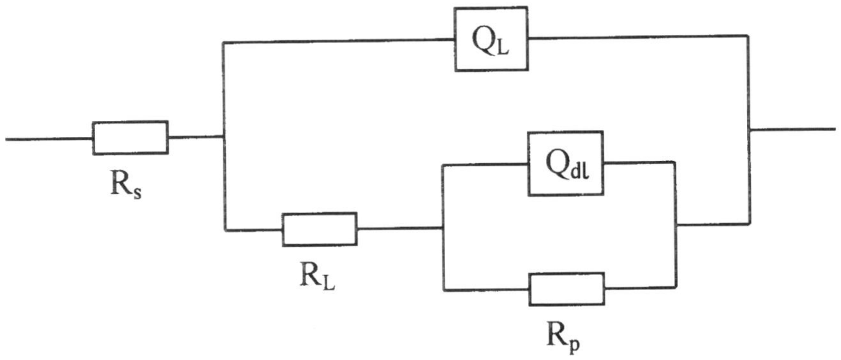

Fig. 3 Equivalent circuit model. R_s: solution resistance, Q_L: constant phase element of surface layer, R_L: layer resistance, R_p: polarisation resistance, Q_{dl}: constant phase element of double layer.

3.3. Concentration Dependence of Corrosion Protection Effect of 1,7-Diphosphono-Heptane

Figure 4 shows the concentration dependence of polarisation resistance of iron corrosion for 1,7-diphosphono-heptane after different exposure times (2 h, 5 h, 1 day). As with the gravimetric measurements, the results show that the inhibition effect of diphosphonates in neutral aqueous solution strongly depends on the solution concentration. At lower concentration ($c \leq 3 \times 10^{-4}$ mol L^{-1}), only a slight protection effect or, in some cases, acceleration of corrosion was observed. A significant increase in inhibition efficiency was obtained near the concentration of 4×10^{-4} mol L^{-1}. On further increase of concentration the inhibitor efficiency did not change significantly. A higher concentration (5×10^{-4} mol L^{-1} or more) was necessary to achieve the same inhibition effect as in the gravimetric tests ($c > 1.5\text{–}2 \times 10^{-4}$ mol L^{-1}). This difference between the concentrations for inhibition obtained by two different techniques can be attributed to the presence of calcium and magnesium ions in the model water used in the gravimetric tests. As a result of the complex formation between phosphonates and bivalent cations, the appropriate protection layer can be formed at lower diphosphonate concentration. Further experiments would be necessary to explore the exact influence of calcium and magnesium ions on the inhibition behaviour of alkane-diphosphonates.

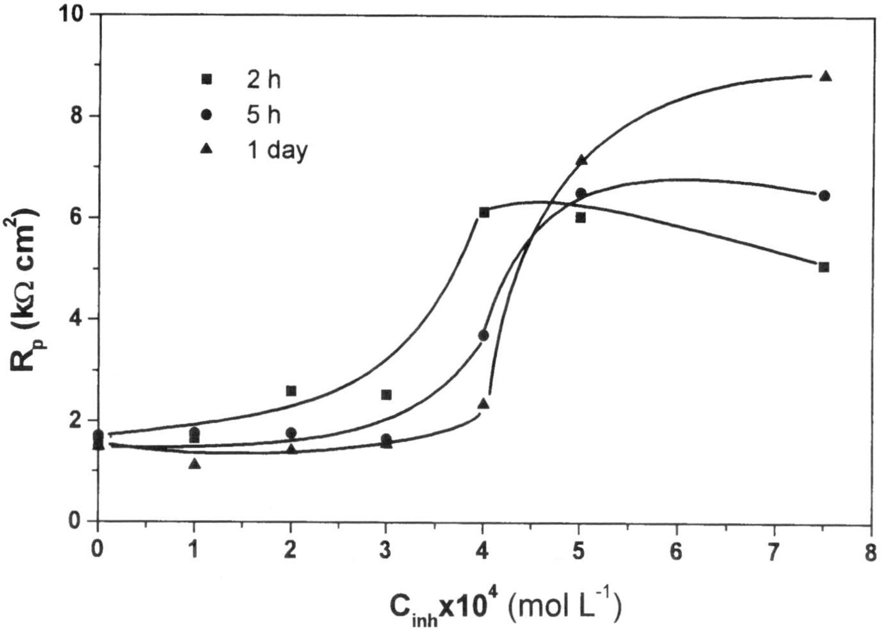

Fig. 4 Concentration dependence of polarisation resistance of Armco iron in 0.1 mol L^{-1} $NaClO_4$ in the presence of 1,7-diphosphono-pentane determined by impedance spectroscopy after different exposure times of 2 h, 5 h and 1 day.

3.4. Time Dependence of Corrosion Protection Effect of 1,7-Diphosphono-Heptane

The time-dependence of the inhibition behaviour of alkane-diphosphonates on Armcoiron corrosion was studied with 1,7-diphosphono-heptane at a concentration of 5×10^{-4} mol L^{-1}. Nyquist and Bode plots measured after different immersion times are depicted in Fig. 5. Although, the polarisation resistance data for iron dissolution does not show any remarkable time dependence, an increase of phase shift in the intermediate frequency range does point to a continuous surface layer formation with a loosely bonded porous structure [16].

4. Conclusions

Corrosion inhibition behaviour of α,ω-diphosphonic acids ($H_2O_3P\text{–}(CH_2)_n\text{–}PO_3H_2$; $n = 5\text{–}10$) was studied in neutral conditions. The inhibition effect of these compounds on iron corrosion in neutral aqueous solution is strongly dependent on the solution concentration. A significant inhibition effect could be observed above a concentration of $2 - 4 \times 10^{-4}$ mol L^{-1}. At lower concentrations, accelerated corrosion was observed. The concentration limit, required for inhibition, was lowered by the presence of calcium and magnesium ions. A correlation was found between the number of methylene groups and the inhibition effect. The better protection effect could be achieved by use of diphosphonates containing an odd number of methylene groups. The most effective compound proved to be the 1,7-diphosphono-heptane (1,7-DPH).

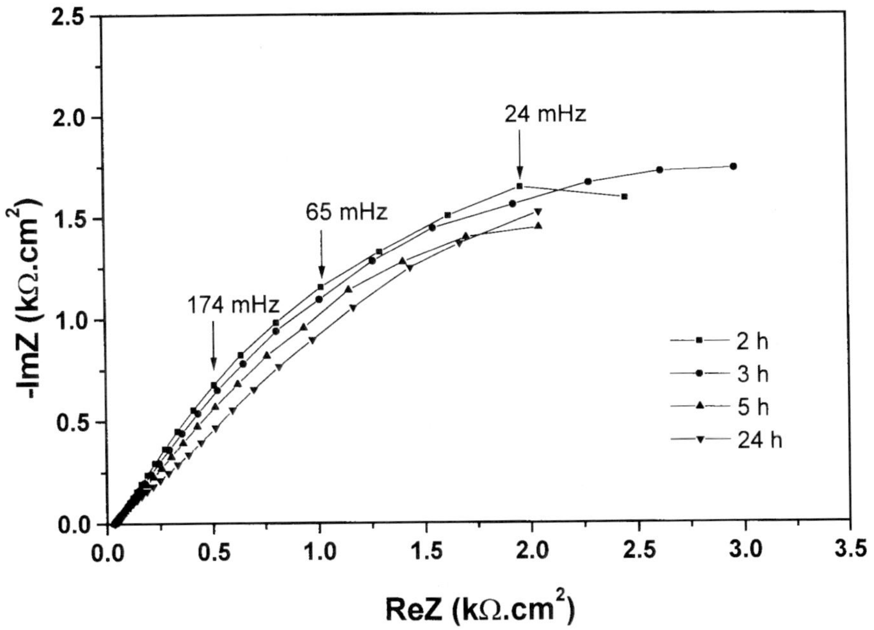

Fig. 5 Impedance spectra of Armco iron in 0.1 mol L^{-1} $NaClO_4$ solution in the presence of 5×10^{-4} mol L^{-1} 1,7-diphosphono-heptane after different exposure times.

5. Acknowledgements

The authors express thanks for the support given by the National Science Foundation (OTKA T022013 and F029709).

References

1. A. Ulman, *Chem. Rev.*, 1996, **96**, 1533.
2. H. Lee, L. J. Kepley, H. G. Hong, S. Akhter and T. E. Mallouk, *J. Phys. Chem.*, 1988, **92**, 2597.
3. M. L. Schilling *et al.*, *Langmuir*, 1993, **9**, 2156.
4. H. C. Yang *et al.*, *J. Am. Chem. Soc.*, 1993, **115**, 11855.
5. R. Feser and T. H. Schmidt-Hansberg, *Proc. Eurocorr '97*, Publ. NTNU Gløshaugen, N-7034, Trondheim, 1997, **2**, 291.
6. I. Maege *et al.*, *Prog. Org. Coatings*, 1998, **34**, 1.
7. A. Ulman, *Ultrathin Organic Films*, Academic Press, New York, 1991.
8. P. E. Laibinis and G. M. Whitesides, *J. Am. Chem. Soc.*, 1992, **114**, 9022.
9. Y. Yamamoto, H. Nishihara and K. Aramaki, *J. Electrochem. Soc.*, 1993, **140**, 436.
10. M. Itoh, H. Nishihara and K. Aramaki, *J. Electrochem. Soc.*, 1994, **141**, 2018.
11. G. K. Jennings and P. E. Laibinis, *Colloid Surfaces A: Physicochem. Eng. Aspects*, 1996, **116**, 105.
12. M. Volmer-Uebing, B. Reynders and M. Stratmann, *Werkst. Korros.*, 1991, **42**, 19.
13. K. Nozawa, H. Nishihara, K. Aramaki, *Corros. Sci.*, 1997, **39**, 1625.
14. B. Ackerman, T.A. Jordan, C.R. Eddy and D. Swern, *J. Am. Chem. Soc.*, 1956, **78**, 4444.
15. E. Kálmán, I. Felhösi, F.H. Kármán, I. Lukovits and J. Telegdi, in *Corrosion and Environmental Degradation*, Wiley-VCH, Frankfurt, Ch 9., 2000.
16. P. Póczik *et al.*, to be published.

2
The Inhibition of the Corrosion of Iron in 1M HCl in the Presence of a New Class of Organic Inhibitors

S. EL HAJJAJI, A. LGAMRI, A. GUENBOUR, A . BEN BACHIR, D. AZIANE*, E. M. ESSASSI* and L. ARIES†

Laboratoire d'électrochimie - corrosion, département de chimie, faculté des sciences, Av Ibn Batouta, B. P. 1014, Rabat, Morocco

* Laboratoire de Chimie Organique Hétérocyclique, Département de Chimie, Faculté des Sciences, Av Ibn Batouta, B. P. 1014, Rabat, Morocco

†LCMIE, Université Paul Sabatier, 118 Route de Narbonne, 31064, Toulouse, France

ABSTRACT

Corrosion inhibitors are applied successfully for prevention of corrosion in acid solution. In this work, a new class of organic molecules containing heteroatoms such as nitrogen and sulfur, is proposed. The effect of addition of these molecules on iron dissolution and surface blistering in HCl (1M) is studied through electrochemical (electrochemical impedance spectroscopy as well as polarisation curves) and scanning electron microscopy methods.

The data obtained showed that the corrosion resistance was greatly enhanced in presence of inhibitor. The maximum protection efficiency exceeded 94%. The corrosion protection could be explained by the adsorption of inhibitor through the S or N-atom and formation of a protective layer attached to the metal surface.

1. Introduction

Most of the effective inhibitors of acid corrosion used in industry are organic compounds which generally contain nitrogen, sulfur and oxygen atoms and multiple bonds in the molecules through which they are adsorbed on the metal surface [1–6]. Generally, a strong co-ordination bond causes higher inhibition efficiency with the inhibition increasing in the sequence O < N < S < P [7].

Despite the large number of organic compounds, the choice for selecting an appropriate inhibitor for a particular system is very limited because of the specificity of the inhibitors and the great variety of corrosion systems. Therefore, there will always exist a need for developing new organic corrosion inhibitors.

It is well known that triazole type organic compounds are good corrosion inhibitors for many metals and alloys in various media [8–10]. The aim of this paper was to synthesise and to examine the inhibitive action of two new triazole type compounds towards the corrosion of iron in HCl solution. These compounds contain different substituent radicals (R) and the effect of R on inhibition was studied.

2. Experimental Procedure

Polarisation and impedance measurements were carried out in a three-electrode electrochemical cell with a platinum counter electrode and a saturated calomel electrode as reference. Samples of iron (Table 1) of total area 1.54 cm^2 were used as working electrodes. Before every run the surface of the electrode was abraded to a mirror finish using emery paper of 1000 grade. The scanning rate was 500 μVs^{-1}. Impedance measurements were performed with the aid of a 6310 EG&G unit. The 1M HCl solution was prepared from analytical grade reagents and doubly distilled water. The additive compounds were synthesised and their structures are given below.

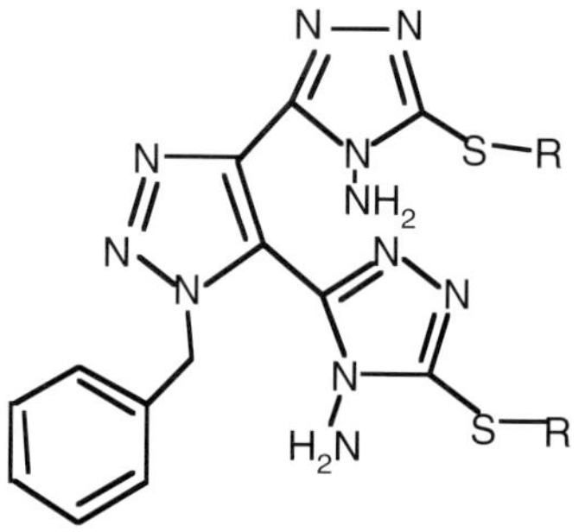

($C_{15}H_{17}N_{11}S_2$), R = CH_3 designated D_{22}
($C_{17}H_{21}N_{11}S_2$), R =CH_2–CH_3 designated D_{24}

Table 1. Composition of samples

Elements	C	P	S	N	Fe
Mass%	0.2	0.055	0.055	0.009	99.7

3. Results

3.1. Electrochemical Testing

The polarisation curves obtained on iron in 1M HCl with and without inhibitors are shown in Figs 1 and 2. The associated corrosion parameters I_{corr}, E_{corr} and corrosion inhibition efficiency (E%) obtained in uninhibited and inhibited solutions are given in Table 2. The inhibitive efficiency is evaluated by the relationship [1]:

$$E = \frac{I_{corr} - I'_{corr}}{I_{corr}} \times 100$$

in which I_{corr} and I'_{corr} represent the current densities without and with inhibitor respectively as determined by extrapolation of cathodic Tafel lines to the corrosion potential.

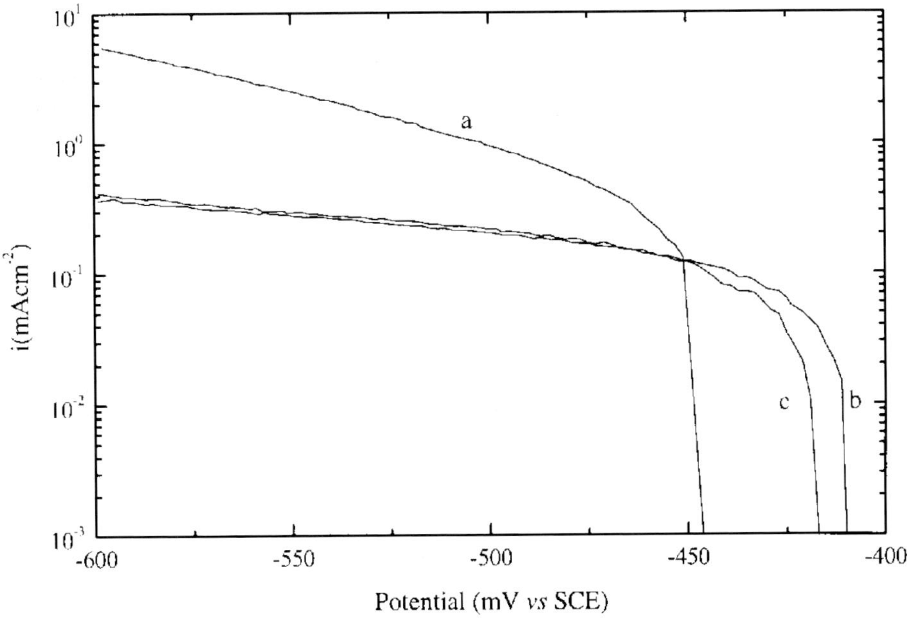

Fig. 1 *Cathodic polarisation curves of iron in 1*M *HCl (a) without inhibitors, (b) with* D_{24} *and (c) with* D_{22}.

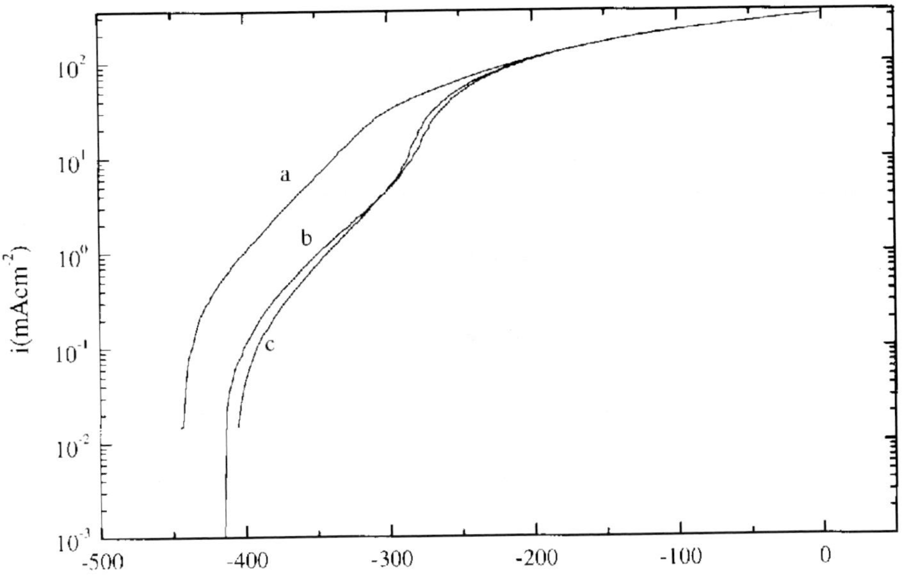

Fig. 2 *Anodic polarisation curves of iron in 1*M *HCl (a) without inhibitors, (b) with* D_{22} *and (c) with* D_{24}.

***Table 2.** Electrochemical parameters (E_{corr} and I_{corr}) of the iron electrode in 1M HCl with and without inhibitors*

Solution	E_{corr} **(mV vs SCE)**	I_{corr} **(mAcm^{-2})**	*E* **(%)**
HCl 1M	–445	93×10^{-3}	—
HCl 1M + 10^{-3} mol L^{-1} of D_{24}	–409	5.1×10^{-3}	94.5
HCl 1M + 10^{-3} mol L^{-1} of D_{22}	–416	6.9×10^{-3}	92.6

As can be seen from Fig. 1, Fig. 2 and Table 2, the addition of inhibitors generally, induced a decrease in both anodic and cathodic currents, indicating that these compounds influence both the anodic and the cathodic processes. The decrease of i_{corr} demonstrates the efficiency of the tested materials as corrosion inhibitors.

The open circuit potential was monitored during 20 min of immersion (Fig. 3). An initial shift shift of E_{corr} towards more positive values with increasing immersion time in presence of inhibitors indicates a deceleration of corrosion.

Nevertheless, after an exposure time of 20 min in the presence and in absence of inhibitors, the potential assumed a value near the corrosion potential as determined by polarisation curves. The immersion time of about 20 min used for this study is too short and other tests must be carried out for a longer period in order to determine

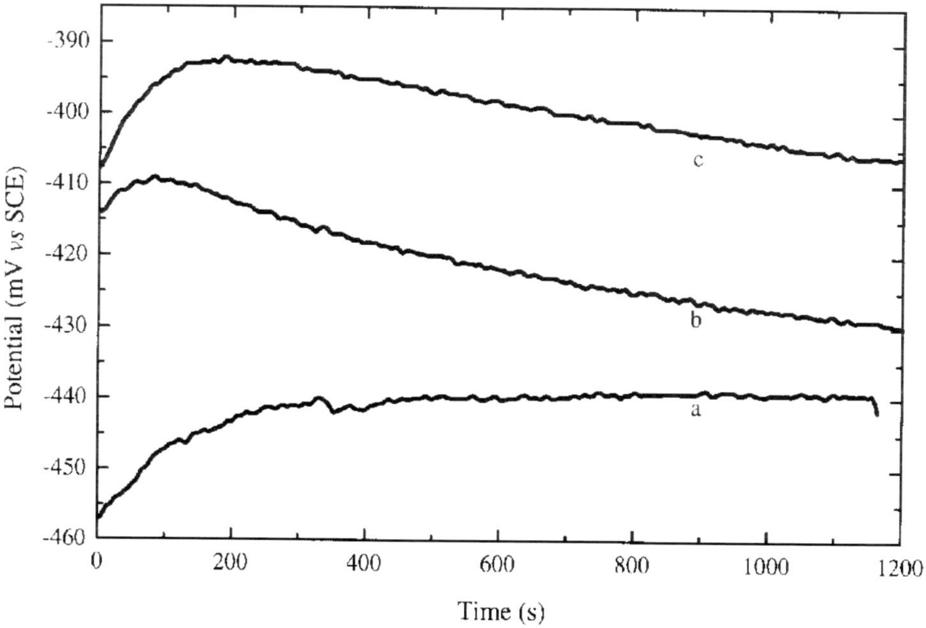

Fig. 3 The open circuit potential of iron in 1M HCl (a) without inhibitors, (b) with D_{22} and (c) with D_{24}.

if the formation of a protective film on the electrode surface is progressive and dependent on the treatment time.

3.2. Gravimetric Measurements

The inhibitor efficiency ($E\%$) was also calculated from mass loss measurements using the relation:

$$E\% = (1 - \frac{W}{W_0}) \times 100$$

where W and W_0 are rate of corrosion ($gcm^{-2}h^{-1}$) with and without inhibitor respectively. The data obtained are given in Table 3, for an immersion period of 58 h.

Table 3. *Corrosion rates, V, in the absence of and in the presence of inhibitor D_{24} in 1M HCl. Inhibitor efficiencies E% determined by gravimetric method (58 h test)*

Solution	**V (g cm^{-2}h^{-1})**	**E (%)**
HCl 1M	6.54×10^{-4}	—
HCl 1M + 10^{-3} mol L^{-1} of D_{24}	3.07×10^{-5}	95.3

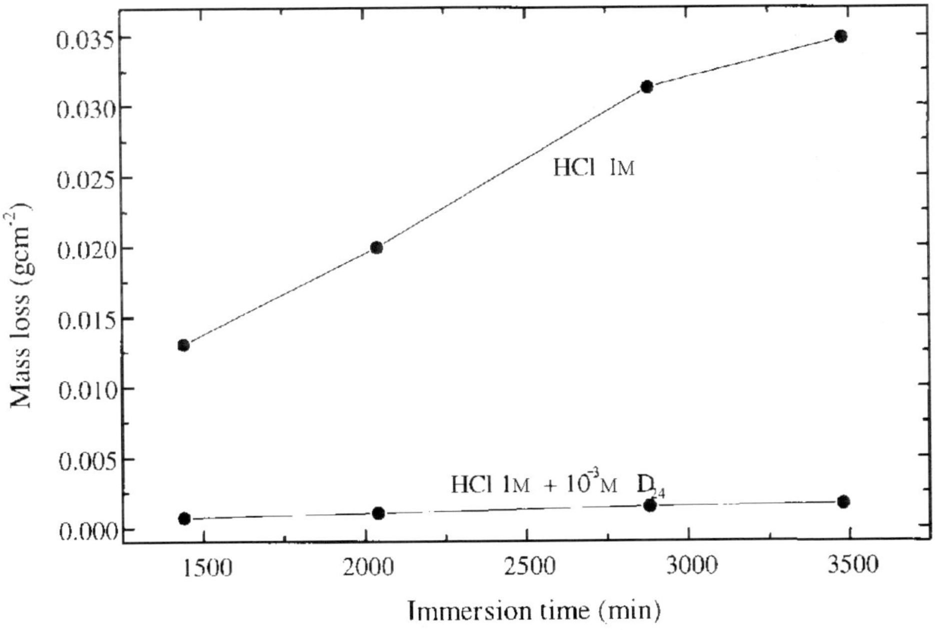

Fig. 4 *Effect of compound D_{24} on the mass loss–time curves for iron in 1M HCl.*

Figure 4 shows mass loss–time curves for the system with and without inhibitor D_{24}. Comparison between *E*% values obtained by two methods indicates good agreement. At 10^{-3}M of each inhibitor tested, the inhibition efficiency reached 92.6 and 94.5 for D_{22} and D_{24} respectively. Thus, we deduce that D_{24} is the better inhibitor.

3.3. Potentiostatic Study

The potentiostatic study involved the polarisation of iron specimens in solutions at –300 mV (SCE) and measuring the resulting current intensity against time.

In the presence of inhibitor D_{24}, the current density decreased rapidly with time to very low stationary values (5 $mAcm^{-2}$) (Fig. 5 curve a). The curve corresponding to iron in absence of inhibitor showed higher currents (Fig. 5 curve b) and the stationary value was 60 $mAcm^{-2}$. Before reaching the stationary state, fluctuations of the current were observed, indicating instability of the film. Thus, the film formed on the iron in presence of inhibitor was more stable and more protective than that on the iron in absence of inhibitor. In absence of inhibitor, iron had a dissolution rate 10 times faster than in presence of inhibitor.

Samples were examined by scanning electron microscopy (SEM) following a 1 h exposure test at –300 mV (SCE) (Fig. 6). In absence of inhibitor, a thick and discontinuous corrosion product layer was observed. In presence of inhibitor D_{24}, the surface was uniform and did not show any corrosion pattern.

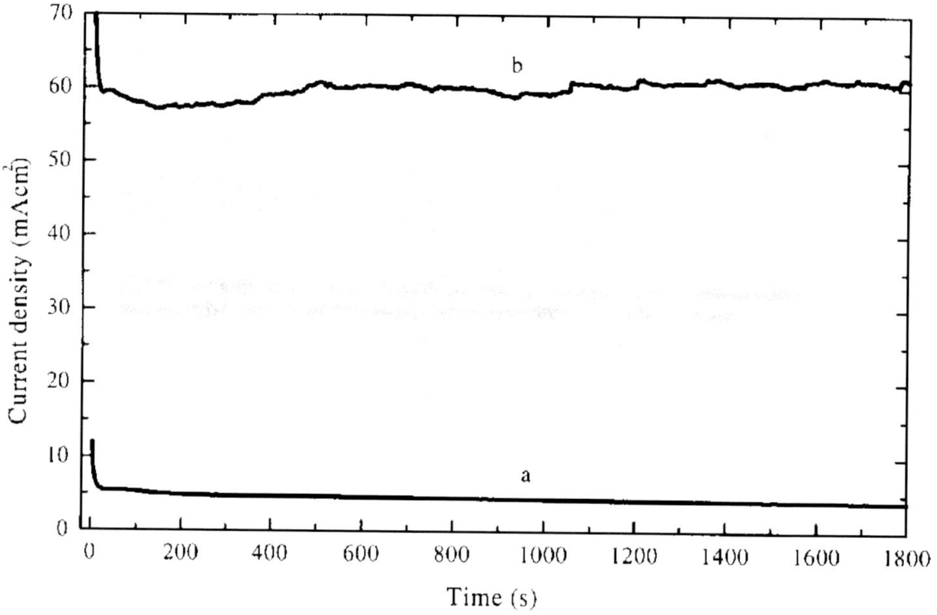

***Fig.** 5 Current transients of iron in HCl (1M), (a) with D_{24} and (b) without inhibitor at –300 mV (SCE).*

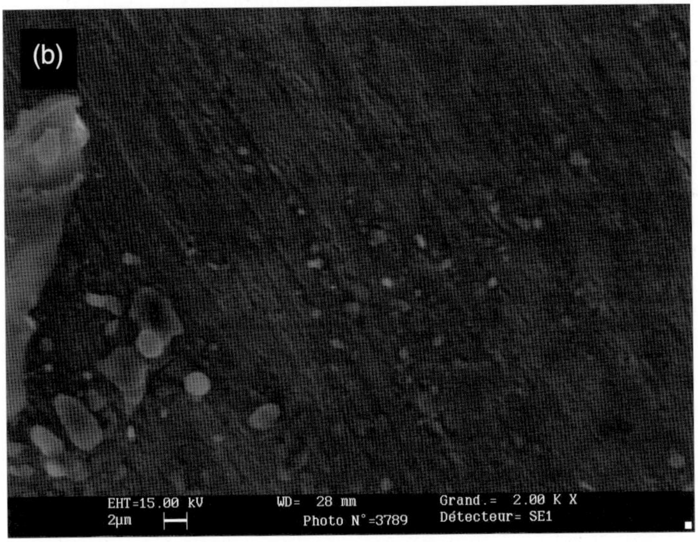

Fig. 6 *Micrographs showing surface of iron electrode after potentiostatic polarisation testing in HCl (1M) solution: (a) in absence of inhibitor; (b) in presence of inhibitor* D_{24}.

3.4. Impedance Measurements

Electrochemical impedance spectroscopy measurements were performed in the potentiostatic mode at the corrosion potential. Typical impedance diagrams obtained for the system iron/HCl solution with or without inhibitor, are presented in Fig. 7.

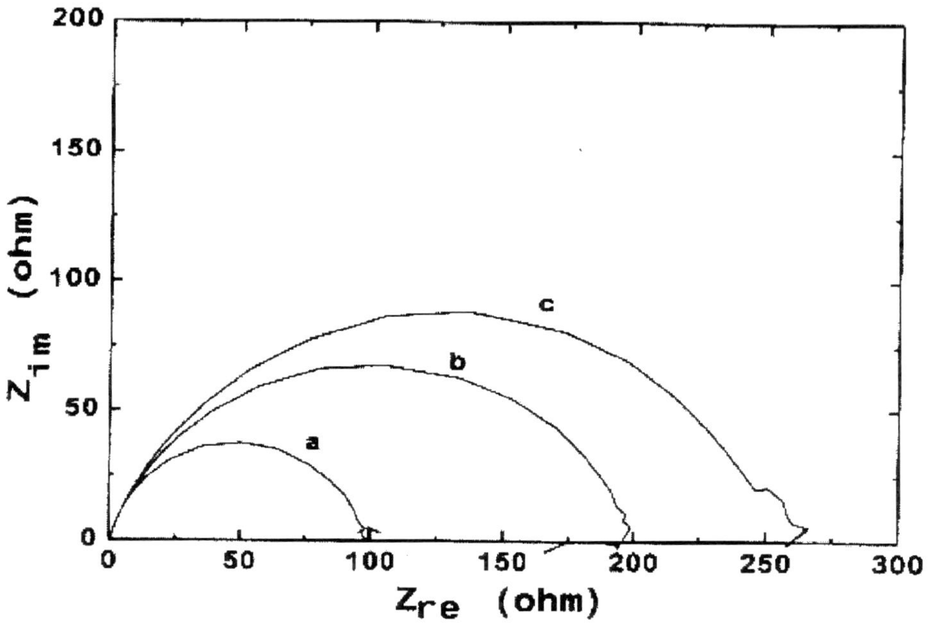

Fig. 7 Nyquist plots for iron at the corrosion potential in HCl (1M), (a) without inhibitor, (b) with 10^{-3} M D_{22} and (c) with 10^{-3} M D_{24}.

The polarisation resistance (R_p) was determined by extrapolating the semicircle toward zero frequency.

In the absence of inhibitors, impedance diagrams showed one capacitive arc. The R_p value determined from the intercept of the semicircle with the real part of the cell impedance at low frequencies was 100 Ωcm^2.

In the presence of the inhibitors, the electrochemical behaviour of the interface was not fundamentally modified. The impedance diagrams again presented a single capacitive arc. Nevertheless, an increase of R_p was observed and confirmed that compounds D_{24} (R_p = 250 Ωcm^2) and D_{22} (R_p = 200 Ωcm^2) act as good inhibitors.

4. Discussion and Conclusions

Polarisation, EIS, mass–loss and SEM data have shown that the two triazole type compounds used in this work and not studied before as inhibitors act as good inhibitors for the corrosion of iron in acid chloride solutions.

The comparative study of corrosion inhibition of 10^{-3} M concentrations of these triazole derivatives indicates that the compound described as D_{24} ($C_{17}H_{21}N_{11}S_2R$ where R = CH_2–CH_3) is the slightly better (*E* 94.5% compared to 92.6% for D_{22} R = CH_3) inhibitor.

The difference of inhibitor effect observed between D_{22} and D_{24} may be attributed to the nature of the radical (R).

The corrosion protection could be interpreted as has been done with other O, N, S, P-containing inhibitors, by adsorption, of inhibitor through the S or N-atom and formation of a protective layer attached to the metal surface. Nevertheless, surface analysis techniques such as Fourier Transform Infrared spectroscopy (FT-IR) must be applied in order to confirm this interpretation.

References

1. G. Benerje and S. N. Malhotra, *Corrosion*, 1992, **48**, 10.
2. K. Pilali and R. Narayan, *Corros. Sci.*, 1983, **23**, 151.
3. A. B. Tadros and B. A. Abdenaby, *J. Electrochem. Soc.*, 1988, **246**, 433.
4. R. Agrawal and T. K. G. Nambododhiri, *J. Appl. Electrochem.*, 1992, **22**, 433.
5. M. El Achouri, S. Kertit, M. Salem, M. S. Hajji and E. M. Essasi, *J. Chem. Phys.*, 1996, **93**, 2011.
6. C. Fiaud, A. Harch, D. Mallouh and M. Tzinmann, *Corros. Sci.*, 1993, **35**, 1437.
7. A. K. Mohamed, A. Al-Nadjm and A. S. Fouda, *J. Chem. Phys.*, 1997, **94**, 1286.
8. A. El Kanouni , Thèse de 3ème cycle (DES), Faculté des Science de Rabat, Maroc (1995).
9. F. El-Taib Heakal and S. Haruyama, *Corros. Sci.*, 1980, **20**, 887.
10. F. Chaouket, A. Sghiri, A. BenBachir and A. Frignani, *Proc. 8th Europ. Symp. on Corrosion Inhibitors*, 8 SEIC Ann. Univ. Ferrara, N. S., Sez, V, 1995. Suppl. No. 10, 1031–1041. Publ. University of Ferrara, Ferrara, Italy, 1995.

3
Imidazole and its Derivatives as Inhibitors for Prevention of Corrosion of Copper in Hydrochloric Acid Solutions

R. GAŠPARAC, E. STUPNIŠEK-LISAC and C. R. MARTIN*
Faculty of Chemical Engineering and Technology, University of Zagreb, Croatia
* Colorado State University, Fort Collins, CO 80523, USA

ABSTRACT
The aim of this work was to investigate the efficiency of imidazole and its derivatives 4-methylimidazole, 4-methyl-5-hydroxymethylimidazole, 1-phenyl-4-methylimidazole, 1-(*p*-tolyl)-4-methylimidazole for the corrosion inhibition of copper in 0.5M hydrochloric acid. Corrosion inhibition was studied using impedance spectroscopy and potentiodynamic methods. Results obtained from the a.c. impedance measurements are in good agreement with those obtained from d.c. polarisation measurements. The copper samples were also analysed by scanning electron microscopy (SEM) and X-ray microanalysis. These studies have shown that 1-(*p*-tolyl)-4-methylimidazole has the highest inhibition efficiency.

1. Introduction

Due to good corrosion resistance, mechanical workability and excellent electrical and thermal conductivity, copper and its alloys are widely used in heating and cooling systems. Such systems are cleaned by acid pickling processes using hydrochloric acid. It is important that the Cu-based systems are resistant to corrosion by HCl. In such applications inhibitors are used to prevent copper dissolution and the destructive effect of HCl on the equipment.

Imidazole derivatives are well known corrosion inhibitors for metals and alloys [1–10]. Imidazole is a heterocyclic organic compound with two nitrogen atoms forming part of a five-membered ring [11] (Fig. 1). It is of particular importance as a corrosion inhibitor because it is environmentally friendly and has relatively low toxicity [12]. In the present study the efficiency of a homologous series of imidazoles for inhibition of copper corrosion in 0.5M HCl was investigated. This series consisted of imidazole, 4-methylimidazole, 4-methyl-5-hydroxyimidazole, 1-phenyl-4-methylimidazole, 1-(*p*-tolyl)-4-methylimidazole (Fig. 1).

Corrosion inhibition was investigated using a.c. impedance spectroscopy and potentiodynamic methods. The copper samples were also analysed by scanning electron microscopy (SEM) and X-ray microanalysis.

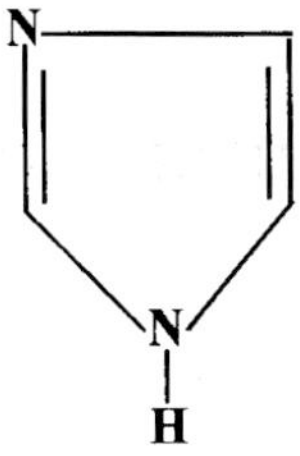

imidazole

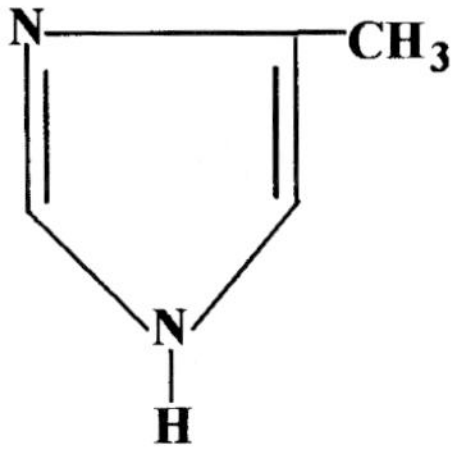

4-methylimidazole

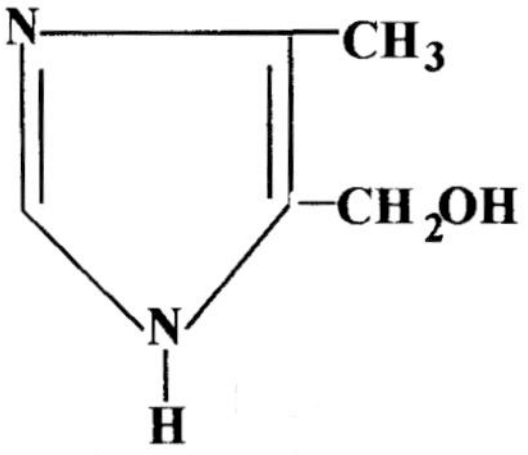

4-methyl-5 hydroxymethylimidazole

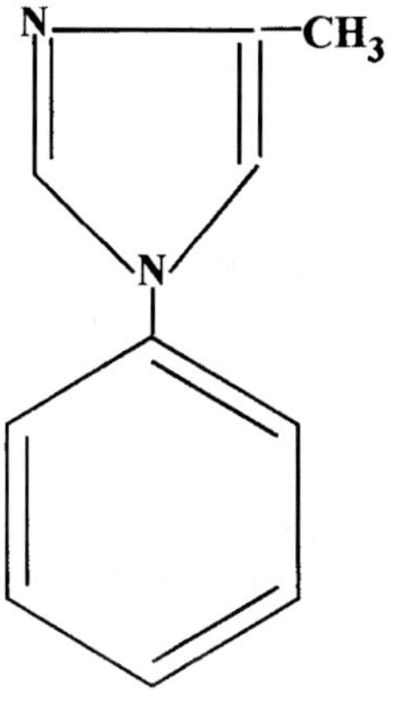

1-phenyl-4-methylimidazole

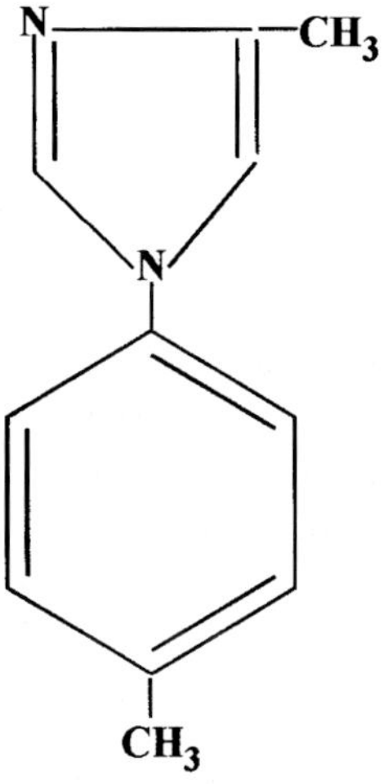

1-(p-tolyl)-4-methylimidazole

Fig. 1 *The inhibitor compounds studied.*

2. Experimental

2.1. Materials

Concentrated HCl (Merck), HNO_3 (Merck), ethanol (Merck) and imidazole (Aldrich) were used as received. The inhibitors 4-methylimidazole, 4-methyl-5-hydroxymethylimidazole, 1-phenyl-4-methylimidazole and 1-(*p*-tolyl)-4-methylimidazole were obtained from Pliva pharmaceutical company (Zagreb, Croatia). Purified water (either doubly distilled or Millipore water) was used to rinse the samples and prepare solutions. The copper working electrode was of 99.98% purity. In addition, some studies were made with Cu composites consisting of two ~18 μm thick Cu films sandwiching an ~2 mm-thick film of epoxy resin. These Cu composites were obtained from Ericsson Nikola Tesla (Zagreb, Croatia).

2.2. Methods

Potentiodynamic experiments were conducted at 20°, 30°, 35° and 45°C using a PARC 263A potentiostat/galvanostat controlled with PARC corrosion analysis software Model 352/252 SoftCorr™. A conventional three-electrode electrochemical cell of volume ~100 mL was used. The working electrode was prepared from a cylindrical copper rod insulated with polytetrafluoroethylene tape such that the area exposed to solution was 0.785 cm^2. A saturated calomel electrode (SCE) was used as the reference; a Pt plate electrode as the counter electrode. All potentials are reported vs SCE.

Prior to use the working electrode was etched for 30 s in 7M HNO_3 and then rinsed in redistilled water and ethanol [13]. The electrode was then immersed into the desired solution and the electrochemical experiment was initiated after 5 min. The supporting electrolyte used for all studies was non-deaerated 0.5M HCl. The investigated inhibitor was added in concentrations ranging from 1×10^{-4}M to 0.15M, and the optimum concentration for each inhibitor was identified. The optimum concentration was determined experimentally by identifying the concentration of inhibitor that maximised the inhibition efficiency (z%). Data are reported here for this optimum concentration and for the concentration 0.01M.

Measurements of a.c. impedance were conducted at 20°C using an ACM Instruments impedance analysis system (Lancashire, UK) and SpectroAnalyser software. An a.c. sinusoid of ±5 mV was applied at the corrosion potential; unless otherwise noted the frequency range was 100 kHz to 0.02 Hz. Instabilities in the impedance data were observed at lower frequencies. These instabilities were also observed when a PARC system was used to obtain analogue data. The a.c. impedance data were obtained at 0.01M and at the optimum concentration. Scanning electron microscopy (SEM) was conducted on the copper composites using a Phillips SEM 505 microscope. The composites were HNO_3-etched and then immersed in 0.5M HCl, with and without addition of the desired inhibitor (0.01M). The samples were immersed for 12 h, rinsed with millipore water and ethanol and sputtered (Hummer 6.2, Anatech, Ltd) with gold (~5 nm) to prevent charging in the microscope. X-ray microanalysis data were obtained using a Kenex Microanalyst 8000.

3. Results and Discussion

3.1. Tafel-Extrapolation Measurements

These measurements were made in the potential region ±200 mV from the corrosion potential (E_{corr}), at a sweep rate of 0.166 mVs^{-1}. The corrosion potential was obtained by measuring the open-circuit potential as a function of time. Typical Tafel plots are shown in Fig. 2. The corrosion current density j_{corr} was determined by extrapolating the Tafel lines. The inhibition efficiency (z) was calculated as follows:

$$z = \frac{100 \cdot (j_{corr,0} - j_{corr,1})}{j_{corr,0}} \tag{1}$$

where $j_{corr,0}$ and $j_{corr,1}$ are the corrosion current densities in the uninhibited and inhibited solutions, respectively.

Table 1 shows the corrosion parameters (E_{corr}, the anodic and cathodic Tafel slopes (b_a and b_c), j_{corr} and degree of protection, (z)) for the five inhibitors investigated at their optimum concentrations and at temperatures ranging from 20° to 45°C. It is interesting that we observe an optimal inhibitor concentration for all five inhibitors (Fig. 3). This has been observed by others [14]; although it is often the case that z simply increases monotonically with concentration. These data show that the

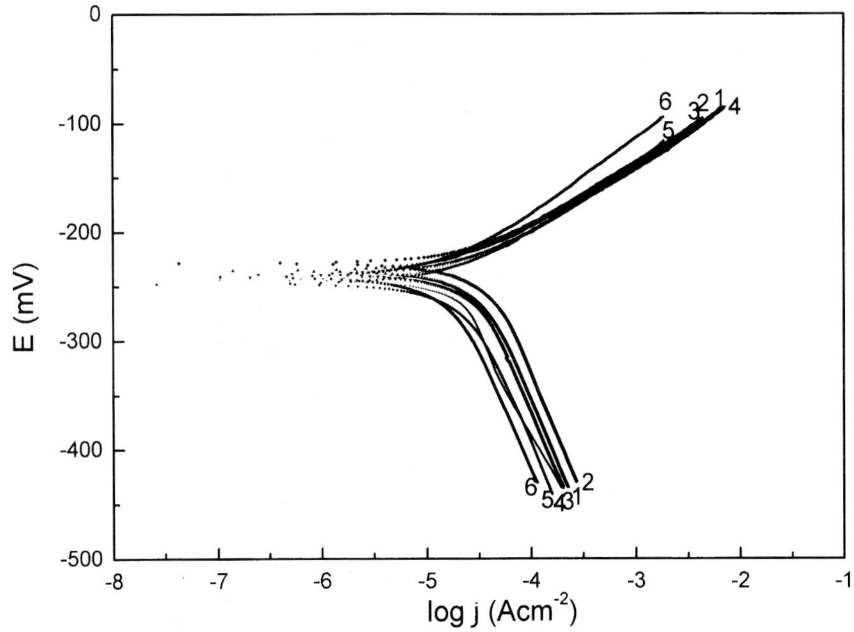

Fig. 2 Anodic and cathodic Tafel lines for copper in 0.5M HCl without (1) and with the addition of 0.01M imidazole (2), 4-methylimidazole (3), 4-methyl-5-hydroxylmethylimidazole (4), 4-methyl-1-phenylimidazole (5) and 1-(p-tolyl)-4-methylimidazole (6) at 30°C.

Table 1. *Corrosion parameters obtained via the Tafel-extrapolation method for copper in 0.5M HCl with and without the addition of the inhibitors at their optimum concentrations*

		Temperature (20°C)			
Solution	E_{corr} **(mV)**	$-b_c$ **(mV/decade)**	b_a **(mV/decade)**	j_{corr} **(μAcm^{-2})**	z **(%)**
0.5M HCl	–235.9	199.2	58.74	14.54	–
Inh 1	237.7	190.4	56.06	9.71	33.22
Inh 2	–248.8	190.8	58.05	10.66	26.69
Inh 3	–238.9	190.0	54.70	11.91	18.09
Inh 4	–256.9	191.4	67.75	8.92	38.65
Inh 5	–237.1	206.5	67.64	6.68	54.06
		Temperature (30°C)			
0.5M HCl	–234.9	220.2	62.80	28.80	–
Inh 1	–234.2	194.7	57.70	18.54	35.63
Inh 2	–243.0	202.1	61.50	18.67	35.17
Inh 3	–240.1	224.2	62.30	24.19	16.01
Inh 4	–246.6	207.3	65.14	18.42	36.04
Inh 5	-250.3	198.7	69.33	9.70	66.32
		Temperature (35°C)			
0.5M HCl	–236.5	219.3	60.30	31.84	–
Inh 1	–245.5	200.1	57.93	20.42	35.87
Inh 2	–244.0	187.0	56.30	22.37	29.74
Inh 3	–248.2	213.7	58.04	25.75	19.13
Inh 4	–251.3	200.6	61.89	20.34	36.12
Inh 5	–261.7	156.4	66.36	10.55	66.87
		Temperature (45°C)			
0.5M HCl	–225.6	240.0	66.20	43.79	–
Inh 1	–256.2	207.2	60.54	36.38	16.92
Inh 2	–253.6	201.3	60.00	37.00	15.51
Inh 3	–252.6	255.5	59.40	41.86	4.41
Inh 4	–258.2	216.0	65.36	31.39	28.32
Inh 5	–275.7	164.0	70.75	18.52	57.71

Inh 1: Imidazole; Inh 2: 4-methylimidazole, 0.0005M; Inh 3: 4-methyl-5-hydroxymethylimidazole, 0.1M; Inh 4: 1-phenyl-4-methylimidazole, 0.005M; Inh 5: 1-(*p*-tolyl)-4-methylimidazole, 0.1M.

(a)

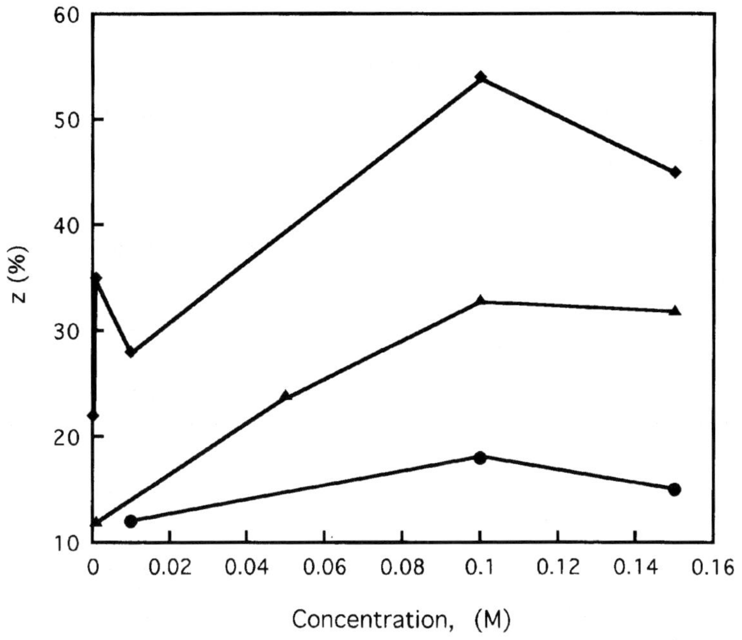

(b)

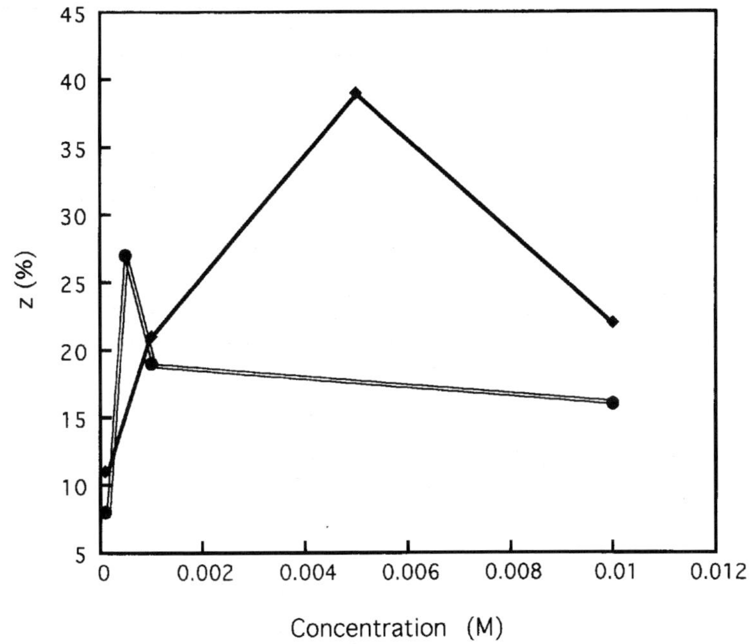

Fig. 3 (a & b) *Plots of inhibitor efficiency (z) vs concentration of inhibitor for (A) 4-methylimidazole (•) and 1-phenyl-4-methylimidazole (♦). (B) imidazole (▲), 4-methyl-5-hydroxylmethylimidazole (•), 1-(p-tolyl)-4-methylimidazole (♦).*

introduction of functional groups to the imidazole ring improves the inhibiting properties. The value of *z* increases in the following order: 4-methyl-5-hydroxymethylimidazole < 4-methylimidazole < imidazole < 1-phenyl-4-methylimidazole < 1-(*p*-tolyl)-4-methylimidazole. 1-(*p*-tolyl)-4-methylimidazole shows the best inhibiting efficiency at all temperatures investigated. The best protection by 1-(*p*-tolyl)-4-methylimidazole is in the range of temperature between 30 to 35°C. Figure 4 shows Tafel plots for copper in 0.5M hydrochloric acid to which was added 1-(*p*-tolyl)-4-methylimidazole at its optimum concentration (0.1M); data at four different temperatures are shown. In general, all inhibitors act as mixed inhibitors shifting E_{corr} to more negative values.

3.2. Linear Polarisation Measurements

These measurements were made over the potential region $E_{corr} \pm 15$ mV at a sweep rate of 0.166 mVs^{-1}. Typical linear polarisation curves are shown in Fig. 5. The polarisation resistance, R_p was obtained from the slopes of such polarisation curves. Figure 6 shows linear–polarisation curves (at four different temperatures) for copper in a solution 0.5M in HCl and 0.01M in 1-(*p*-tolyl)-4-methylimidazole at different temperatures. The variation of inhibitor efficiency (*z*) with molecular structure is identical to that observed from the Tafel-extrapolation method. Finally, as clearly shown by the lower slope in Fig. 5, these measurements confirm that 1-(*p*-tolyl)-4-methylimidazole is the best inhibitor.

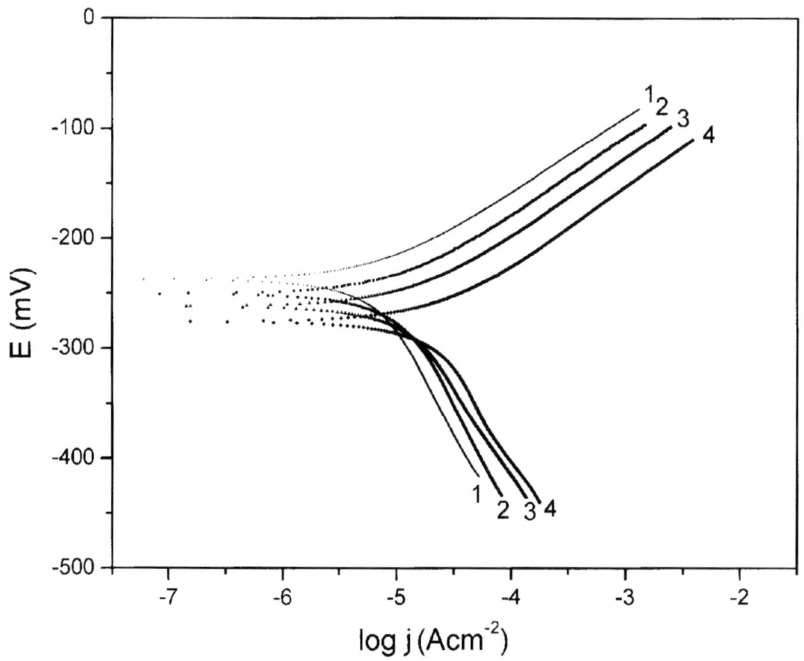

Fig. 4 *Anodic and cathodic Tafel lines for copper in 0.5M HCl with the addition of 0.1M 1-(p-tolyl)-4-methylimidazole at 20°C (1), 30°C (2), 35°C (3) and 45°C (4).*

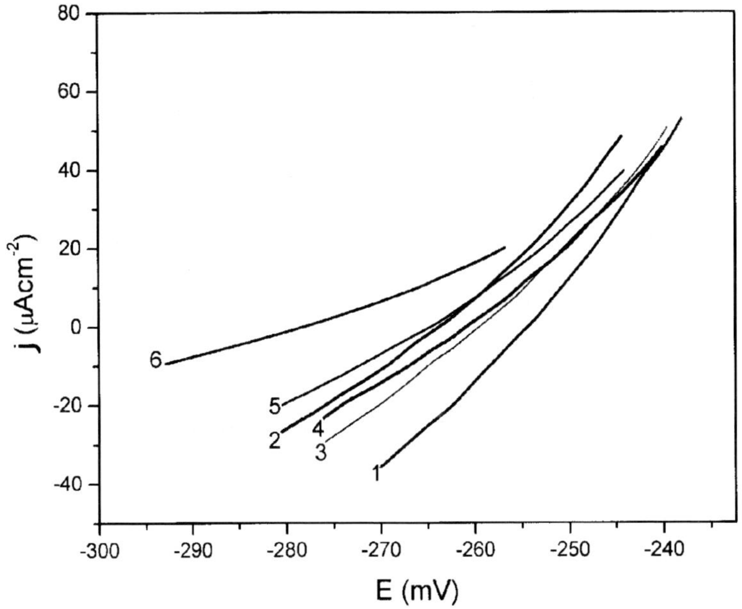

Fig. 5 *Linear-polarisation curves for copper in uninhibited 0.5*M *HCl (1) and with the addition of 0.1*M *imidazole (2), 0.0005*M *4-methylimidazole (3), 0.1*M *4-methyl-5-hydroxymethylimidazole (4), 0.005*M *4-methyl-1-phenylimidazole (5) and 0.1*M *1-(p-tolyl)-4-methylimidazole (6) at 45°C.*

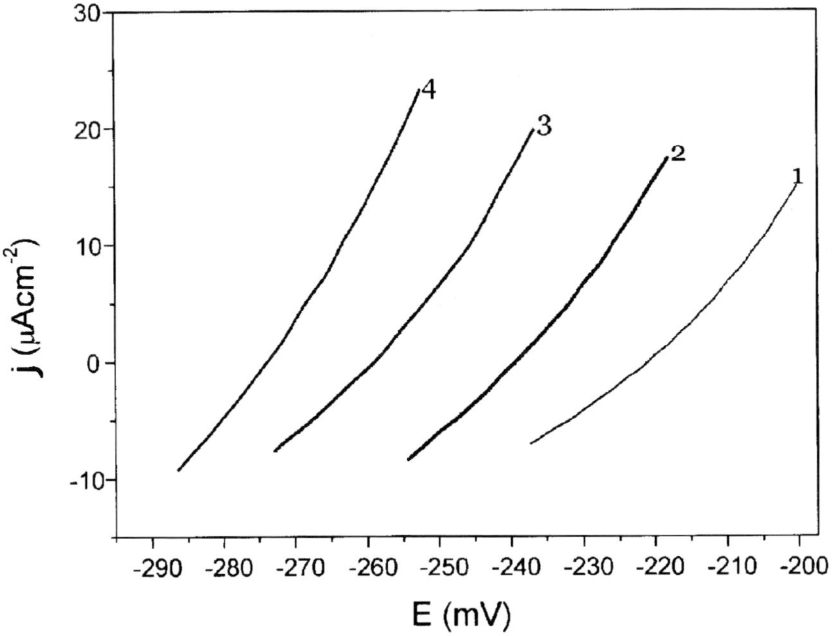

Fig. 6 *Linear-polarisation curves for copper in 0.5*M *HCl with the addition of 0.01*M *1-phenyl-4-methylimidazole at 20°C (1), 30°C (2), 35°C (3) and 45°C (4).*

3.3. Alternating Current–Impedance Measurements

Figure 7(a–d) shows Nyquist plots for all inhibitors at a concentration of 0.01M. The polarisation resistance is given by [16]

$$R_p = \lim_{\omega \to 0} \mathrm{Re}\left\{Z_f\right\}_{E=E_{corr}} \qquad (2)$$

where Re{Z_f} represents the real part of the complex faradaic impedance Z_f and ω coresponds to the angular velocity of the a.c. signal ($\omega = 2\pi f$, where f is frequency, Hz). The R_p values were obtained by fitting the experimental Nyquist data to a simple semicircle and extrapolating to $Z(im) = 0$. The effect of molecule structure on the inhibition efficiency is the same as was observed for the other two electrochemical methods. Hence, these three very different methods are telling a consistent story about the efficiency of these molecules in the inhibition of copper corrosion.

3.4. Scanning Electron Microscopy

Figure 8 (p.31) shows scanning electron microscopic images of a copper composite sample after immersion in 0.5M HCl without inhibitor for 12 h. Figure 9 (p.32) shows analogous images for a sample that had been immersed in a solution 0.5M in HCl and 0.01M in 1-(*p*-tolyl)-4 methylimidazole. Looking first at the low magnification images (Figs 8b and 9b) we see that the surface that had not been exposed to the inhibitor solution shows crystalline deposits; such deposits are not observed when the inhibitor solution is used. X-ray microanalysis was used to obtain qualitative information about the chemical make up of these crystals. Figure 10 (p.34) shows X-ray data obtained from one of the crystals from Fig. 8(b). A strong Cl line is observed at 2.640 keV. In addition, this crystals shows Cu lines at 0.960, 8.080 and 8.920 keV. Emission lines due to gold are also observed from the Au that was sputtered onto the samples.

Figure 11 (p.34) shows X-ray data obtained from the surface from Fig. 9(a); the Cl line is not observed. These data suggest that the crystals observed from the uninhibited surface are either CuCl or $CuCl_2$. However, quantification from the relative intensities of the Cu and Cl signals in Fig. 10 gave superstoichiometric amounts of Cu (i.e. more moles of Cu than moles of Cl). This is undoubtedly because the X-ray data are obtained from both the crystals and the underlying Cu surface.

The kinetics and mechanism of the anodic dissolution of copper in HCl solution have been studied by many authors [17–19]. At concentrations lower than 1M HCl the mechanism of dissolution is [20]

$$\mathrm{Cu} + \mathrm{Cl}^- \Leftrightarrow \mathrm{CuCl}_{ads} + \mathrm{e}^- \qquad (3)$$

$$\mathrm{CuCl}_{ads} + \mathrm{Cl}^- \Leftrightarrow \mathrm{CuCl}_2^- \qquad (4)$$

This mechanism would support the notion that these crystals on the surface of the uninhibited sample are CuCl. If this interpretation is correct, it would not be surprising that such crystals are not observed on the surface when inhibitor is added,

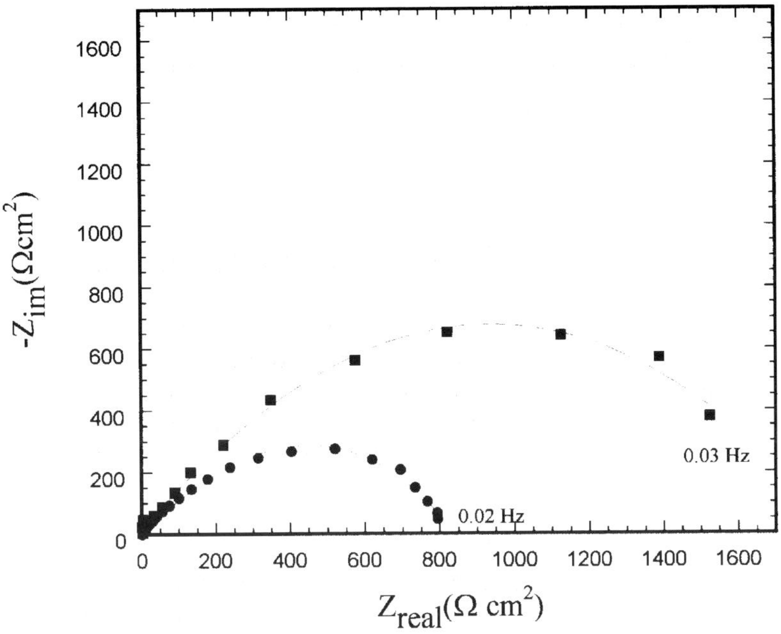

Fig. 7(a) *Nyquist plot for copper in 0.5M HCl without (•) and with the addition of 0.01M imidazole (■).*

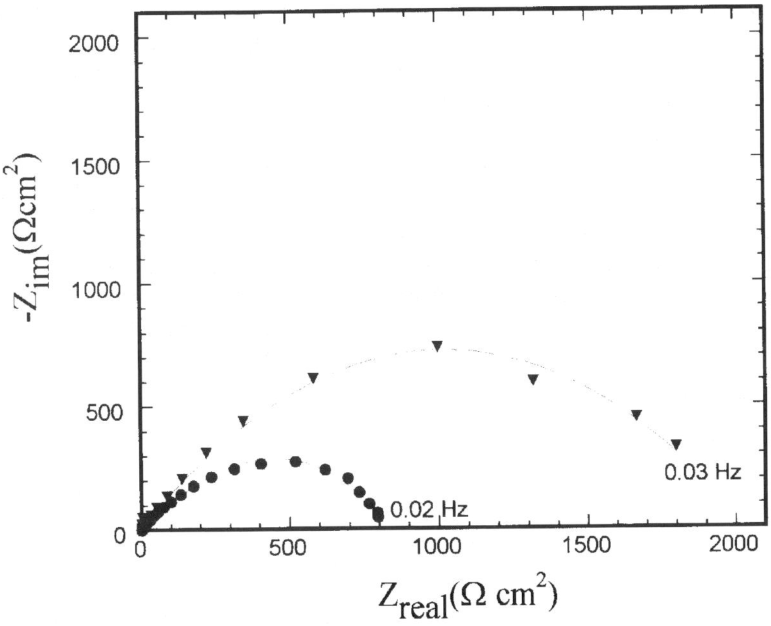

Fig. 7(b) *Nyquist plot for copper in 0.5M HCl without (•) and with the addition of 0.01M 4-methylimidazole (▾) at 20(C.*

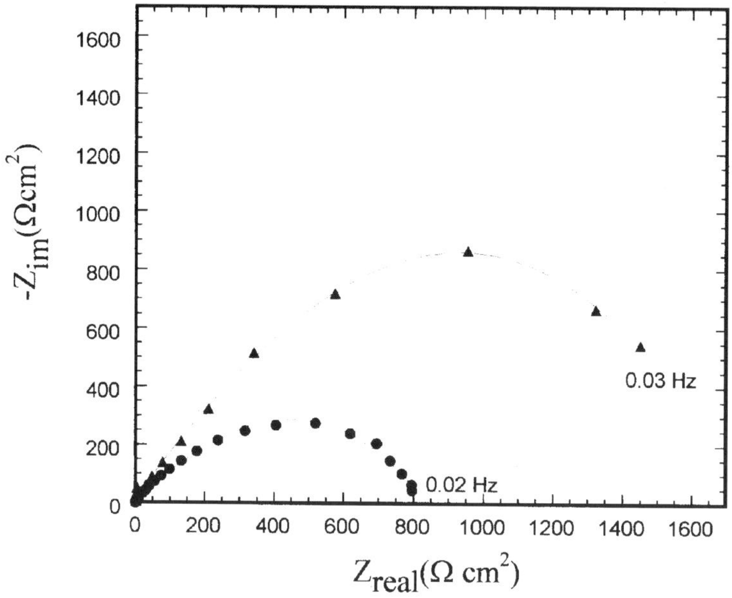

Fig. 7(c) *Nyquist plot for copper in 0.5M HCl without (●) and with the addition of 0.1M 4-methyl-5-hydroxymethylimidazole (▲) at 20°C.*

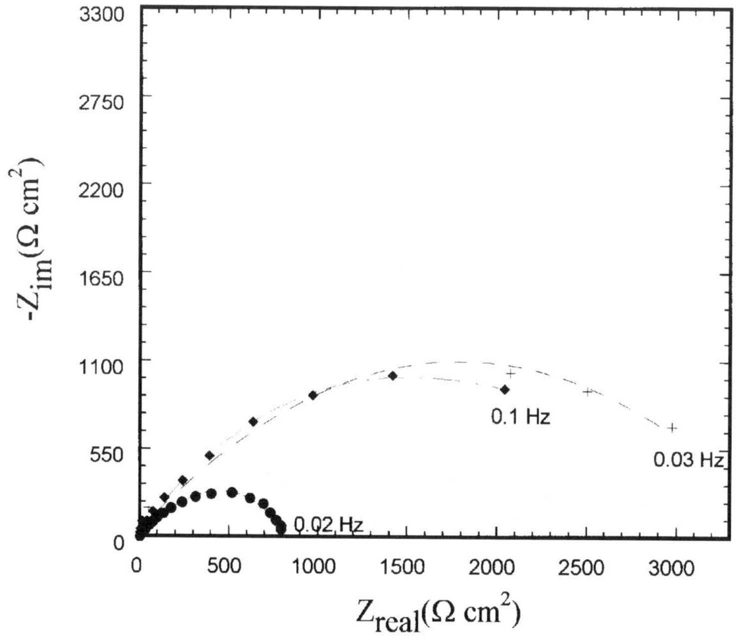

Fig. 7(d) *Nyquist plot for copper in 0.5M HCl without (●) and with the addition of 0.01M 1-phenyl-4-methylimidazole (♦) and 0.01M 1-(p-tolyl)-4-methylimidazole (+) at 20°C.*

Fig. 8 *Scanning electron microscopic images of the surface of a copper sample after immersion for 12 h in 0.5M HCl. Magnification 5000× (a) and 625× (b).*

Fig. 9 *Scanning electron microscopic images of the surface of a copper sample after immersion for 12 h in a solution 0.5M in HCl and 0.01M in 1-(p-tolyl)-4-methylimidazole. Magnification 5000× (a) and 625× (b). X-ray microanalysis shows that the small bright white spots are Au from the sputtered Au film.*

because the inhibitor lowers the rate of corrosion. While this is in agreement with all the electrochemical data, it is clear that chemical analysis of these crystals must be carried out before definite conclusions can be made.

It is also of interest to consider the high magnification images shown in Figs 8(a) and 9(a). The faceting seen in Fig. 9(a) is a result of the differences in HNO_3-acid etching rates for the different crystal faces of copper [21]. That this faceted surface is preserved after exposure to the HCl-plus-inhibitor solution indicates that the Cu surface has not been significantly corroded by the HCl. This may be contrasted to the Cu surface obtained after exposure to HCl solution that contained no inhibitor (Fig. 8a). Note that the faceting observed in Fig. 9(a) is gone and that the surface is much rougher than when inhibitor is present. These images clearly show that corrosion is strongly inhibited when the best inhibitor is present in the HCl solution.

The results obtained here clearly show that adding a phenyl or *p*-tolyl group to the imidazole ring improves the ability of the molecule to inhibit Cu corrosion in 0.5M HCl solution. It is of interest to consider the mechanism of this inhibitory effect. Three possible mechanisms can be proposed — inhibition based on chemisorption [3,7,12,22], inhibition based on the formation of a Cu(II)-based polymer film [21,23–27], and inhibition based on physisorption. We shall discuss each of these possibilities.

3.5. Chemisorption

Thierry and Leygraf have used surface-enhanced Raman spectroscopy to show that imidazole chemisorbs on Cu from 0.1M aqueous NaCl solution [22]. Chemisorption has been proposed as a possible mechanism of corrosion inhibition for steel in 1M H_2SO_4 solution [3]. However, it seems unlikely that chemisorption of imidazole and its derviatives would occur in acidic solutions such as 1M H_2SO_4 or the 0.5M HCl used here. This is because imidazole is a weak base and is present as the protonated, cationic form in such highly acidic solutions [11]. Protonation decreases the number of lone pairs available for chemisorption from two to one, and would clearly lower the basicity of the remaining lone pair. Finally, because the Cu atom has 10 d electrons, chemisorption (as opposed to physisorption, see below) involving simple donation of electron density from the imidazole to an empty d orbital on the metal seems impossible [28]. The only possibility would seem to be oxidative chemistry to form Cu(I) or Cu(II), followed by formation of a coordination complex between the inhibitor and the Cu ion. This possibility is discussed in the next section.

3.6. Cu(II)-Based Films

It is well known that metallic Cu reacts with imidazole and its derivatives to form a polymeric film based on Cu(II) and the imidazolato ligand [23]. However, such films are typically prepared by applying imidazole to the metal surface and allowing the film-formation reaction to occur in air. The mechanism of film formation entails loss of the proton from the pyrrole nitrogen of the imidazole. Loss of this proton seems problematic in strongly acidic solution. Indeed, when such complexes are prepared from Cu^{2+} and imidazole, the solution has to be strongly basic [23,24]. In addition, in our best inhibitors (1-phenyl-4-methylimidazole and 1-(*p*-tolyl)-4-methylimidazole) the pyrrole proton is replaced by the phenyl or *p*-tolyl group. Therefore, if film

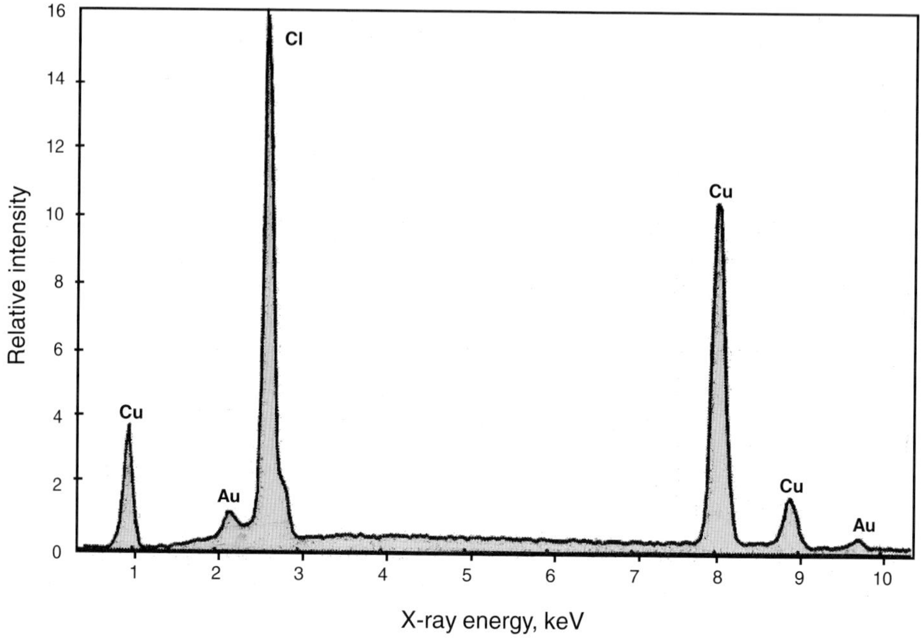

Fig. 10 X-Ray microanalysis of a copper sample exposed to 0.5M HCl for 12 h.

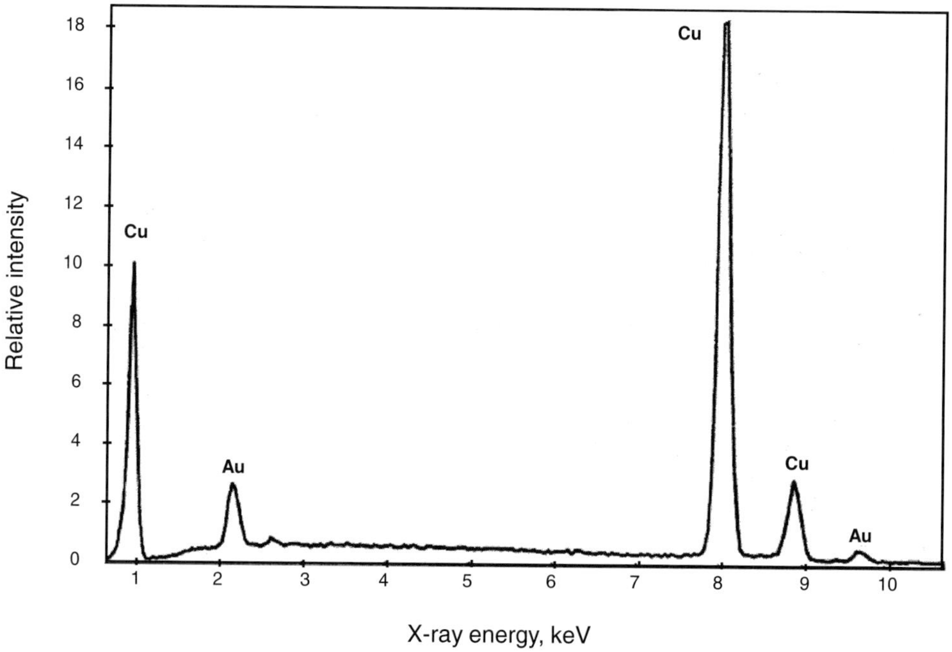

Fig. 11 X-Ray microanalysis of a copper sample exposed to 0.5M HCl plus 0.01M 1-(p-tolyl)-4-methylimidazole for 12 h.

formation does occur, the mechanism and stoichiometry would have to be very different from that for films based on imidazole. Finally, the first step in the film formation process on Cu metal is chemisorption which is critical to the process because this increases the acidity of the pyrrole hydrogen [23,27].

3.7. Physisorption

Very early in corrosion research it was discovered that the extent of inhibition by the members of a homologous series of inhibitors increased with the molecular weight of the molecules in the series [29]. Hackerman showed, in studies with steel electrodes, that this increase is caused by an increase in physisorption of the higher molecular weight members of the homologous series of inhibitor molecules [29]. The results obtained here, which show that the higher molecular weight members of the imidazole series show better corrosion inhibition, are in agreement with this physisorption model for corrosion inhibition. Surface analytical data would be particularly helpful to explore the mechanism for these inhibitors.

4. Conclusions

The influence of methyl, hydroxymethyl, phenyl and tolyl substituents in the imidazole molecule used as an inhibitor of the corrosion of copper in 0.5M HCl was studied.

- The value of the inhibition efficiency *z* increases in the following order: hydroxymethyl < methyl < imidazole < phenyl < tolyl at optimum concentrations at temperatures ranging from 20 to 45°C.

- a.c. impedance results at 20°C are in good qualitative agreement with those obtained by potentiodynamic measurements for inhibition of copper corrosion in 0.5M HCl. These investigations have also shown that 1-(*p*-tolyl)-4-methylimidazole has the best inhibiting efficiency.

- Images obtained with SEM show the formation of crystalline deposits on the copper surface in 0.5M HCl and greater surface roughness. With the addition of 0.01M 1-(*p*-tolyl)-4-methylimidazole the surface roughness is considerably lower and facets are observed.

- Results obtained by X-ray microanalysis show that CuCl crystals are not formed with the addition of the best inhibitor to the 0.5M HCl solution.

5. Acknowledgements

The financial support was obtained from the Ministry of Science and Technology of the Republic of Croatia under Project 125012. Charles R. Martin acknowledges support from the US office of Naval Research.

References

1. Y. I. Kuznetsov, *Organic Inhibitors of Corrosion of Metals*. Plenum Press, New York, NY, USA, 1996.
2. E. Stupnišek-Lisac and Z. Ademovic, *Proc. 8th SEIC*, Ann. Univ. Ferrara, N. S., Sez. V, 1995. Suppl. No 10, 257–265, Publ. University of Ferrara, Ferrara, Italy, 1995.
3. M. H. Wahdan and G. K. Gomma, *Mater. Chem. & Phys.*, 1997, **47**, 176–183.
4. E. Stupnišek-Lisac, N. Galic and R. Gašparac, Meeting Abst. of the Joint Intnl Meeting of The Electrochemical Society and The International Society of Electrochemistry, 97–2, 673, Paris, France, 1997.
5. N. K. Patel, J. Franco and I. S. Patel, *J. Indian Chem. Soc.*, 1997, **54**, 815–816.
6. S. N. Raicheva, B. V. Aleksiev and E. I. Sokolova, *Corros. Sci.*, 1993, **34**, 343–350.
7. E. Stupnišek-Lisac, D. Kasunic and J. Vorkapic-Furac, *Corrosion*, 1995, **51**, 767–771.
8. G. Lewis, *Corrosion*, 1982, *38*, 60–62.
9. J. Jang and H. Ishida, *Corros. Sci.*, 1992, **33**, 1053–1066.
10. S. Yoshida, H. Ishida, *Appl. Surf. Sci.*, 1995, **89**, 39–47.
11. K. Hofmann, *Imidazole and its Derivatives*. Interscience Publishers, Inc., New York, NY, USA, 1953.
12. E. Stupnišek-Lisac, A. Loncaric-Božic and I. Cafuk, *Corrosion*, 1998, **54**, 713–720.
13. Y. Feng, W. K. Teo, K. S. Siow, Z. Gao, K. L. Tau and A. K. Hsieh, *J. Electrochem. Soc.*, 1997, **144**, 55–59.
14. M. Metikoš-Hukovic, R. Babic, Z. Grubač and S. Brinić, *J. Appl. Electrochem.*, 1996, **26**, 443–448.
15. M. Stearn and A. L. Geary, *J. Electrochem. Soc.*, 1957, **104**, 56–60.
16. D. J. Lorenz and F. Mansfeld, *Corros. Sci.*, 1981, **21**, 647–672.
17. O. E. Barcia, O. R. Mattos, N. Pebere and B. Tribollet, *J. Electrochem. Soc.*, 1993, **140**, 2825–2832.
18. J.-P. Diard, J.-M. Le Canut, B. Le Gorrec and C. Montella, *Electrochim. Acta*, 1998, **43**, 2469–2483.
19. J.-P. Diard, J.-M. Le Canut, B. Le Gorrec and C. Montella, *Electrochim. Acta*, 1998, **43**, 2485–2500.
20. F. K. Crundwell, *Electrochim. Acta*, 1992, **37**, 2707–2714.
21. M. L. Lewis, G. Xue and K. Carron, *Polymer Mater. Sci. Eng.*, 1991, **64**, 199–200.
22. D. Thierry and C. Leygraf, *J. Electrochem. Soc.*, 1986, **133**, 2236–2239.
23. G. Xue, S. Jiang, X. Huang and G. Shi, *J. Chem. Soc. Dalton Trans.*, 1998, 1487–1488.
24. S. Yoshida and H. Ishida, *Appl. Surf. Sci.*, 1985, **20**, 497–511.
25. S. Yoshida and H. Ishida, *J. Chem. Phys.*, 1983, **78**, 6960–6969.
26. S. Yoshida and H. Ishida, *Appl. Surf. Sci.*, 1995, **89**, 39–47.
27. K. T. Carron, G. Xue and M. L. Lewis, *Langmuir*, 1991, **7**, 2–4.
28. J. J. Bordeaux and N. Hackerman, *J. Phys. Chem.*, 1957, **61**, 1323–1327.
29. E. L. Cook and N. Hackerman, *J. Phys. Chem.*, 1951, **55**, 549–557.

4
In situ Surface Raman Spectroscopic Studies on Benzotriazole and Thiourea as Corrosion Inhibitors for Bare Iron Surfaces

J. L. YAO, H. C. LIU, P. G. CAO*, B. REN, B. W. MAO, R. A. GU* and Z. Q. TIAN†

State Key Laboratory for Physical Chemistry of Solid Surfaces, Institute of Physical Chemistry and Department of Chemistry, Xiamen University, Xiamen 361005, China

* Department of Chemistry, Suzhou University, Suzhou 216006, China

†Author for correspondence

ABSTRACT

Surface enhanced Raman spectra from benzotriazole (BTA) and thiourea (TU) adsorbed at roughened bare Fe electrodes were obtained for the first time in electrochemical environments. A set of special surface pretreatment procedures for Fe has been developed to enable high-quality surface Raman spectra to be obtained. The potential-dependent Raman spectra reveal that the inhibition process of BTA involves the formation of surface complexes and a compact polymer film in two different potential regions. It has been shown that TU can strongly interact with an Fe surface via its S atom and form only an adsorbed layer. Higher inhibition efficiencies for BTA than for TU for iron corrosion are mainly attributed to the difference in the mechanisms for the film or adlayer formation on the Fe surface.

1. Introduction

Iron is probably one of the most important and widely used materials, and plays a central role as basic material for the metallurgical process industries. However, the corrosion and passivation behaviour are quite general and important phenomena in the chemical engineering industry and our daily life. Therefore, there is great economic incentive in developing methods and materials to alleviate corrosion [1], which can come only from a good understanding of the mechanisms and processes involved in this complex phenomenon.

A wide variety of electrochemical and spectroscopic techniques have been employed in investigating corrosion and inhibition processes on Fe surfaces. However, conventional electrochemical techniques can only provide macroscopic information on the metal/solution interface. On the other hand, spectroscopic techniques are capable of investigating the interface microscopically at the molecular (atomic) level although few of these have the required sensitivity to investigate directly the interaction of inhibitors with a metal surface. There are some *ex situ* spectroscopic investigations that have been reported, including infrared [2] and X-ray photoelectron spectroscopy (XPS) [3,4], for probing the corrosion mechanism.

The XPS measurement has to be conducted in an ultra high vacuum (UHV) environment, which will inevitably lead to some changes in the surface states of the metal electrodes. Recently, there have been significant developments in vibrational spectroscopies which have some advantages over the conventional techniques in monitoring inhibition mechanisms at the molecular level. Among these techniques, surface enhanced Raman spectroscopy (SERS) has extremely high surface sensitivity which makes it feasible to investigate *in situ* the monolayer adsorption on metal electrodes [5–7]. Unfortunately, in the past two decades, the application of this technique has been restricted almost entirely to Ag, Au and Cu, on which the most pronounced SERS effect exists [8–11]. This has impeded the wider application of Raman spectroscopy in surface science and electrochemistry of Fe surfaces. Corrosion specialists have never been able to study the inhibition processes by *in situ* Raman spectroscopy directly on a bare Fe electrode.

All previous Raman studies on the corrosion, inhibition and passivation of Fe have been carried out through one of the following three approaches. The first involves coating a SERS active Ag or Au electrode with a very thin overlayer of Fe by electrochemical deposition [12–14]. However, it is extremely difficult to eliminate entirely the possibility that the adsorbate is bound to the exposed rough substrate rather than to the overlayer sites because the ultra thin film is only one to five atomic monolayers in thickness. For the second approach, SERS active Ag micro islands were deposited on Fe substrates, but this is appropriate only for studying the passive film formed at the Fe substrate [15–17]. Recently our laboratory reported a third approach of electrochemically depositing a thick Fe film on a non-SERS active substrate such as glassy carbon [18]. It should be pointed out that the crystalline structure and properties of these deposited Fe materials may not be exactly the same as that of the Fe bulk phase formed by metallurgy. Therefore, it is highly desirable to extend SERS studies directly to the processes of corrosion and inhibition on bulk Fe. To the best of our knowledge, there are no reports of surface Raman spectroscopic studies of electrosorption and corrosion on bare bulk Fe electrodes.

The most recent progress in surface Raman spectroscopy, especially the very high sensitivity it provides, allows one to gain vibrational information reflecting the adsorbate and substrate interaction from bare transition metal surfaces [18–24]. From the viewpoint of the special interest in corrosion and inhibition of Fe, it is well known that benzotriazole (BTA) and thiourea (TU) are effective inhibitors for Fe corrosion. Although the remarkable efficiency of BTA and TU as corrosion inhibitors for Cu and its alloys have been well-documented [25–27], little information has been obtained for Fe surfaces. In this study, with the appropriate roughening procedure developed for Fe surfaces, we are able to obtain the surface Raman spectra of BTA and TU adsorbed on bare Fe electrodes and to investigate their inhibition mechanisms.

2. Experimental

Raman spectra were acquired on a confocal microprobe Raman system (LabRam I from Dilor). The exciting wavelength was 632.8 nm from an internal He–Ne laser. The detailed description of the Raman system has been given elsewhere [19–23]. The working electrode was a polycrystalline Fe rod embedded in a Teflon sheath with a geometric surface area of 0.1 cm^2. Before the roughening pretreatment, the Fe electrode was first mechanically polished successively with 0.3 and 0.05 μm alumina powder to a mirror finish followed by ultrasonic cleaning with Milli-Q water. All the potentials were quoted versus saturated calomel electrode (SCE). All the chemicals used were of analytical reagent and the solutions were prepared using Milli-Q water.

3. Results and Discussion

As is well known from previous SERS studies on Au, Ag and Cu, a necessary, but not sufficient, requirement for the presence of significant surface enhancement is some form of surface roughness. A proper surface roughness can at least raise the surface Raman signal by increasing the number of adsorptive sites. We have examined various surface-roughening procedures for Fe electrodes so as to obtain surface Raman spectra with better signal-to-noise ratio (S/N). Four types of Fe surfaces have been studied: (i) mechanically polished surfaces, (ii) chemically etched surfaces, (iii) *ex situ* oxidation–reduction cycle (ORC) electrochemically roughened surfaces and (iv) *in situ* ORC roughened surfaces. On comparing the spectra of adsorbed pyridine (a model molecule in SERS studies) from these surfaces, two interesting differences arise. The first difference is the band frequencies on smooth and roughened Fe surfaces. In the former case, the frequency of the ring stretching vibration (ν_1) is very close to that of pyridine in aqueous solution. Therefore, the Raman signal detected comes, in fact, mostly from the bulk solution but it may also infer that the surface Raman signal of pyridine from the smooth surface is too weak to be detected. This assumption is supported by the large shift of the vibrational frequency of ν_1 mode observed from the roughened Fe surfaces. The second difference is the band intensities. Raman signals from the mechanically polished Fe surface were extremely weak whilst those from roughened surfaces could be clearly observed. It should be noted that the surface Raman intensities are not simply proportional to the surface roughness factor. The intensity increased significantly if the smooth Fe electrode was chemically etched in H_2SO_4 solution, although the surface roughness factor only increased by a factor of about 3. The intensity of the strongest band (ν_1) of pyridine doubled if the electrode was further treated in the same solution by an ORC procedure, which is called *ex situ* ORC. The band intensity increased dramatically when the additional ORC was performed *in situ* in the spectroelectrochemical cell containing 0.01M pyridine and 0.1M KCl prior to the measurement. The surface roughness factor remained at about 3. This illustrates that the proper surface roughening procedure is vitally important to obtain reasonably good spectra for detailed investigation. Using a unique method for calculation of the surface enhancement factor (SEF) based on the confocal Raman system developed in our lab [23], we were able to estimate the SEF for the Fe surface, which amounted to 10^2 to 10^3 for pyridine adsorbed on roughened Fe surfaces, depending on the surface roughening procedures [24]. It is evident from the above result that surface Raman spectroscopy can be employed for obtaining detailed information of the solid/liquid interface and this stimulated us to perform *in situ* studies of the corrosion and inhibition processes on bare Fe electrode surfaces.

Figure 1 compares a normal solution Raman spectrum of BTA with SERS spectra of BTA adsorbed on ORC roughened Cu and Fe electrodes respectively. Although the SERS intensity of Fe is about one order smaller than that of Cu, the spectral quality is good enough for detailed analysis. After examining carefully the solution and surface Raman spectra, it was found that three weak bands, at 1100, 1126, and 1148 cm^{-1} respectively assigned to NH in-plane deformation of BTA in the solution spectrum, decrease in intensity or even disappear after BTA adsorbs onto the metal surfaces. Moreover, the 1219 cm^{-1} band attributed to the triazole ring-breathing mode red-shifts to 1191 cm^{-1} and 1195 cm^{-1} for Cu and Fe, respectively. It was also found that the spectral features from both electrodes were almost identical. The high similarity of the Raman spectra of BTA on Cu and Fe surfaces could provide evidence that BTA may coordinate with Cu and Fe surfaces in a similar configuration. Rubim *et al.* proposed that the protective film on Cu in the presence of BTA is copper(I)BTA, i.e. in which BTA could coordinate with Cu through two of the triazole nitrogen atoms to form a compact resistant film [28]. Accordingly, it is reasonable to assume that BTA could also coordinate with the Fe electrode through two of the triazole nitrogen atoms to form a compact resistant film. In the case of the Cu system in a different potential region [29], the interaction of BTA with Cu involves a de-protonation process before interacting with the surface. Thus, potential dependent SERS studies with BTA on Fe were performed and are discussed in detail in the following section.

Figure 2 shows the cyclic voltammogram of the roughened Fe in 0.1M KCl without and with 0.01M BTA. A rapid increase in anodic current occurred at about –0.58 V without the BTA. After the addition of BTA, the initial potential of the anodic peak shifted in the positive direction and the oxidation current decreased significantly, indicating that the BTA can inhibit the oxidation of the Fe. Therefore, it is necessary to use *in situ* surface Raman spectroscopy to investigate this process at the molecular level.

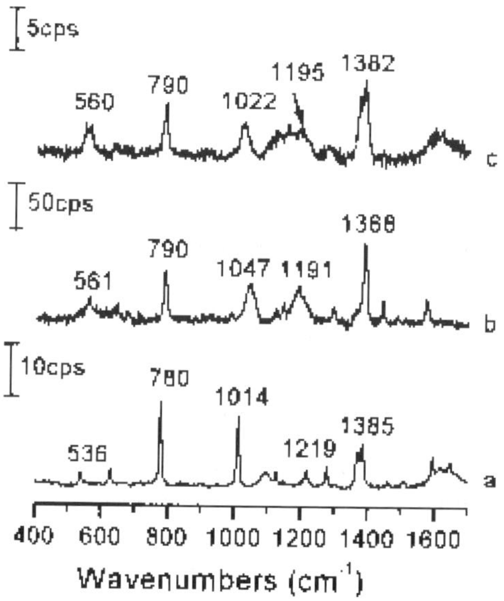

Fig. 1 Raman spectrum of a BTA solution (a), SERS spectra of BTA adsorbed on Cu electrode at –0.5 V (b) and Fe electrode at –0.6 V (c).

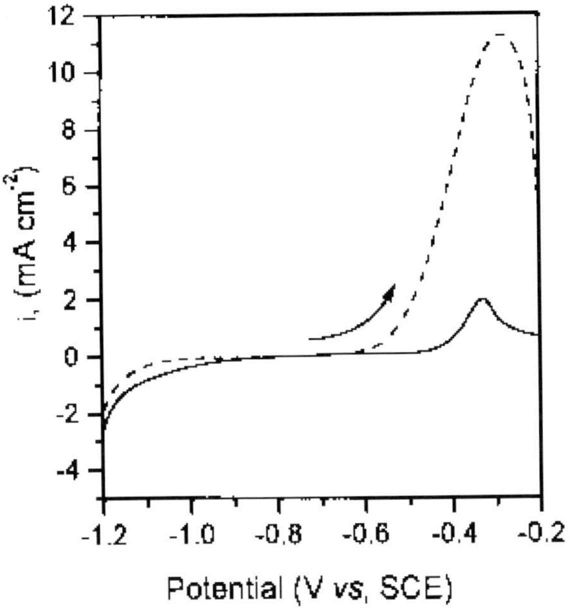

*Fig. 2 Cyclic voltammograms of a roughened Fe electrode in 0.1*M *KCl solution without (- - -) and with 0.01*M *BTA (—). The scan rate was 20 mVs^{-1}.*

Figure 3 presents a set of the surface Raman spectra of BTA adsorbed on the roughened Fe surface at different potentials from –0.3 V to –1.2 V in a BTA solution containing Cl^- ion. The Fe electrode was immersed into the solution at a negative potential of

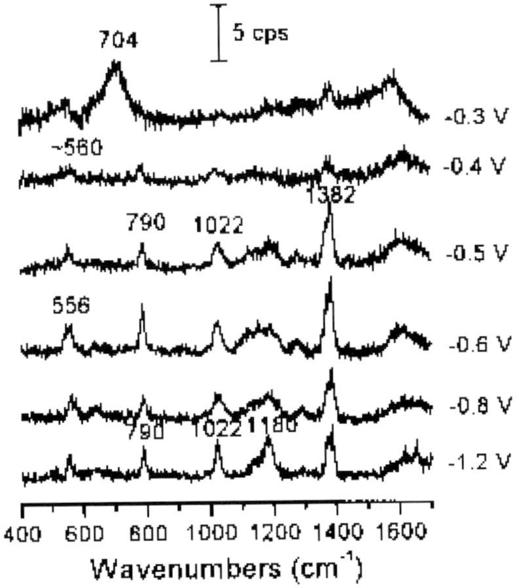

*Fig. 3 Potential-dependent surface Raman spectra of BTA adsorbed at Fe electrodes. The solution was 0.01*M *BTA + 0.1*M *KCl.*

–1.2 V in order to investigate the initial stage of inhibition prior to the formation of the protective inhibition film. The potential-dependent spectra were recorded stepwise from the negative limit to more positive potentials. It can be seen that the SERS intensities change significantly with the applied potentials. This could serve as evidence that the signal is exactly from the Fe surface rather than from the bulk solution. In contrast to the Raman spectrum of BTA in aqueous solution, the relative intensities and frequencies of those bands corresponding to the triazole ring and the N-H vibrational modes change distinctively. Therefore, it is reasonable to assume that the BTA molecule very probably interacts with the Fe surface through its triazole ring. As mentioned above, Rubim *et al.* reported that the BTA molecule adsorbs at Cu surfaces and forms a surface complex as $[Cu(I)Cl(BTA)]_n$ [28]. In the present study, the spectral feature of BTA on an Fe surface is similar to that obtained on Cu surfaces in the more negative potential region. Accordingly, the surface complex $[Fe_n(Cl)_p(BTA)]_m$ could be formed by participation of chloride ion and that the adsorption may occur by coordination through the lone pair electrons on a nitrogen atom in the triazole ring with Fe. When the potential was moved to more positive values until the current polarisation changed from cathodic to anodic, the frequencies and shapes of bands corresponding to the N-H vibrational modes noticeably changed, thus, indicating that Fe could very easily be oxidised in this region. Therefore, the band intensities of the adsorbed BTA decreased dramatically with the positive shift of the potential. In the anodic polarisation region (about –0.4 V), a broad band centred at ~ 560 cm^{-1} was observed which differs from the narrow band of BTA located at 556 cm^{-1}. The band feature of this is very like a surface oxide and in this case, could be attributed to FeOOH forming in the oxidation process taking place in this potential region. Due to the dissolution and oxidation of Fe in this potential region, the Raman signal of BTA disappeared and a band at about 704 cm^{-1} appeared, which could be unambiguously assigned to the vibrational mode of $Fe(OH)_3$.

After overviewing all the Raman features of BTA adsorbed at the Fe surface, a clearer picture of the inhibition processes of BTA on the Fe surface is obtained. At negative potentials, the surface complexes formed were not sufficiently stable for efficient inhibition. On the other hand, after the electrode had been held at the open circuit potential in the solution in the presence of BTA and KCl for several hours, a compact and stable layer was formed at the surface which inhibited the corrosion effectively. It is of interest that the potential for the appearance of Fe dissolution and formation of Fe oxides was moved about 200 mV positively, and moreover, that the process for the formation of the inhibition layer depends on the immersion time. The Raman signal of BTA from the Fe surface could hardly be detected in the first 10 min. But when the immersion time was prolonged to about 30 min, the BTA signal becomes discernible. It is evident that the intensities of the major bands of BTA increase with the length of the immersion time. When the electrode was immersed in the solution for about 2 h, the Raman signal intensity of the polymer layer remained unchanged with time.

Figure 4 shows a set of potential-dependent surface Raman spectra of the BTA film formed on Fe by immersing the electrode in the BTA solution for about two hours. Some spectral features should be noted in Figs 3 and 4. No significant changes in frequencies could be observed for those bands related to benzene ring vibration, for instance, the frequency of benzene ring stretching mode (~ 790 cm^{-1}) remained

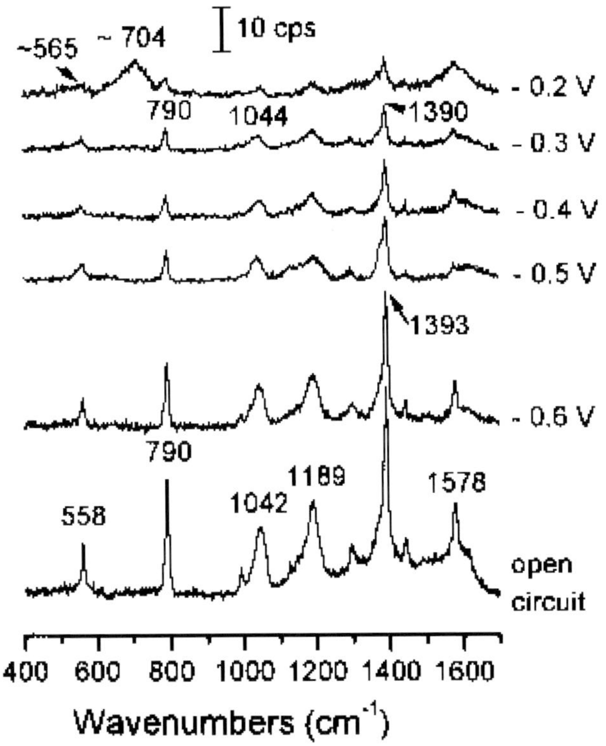

Fig. 4 *Potential-dependent surface Raman spectra of BTA adsorbed at Fe electrodes coated with a BTA polymer film. The solution was 0.01*M *BTA + 0.1*M *KCl.*

almost unchanged. In contrast, the 1022 cm^{-1} band assigned to the triazole ring of BTA in solution (see Fig. 3) shifts up to 1042 cm^{-1} for the BTA film (Fig. 4). This significant frequency shift for the triazole ring mode could be due to the de-protonation of BTA during the immersion. So the interaction of BTA^- ion with the Fe surface is stronger than that of Cl^- ion. Therefore, BTA^- ion can replace the Cl^- already existing in the surface complex $[Fe_n(Cl)_p(BTA)]_m$. It is known that BTA could interact with Cu(I) to form a compact polymer film $[Cu(BTA)]_n$ in a relatively positive potential region [29]. On grounds of the similar spectral features of BTA adsorbed at the Cu and Fe surfaces, it is reasonable to propose that the BTA is de-protonated and coordinated with Fe ion. The composition of the surface complex polymer might be $[Fe_n(BTA)_p]_m$. The formation of this surface polymer film may suppress the dissolution and oxidation of Fe effectively. In addition, the adsorption behaviour of BTA has been investigated in acid solution in order to examine the influence of the pH on the inhibition efficiency. It was found that the adsorption configuration changes with the pH. With decrease of the pH, the intensity and frequency of some vibrational modes change noticeably. For example, the intensities of the triazole ring breathing vibration mode decreased rapidly and disappeared at about pH 2.0. In contrast, the intensities of NH in-plane deformation modes increased significantly in acid solution. As the BTA molecule can interact with H^+ and be protonated easily, the Raman spectral features are found to be similar to that of the normal solution spectra. The implication is that the Raman signal is from the neutral BTA molecules or the protonated BTA cation rather than from the polymer film formed in the neutral solution. Therefore, it is rather difficult to form the surface complex or polymer film between the BTA molecule and the Fe surface in the solution with pH below 2.0. In consequence, the inhibition efficiency of BTA is low in acid solution and reduces with the pH value. The detailed discussion on this aspect will be given elsewhere [30].

In addition to the inhibition by BTA, the adsorption of, and inhibition by thiourea (TU) have been investigated at the ORC roughened Fe electrode surface, since TU has long been known as an inhibitor and widely used as an additive in electroplating and as a catalyst for some electrode reactions [31]. Although the interaction between TU and Fe has been investigated extensively by conventional electrochemical techniques [32,33], as mentioned in the introduction, the detailed mechanisms involved remain unclear. Therefore, it might be helpful to get a deeper insight into the mechanisms of the inhibition processes using Raman spectroscopy. The surface Raman measurements were carried out from the negative extremity of the potential range studied towards more positive potentials so as to avoid the oxidation of TU and the formation of Fe oxides. In contrast to the case of the Raman spectrum from the bulk solution of TU in which the stretching vibrational mode of C=S was about 734 cm^{-1}, the frequency of the CS vibration for the adsorbed TU down-shifts about 25 cm^{-1}. This significant change cannot be the result of the protonation of the S atom in acid solution since the SH stretching band was not observed throughout the whole potential range studied. Alternatively, the large shift in frequency indicates that the TU interacts with the Fe surface through its S atom. Furthermore, it was found that with change in potential all the intensities of Raman bands changed significantly, while the frequencies changed only slightly. At potentials positive to –0.6 V, the surface Raman signals almost disappeared, which was similar to the case of BTA adsorbed at Fe surfaces. Even after the Fe electrode was immersed in the TU solution for several hours, the SERS spectra show that a TU polymer film did not form on the Fe surface in contrast to that of BTA. It may be due to the configuration of the complex of TU and Fe ion not being sufficiently suitable to grow a compact film. This difference between BTA and TU gives an indication that the inhibition efficiency of BTA is higher than that of TU. Previous studies in our laboratory have revealed that the adsorption behaviour of TU is more like a surface-active anion rather than a neutral organic molecule [34]. Therefore, we may expect the surface adsorption process of TU to be promoted when the electrode potential becomes more positive than the potential of zero charge (PZC). The maximum intensity observed at –0.6 V seems to indicate that the variation of the surface coverage of adsorbed TU could explain the surface Raman intensity change with the potential in the present system. Consequently, the adsorbed layer, which suppresses the corrosion of Fe was formed at the relatively positively charged surface. This information may be helpful to explain, at the molecular level, the reasons why TU can be used as corrosion inhibitor.

4. Conclusions

Proper surface roughening procedures for Fe electrodes assisted by a highly sensitive confocal microprobe Raman system have enabled us for the first time to obtain high-quality surface Raman spectra of pyridine, BTA and TU adsorbed at bare roughened Fe electrodes over a wide applied potential range. The SERS spectra indicate that the complex polymer film of BTA and the adsorbed layer of TU play an important role in the inhibition processes of Fe. Higher inhibition efficiencies of BTA than TU for Fe corrosion are mainly attributed to the difference in the mechanisms for the film formation on the surfaces. This preliminary result may provide a good reason for optimism that confocal Raman microscopy will be an advantageous tool for *in situ* investigation of inhibition processes and mechanisms of the corrosion of Fe-group or Fe-alloy surfaces.

5. Acknowledgements

This work was supported by the Natural Science Foundation of China and the Ministry of Education of China.

References

1. C. A. Melendres, N. Camillone and T. Tipton, *Electrochim. Acta*, 1989, **34**, 281–286.
2. J. O'M. Bockris, M. A. Hobib and J. L. Carbajal, *J. Electrochem. Soc.*, 1984, **131**, 3032–3033.
3. D. Chadwick and T. Hashemi, *Corros. Sci.*, 1978, **18**, 39–51.
4. Y. Yamamoto, H. Nishihara and K. Aramaki, *Corrosion*, 1992, **48**, 641–648 .
5. L. W. H. Leung and M. J. Weaver, *J. Electroanal. Chem.*, 1987, **217**, 367–384.
6. M. Fleischmann, Z. Q. Tian and L. J. Li, *J. Electroanal. Chem.*, 1987, **217**, 397–410.
7. Y. Zhang, X. Gao and M. J. Weaver, *J. Phys. Chem.*, 1993, **97**, 8656–8663.
8. B. Pettinger and H. Wetzel, Organic and Inorganic Species at Ag, Cu, and Au electrodes, in *Surface Enhanced Raman Scattering*, pp.293–314 (R. K. Chang and T. E. Furtak, Eds). Plenum Press, New York, USA, 1982.
9. R. K. Chang, Raman Spectroscopic Techniques in Interfacial Electrochemistry, in *Spectroscopic and Diffraction Techniques in Interfacial Electrochemistry*, pp.155–180 (C. Gutierrez and C. Melendres, Eds). Kluwer Academic, Drodrecht, Netherlands, 1990.
10. J. Lipkowski and L. Stolberg, Molecular Adsorption at Gold and Silver Electrodes, in *Adsorption of Molecules at Metal Electrodes*, Ch. 4 (J. Lipkowski and P. N. Ross, Eds). VCH: New York, USA, 1992.
11. L. W. H. Leung and M. J. Weaver, *J. Am. Chem. Soc.*, 1987, **109**, 5113–5119.
12. G. Mengoli, M. M. Musiani, M. Fleischmann, B. W. Mao and Z. Q. Tian, *Electrochim. Acta*, 1987, **32**, 1239–1245.
13. J. Uehara and K. Aramaki, *J. Electrochem. Soc.*, 1991, **138**, 3245–3251.
14. K. Aramaki, M. Ohi and J. Uehara, *J. Electrochem. Soc.*, 1992, **139**, 1525–1529.
15. J. Gui and T. M. Devine, *J. Electrochem. Soc.*, 1991, **138**, 1376–1384.
16. C. A. Melendres, M. Pankuch, Y. S. Li and R. L. Knight, *Electrochim. Acta*, 1992, **37**, 2747–2754.
17. J. C. Rubim and J. Dunnwald, *J. Electroanal. Chem.*, 1989, **258**, 327–344.
18. Z. Q. Tian, *et al.*, *J. Raman Spectrosc.*, 1998, **29**, 703–711.
19. B. Ren, *et al.*, *J. Electroanal. Chem.*, 1996, **415**, 175–178.
20. Z. Q. Tian, B. Ren and B. W. Mao, *J. Phys. Chem. B*, 1997, **101**, 1338–1346.
21. J. S. Gao and Z. Q. Tian, *Spectrochimica Acta A*, 1997, **53**, 1595–1600.
22. Q. J. Huang, J. L. Yao, R. A. Gu and Z. Q. Tian, *Chem. Phys. Lett.*, 1997, **271**, 101–106.

23. W. B. Cai, *et al.*, *Surf. Sci.*, 1998, **406**, 9–22.
24. P. G. Cao, J. L. Yao, B. Ren, R. A. Gu and Z. Q. Tian, *Chem. Phys. Lett.*, 2000, **316**, 1–5.
25. J. J. Kester, T. E. Furtak and A. J. Bevolo, *J. Electrochem. Soc.*, 1982, **129**, 1716–1719.
26. M. Fleischmann, G. Mengoli, M. M. Musiani and C. Pagura, *Electrochim. Acta*, 1985, **30**, 1591–1602.
27. G. A. Hope, D. P. Schweinsberg and P. M. Fredericks, *Spectrochim. Acta A*, 1994, **50**, 2019–2026.
28. J. Rubim, I. G. R. Gutz, O. Sala and W. J. Orville–Thomas, *J. Mol. Struct.*, 1983, **100**, 571–583.
29. R. Youda, H. Nishihara and K. Aramaki, *Corros. Sci.*, 1988, **28**(1), 87–96
30. P. G. Cao, J. L. Yao, B. Ren, R. A. Gu and Z. Q. Tian, to be submitted.
31. B. Ke, J. J. Hoekstra, B. C. Sison and D. Trivich, *J. Electrochem. Soc.*, 1959, **106**, 382–388.
32. R. Parsons and P. C. Symons, J. Chem. Soc., *Trans. Faraday Soc.*, 1968, **64**, 1077–1092.
33. C. Yarnizky and R. Schreiber-Stanger, *J. Electroanal. Chem.*, 1986, **214**, 65–78.
34. Z. Q. Tian, W. H. Li, B. W. Mao and J. S. Gao, *J. Electroanal. Chem.*, 1994, **379**, 271–279.

Part 2

Organic Coatings

5
Electrochemical Methods for the Study of Protective Organic Coatings after Mechanical Deformation

F. DEFLORIAN, L. FEDRIZZI*, S. ROSSI and P. L. BONORA
Materials Engineering Department, University of Trento, Trento, Italy
*Department ICMMPM, University of Rome "La Sapienza", Rome, Italy

ABSTRACT

The possible presence of defects in the coatings that could be introduced in the shaping and forming of coated metal sheets could decrease the performance of coil coated products, thus reducing the advantages of this technology, which are significant from an economic as well as an environmental point of view. Different coating products on galvanised steel, containing environmentally friendly pigments (Ion Exchange Pigments IEP), deformed to simulate actual applications, for example, bending and deep drawing, were studied. After calculation of the strain induced by deformation by finite element analysis, the samples were characterised by Electrochemical Impedance Spectroscopy (EIS) to study the protective properties of the organic coatings as a function of the imposed deformation. For the electrochemical data analysis an equivalent electrical circuit modelling the electrochemical behaviour of the painted sheets was developed. In this way it was possible to compare the different coating systems and to identify a maximum acceptable strain threshold for every material. The coating performance was studied by measuring the ionic and water permeability of the coating after deformation, the corrosion rate of the substrate and the loss of adhesion as evaluated by EIS measurements. In order to optimise the experimental approach, the data obtained from different deformation methods (bending test and cupping test) but with a similar strain, were compared.

The results showed that the effects of the different deformation procedures on barrier properties (the ionic resistance of the coating R_p) were similar. Moreover deformation by deep drawing (cupping test) seems to be the simplest and most useful way for an electrochemical study of organic coated metal sheets after mechanical deformation. The cupping test together with an EIS characterisation is a valid method for the evaluation of the protective behaviour of coil coating products after deformation.

1. Introduction

The coil coating technology is a widely adopted method for protecting metal sheets with organic coatings [1]. In addition to economic advantages and good quality control, this technique can limit the environmental impact in comparison to "in field" coating application technologies.

However, the post-coating shaping can be a source of some problems. Bending or forming the metal sheets can produce defects or can induce internal stresses in the coating affecting the protection properties of the paint [2]. In the coil coating industries, an optical observation and simple mechanical tests are used to check the coating integrity. However it is possible that small defects, which are impossible to detect visually, can reduce the protective properties of the coating, without any significant modification in the paint appearance.

Moreover, during the last few years the coil coating industry has been working on the production of new surface pre-treatments and painting cycles with better environmental compatibility [3]. Toxic pigments, for example chromates, used in organic coating formulation will have to be replaced in the near future by pigments which exhibit sufficient corrosion protection properties without an adverse impact on the environment. A number of different solutions have been proposed, for example pigments based on phosphates, zinc compounds, or barrier coatings without inhibitive pigments [4]. An interesting new class of pigments, proposed in the last 10 years, is that of ion exchange pigments (IEP), produced under the trade name Shieldex" [5,6]. This pigment (calcium/silica) is the result of exchanging calcium ions onto microporous silica gel. The protective mechanism is related to the ion-exchange reaction: the calcium ions are released by exchange with cations (Na^+, H^+) and there is a dissolution of polysilicate anions. A deposition of silica species and calcium ions occurs on the metal surface forming a protective layer. The behaviour is similar to that observed with calcium silicate.

The aim of the present work is the characterisation of coil coating products containing environmentally friendly active pigments after mechanical deformation, that is, a condition similar to the real application, and at the same time, to contribute to a development of a new experimental approach for evaluating the corrosion resistance properties of deformed metal sheets. Two different mechanical deformation techniques, representative of two different ways of shaping coil coated sheets, were used: bending and deep drawing. The strain was introduced in the test samples following the specific standards: ISO 1519 (mandrel test) for bending and ISO 1520 for deep drawing (cupping test).

In order to be able to compare the different mechanical deformations it is essential to know the strain of the metal and the coating at every point of the sample. This information can be calculated simply in the case of the bending test on cylindrical tools (mandrel test) because the strain is constant and depends only on the diameter of the cylinder. The calculation in the case of samples deep drawn by the cupping test is more complex: the deformation is biaxial (i.e. in two directions) and differs from point to point. A finite element analysis was carried out in this case to model the deformation of the samples. The corrosion protection properties were characterised by Electrochemical Impedance Spectroscopy (EIS).

2. Experimental

Three different kinds of coil coating products on steel were studied; the details are shown in Table 1. Sample C is a standard product and comes from an industrial production, while samples P_a and P_b were prepared in the laboratory and attempt to

Table 1. Materials

Symbol	Metallic coating		Polyester primer +5% IEP	Polyester top coat
	Composition	Thickness (μm)	Thickness (μm)	Thickness (μm)
C	Zn	12	2	18
P_a	Zn + 1–2% Al	7	5	18
P_b	Zn + 1–2% Al	7	5	25

improve the formability of the coil coating. The samples differ in respect of chemical composition and thickness of the metallic and organic layers. The metallic coating was hot dipped zinc (thickness about 12 μm) for the *C* samples and hot dipped zinc + 1–2% of Al for *P* samples. The thickness for this class of samples is lower then in the case of *C* samples and is about 7 μm.

Before painting, the substrate surfaces were pretreated by using a chemical conversion bath based on titanium and fluoride salts (chromium-free).

All the primers were based on polyester polymers (with slightly different formulations) and contained about 5% of ion exchange inhibitive pigments, but differed in thickness, i.e. low for standard *C* samples (about 2 μm) and high (5 μm) for the *P* samples. All the top coats were polyester of 18 μm thickness; the top coat of samples P_b was slightly thicker at 25 μm.

The samples were deformed following the ISO 1519 standard by using a cylindrical mandrel with three different diameters: 13, 6 and 2 mm (the corresponding strains ε are 5, 10 and 25%) and were deep drawn with 3 different indentations: 2, 4 and 6 mm (cupping test ISO 1520). The value of ε for the bend test samples and the values in millimetres of depth of indentation for the cupping test samples are reported with the letter characteristic of the coating for identifying the samples (Table 2).

After deformation the samples were prepared for electrochemical testing. A polymeric tube containing the electrolyte was fixed on each of the undeformed samples and the samples after the cupping test. The diameter of this tube corresponded to the diameter of the deformed area after deep drawing, that is, a test area of about 15 cm^2. The bent samples were embedded in a polyester resin leaving only the surface with the radius of curvature of the mandrel in contact with the electrolyte. In this way the area of testing was quite small, about 1–2 cm^2. It is important to check that adhesion of the resin to the coating is good in order to avoid electrolyte penetration in the interface between the sample and the resin which would lead to a contact between the electrolyte and the unprotected edges of the sample. The embedded sample, as working electrode, was put in the testing solution together with a platinum counter electrode and a reference electrode.

The EIS measurements were performed in the 100 kHz–1 mHz frequency range using a Solartron 1255 Frequency Response Analyser connected to a 273 PARC potentiostat. The amplitude of the sinusoidal voltage signal was 10 mV. The

Table 2. *Levels of deformation and symbols of samples*

Deformation tests		**Symbols**	
	C	P_a	P_b
Cupping test, indentation 2 mm	C2	P_a2	P_b2
Cupping test, indentation 4 mm	C4	P_a4	P_b4
Cupping test, indentation 6 mm	C6	P_a6	P_b6
Bending test, ε = 5%	C5	P_a5	P_b5
Bending test, ε = 10%	C10	P_a10	P_b10
Bending test, ε = 25%	C25	P_a25	P_b25

electrochemical tests were carried out in a 0.3% quiescent aerated sodium sulfate aqueous solution. An Ag/AgCl electrode (+207 mV vs SHE) as reference and a platinum counter electrode were employed. All the impedance measurements were carried out in a Faraday cage in order to minimise external interference on the studied system; the experimental impedance spectra were interpreted on the basis of equivalent electrical circuits using the fitting software Equivcrt [7].

The measurement of the adhesion strength between the organic coating and the substrate is quite difficult, particularly in wet conditions where the information is more useful [8]. An interesting method which is also suitable for non-flat geometry of the samples is the NMP (N-methyl-pyrrolidone) test developed by van Ooij [9]. The NMP liquid (C_5H_9NO) is a polar solvent able to swell the polymeric matrix of organic coatings without dissolving the polymer and thereby maintaining the consistency of the film. The increase in volume of the coating induces interfacial stresses causing a delamination proportional to the energy of adhesion. The adhesion evaluation can be made by comparing the time for complete detachment of the coating, or by comparing the amount of delamination at a constant time of immersion in the solvent. The measurement is made at a temperature of 60°C in order to increase the swelling rate. This kind of measurement can only be used to compare the adhesion for similar coatings (for example on different substrates or pretreatments), because it is possible that the chemical nature of the polymer could have an influence on the swelling process. The values give only qualitative information, but the technique is a useful way to rank the adhesion of different materials.

The materials studied were immersed for 1 h in the NMP solvent at 60°C. The samples were flat sheets with a surface area of about 5 cm^2. It is not useful to test samples after a cupping test because the presence of the solvent changes the viscoelastic properties of the polymer by relaxing the stresses completely, thus making the interfacial residual stresses insignificant. For these reasons the same results would be obtained with deformed or undeformed materials.

3. Results and Discussion

The mandrel test and the cupping test induce completely different deformations. The bending of the sheets causes a uniaxial strain that is about constant on the deformed area, while deep drawing causes a biaxial strain that is variable depending on the point at which it is measured. To compare the different testing procedures it is essential to know precisely the strain at every point of the sample for both kinds of deformation. Calculation of the uniaxial deformation for the samples after bending assumes that a rectangular portion of the material deforms, taking the shape of an arc of a ring, and that the neutral axis coincides with the centre line. The deformation of the coating is calculated as

$$\varepsilon = \frac{l - l_0}{l_0} = \frac{s/2}{R + s/2} \tag{1}$$

where l_0 is the initial length, l the final length, R the diameter of the mandrel used and s is the thickness of the steel sheet. Three different deformation were produced: 5, 10 and 25%.

The field of deformation caused by the cupping test is not simple and is not equal at every point of the sample (as in the case of the bending test), and in order to know the strain distribution for samples tested using the cupping test it was necessary to develop a numerical finite element model. The contribution to the mechanical properties due to the polymeric coatings was neglected and the stress–strain behaviour in the elastic and plastic regions was found from the literature for a mild steel with a similar composition. It was necessary to transform the nominal value of stress and strain (σ_n and ε_n) into the real values (σ_r and ε_r) by using the following equations:

$$\sigma_r = \sigma_n\,(\varepsilon_n + 1) \tag{2}$$

$$\varepsilon_r = \ln\,(\varepsilon_n + 1) \tag{3}$$

The deformation was calculated in the radial cross section starting from the centre of the sample to the external circumference, since the system was radial–symmetrical. The deformation was split in two directions: the radial direction (parallel to the cross-section) and a circumferential direction (perpendicular to the cross-section) as shown in Fig. 1. For the calculation of the permanent deformation, the load was assumed to be applied in three different steps: the first step inducing elastic deformations, the second step causing plastic deformations and, finally, the unloading. The stresses due to wear in the surface between the metal sheet and the indenter were calculated, taking into account the fact that the contact area increases during indentation.

An example of the results obtained is shown in Fig. 2 for the samples deformed with 2 mm depth of indentation. The maximum deformation occurs at a certain distance from the centre of the sample which was also the load application centre. The maximum and average values of circumferential and radial strain and the radius of the area deformed were calculated for every deformation; the results are reported in Table 3. The deformed area is defined as the sample area in which the radial and circumferential strain is greater than 1%. From this analysis it is possible to recognise the complexity of the deformation conditions in metal sheets after the cupping test.

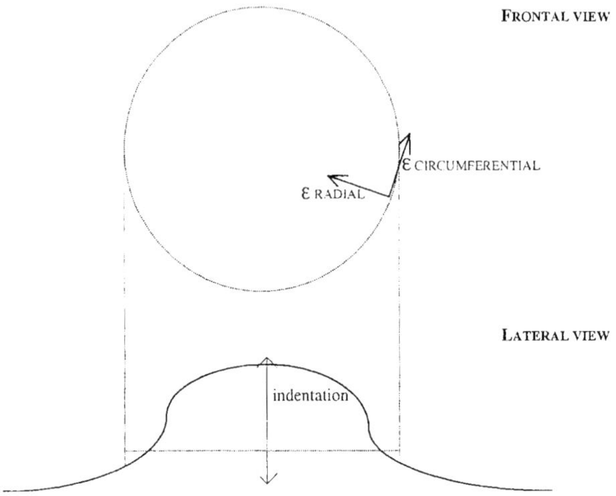

Fig. 1 Strain components in samples deformed with the cupping test.

The electrochemical behaviour of the undeformed samples was studied in previous work [10] in which the inhibitive behaviour of the ion exchange pigments in the primer was investigated. The typical impedance plot is shown in Fig. 3 and the equivalent electrical circuit used to model the impedance results is shown in Fig. 4.

The elements forming the circuit at the highest frequencies are the non ideal capacitance Q_c, mathematically described as a constant phase element, and which is related to the dielectric properties of the coating, and the resistance R_p, that is, the ionic resistance through the coating (and therefore dependent on the presence of defects or pores) whereas at the lowest frequencies there is the non ideal capacitance Q_{dl} and the charge transfer resistance R_{ct}, related to the faradaic reaction on the metal surface. These last two elements are considerable simplifications of the real

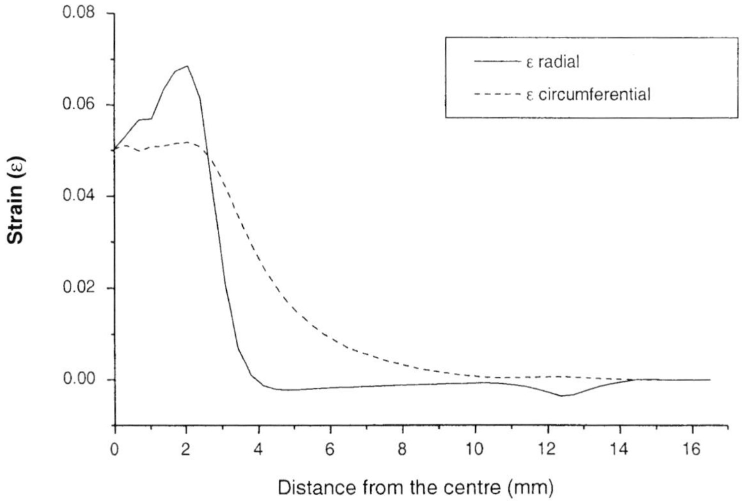

Fig. 2 Strain distribution in samples with 2 mm of indentation depth.

Table 3. Deformation parameters

Depth of indentation (mm)	Circumferential strain (ε%)		Radial strain (ε%)		Radius of the deformed area (mm)
	Max.	Mean	Max.	Mean	
2	5	5	7	6	5
4	9	9	12	10	8
6	16	15	23	20	10

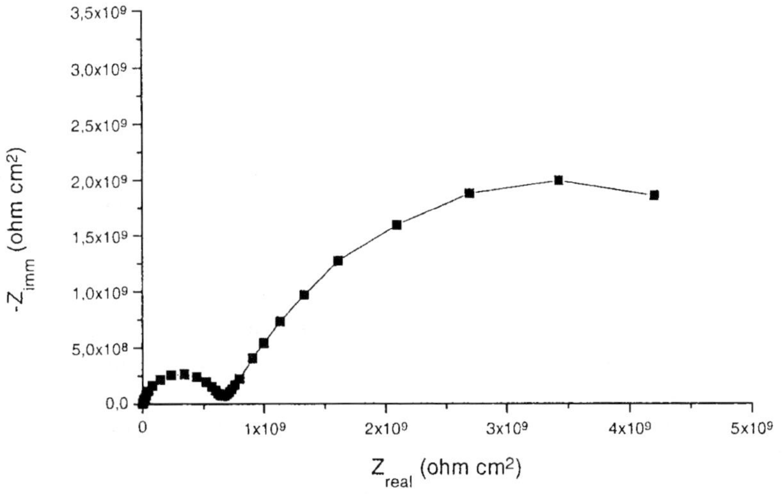

Fig. 3 Typical impedance plot in Nyquist representation after 24 h of immersion.

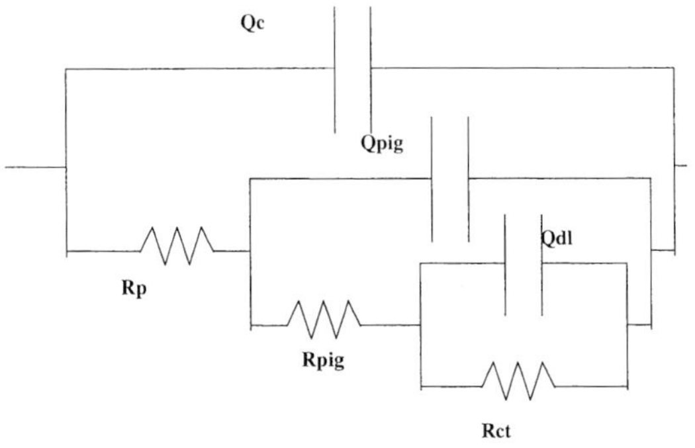

Fig. 4 Equivalent electrical circuit.

electrochemical situation on zinc, the corrosion reaction in these conditions being quite complex and involving different steps which determine the total impedance [11,12]; moreover, the possible presence of corrosion products influences the apparent value of the double layer capacitance. However, considering the aim of this work, we have preferred, as in similar cases, to avoid a further circuit complication.

More difficult is the interpretation of the elements seen at intermediate frequencies. These parameters have been interpreted as a contribution to the total impedance due to the presence of the pigments which can modify their capacitance and resistance as a function of the hydration mechanism following the water penetration [13]. In general, following the water uptake phenomenon, the values of Q_{pig} increase and the R_{pig} values decrease.

Before characterising the samples by electrochemical methods, an optical observation (50×) of the surface of the coating was carried out to check for the presence of macro-defects and cracks in the coatings produced by the deformation procedure. No significant defects or cracks could be seen in coatings subjected to deformations lower or equal than 25%, while a large number of defects are present for higher deformations.

In analysing the electrochemical results obtained by EIS, which means discussing the trend with time of immersion in the electrolyte of the main parameters of the circuit in Fig. 4, the first comparison made involved the undeformed sample and samples after bending with the minimum strain (5%). No significant differences were found for all the kinds of samples (*C*, P_a and P_b) on comparing these two levels of deformations (0% and 5%), from which it seems that the minimum strain studied (5% after bending) does not cause a significant deterioration of the protective properties of the coating. For this reason only the EIS results of the deformed materials will be presented and discussed in this paper.

The general trend of some parameters of the equivalent electrical circuit differs depending on the kind of coating. The first example to be described will be the standard material, sample *C*, after the bending test. The evolution of the coating capacitance Q_c vs the time of testing is shown in Fig. 5. The first result that is clear from this Figure is the influence of the level of deformation on the capacitance variation: the higher the strain, the higher the capacitance increase.

The values of Q_c as a function of strain are not very different during the first 200 h of immersion, although there are small differences in the starting value, this being higher for the C25 sample. In this case the deformation slightly damaged the coating

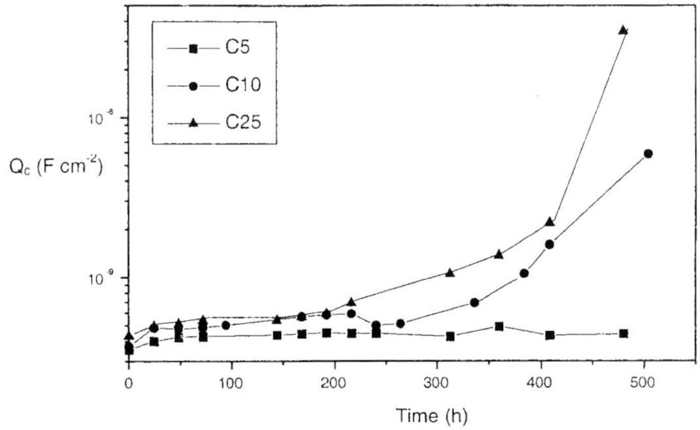

***Fig.* 5** *Coating capacitance* (Q_c) *vs the immersion time for the* C *samples after the bending test.*

allowing an instantaneous higher water penetration into the coating. This increase of the coating capacitance is due initially to the water uptake into the coating [14], but the very big increase of Q_c (one or more orders of magnitude) as in the case of samples C10 and C25 after 250 h of immersion cannot be simply explained by water diffusion into the polymer matrix of the coating, but rather as water accumulation at the metal–coating interface [15]. This hypothesis is proved by the presence in both the samples of small blisters after long times of immersion.

The trend of the R_p parameter can be associated with the evolution of the ion barrier properties of the coating [16]. Despite the fact that this kind of coating contains anti-corrosive active pigments, the barrier properties of the coating towards ions play a very important role in the performance of the protective system. For this reason the pore resistance parameter is probably one of the most significant pieces of information for ranking the different samples as a function of the protective properties. In Fig. 6 it is possible to see that the starting values are close to each other for all levels of deformation which confirms the observations that indicate the absence in the coating of macro-defects since these would have caused an instantaneous decrease in the R_p values.

The sample with 5% strain (C5) shows a gradual and limited decrease of R_p. A different behaviour is shown by samples C10 and C25; the first has a sharp decrease of R_p after 200 h, while C25 drops after only about 100 h of testing. The mechanical deformation, even if it does not induce visible defects, causes a degradation of the ion barrier properties. It is interesting to note that the R_p decrease, that is without doubt related to the water uptake process which would favour ion diffusion, occurs before the huge increase in Q_c due to blister formation. Moreover when the Q_c values become very large after 250 h of immersion the pore resistance is lower than 10^6 ohm cm^2 and the values are very similar for C10 and C25 samples. For these two samples it is possible to suppose that the delamination process, with water accumulation at the interface, can cause further stresses in the coating (due to the formations of blisters) thereby increasing the concentration of defects, so making the ion penetration easier.

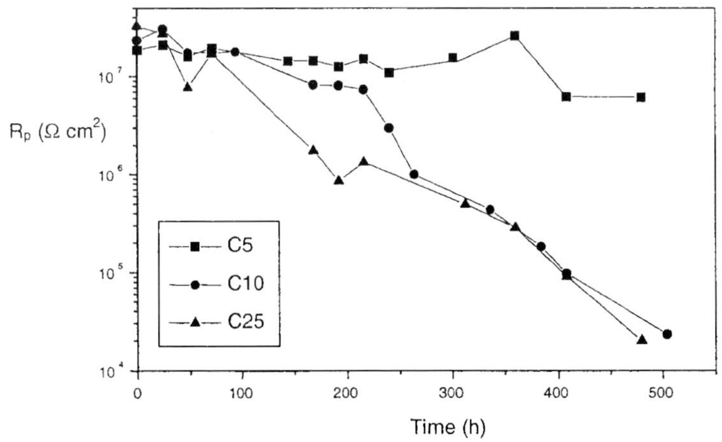

Fig. 6 *Pore resistance* (R_p) *vs the immersion time for the* C *samples after the bending test.*

For the standard samples (C) it is possible to conclude from these results that the maximum acceptable strain after bending is 5%, because ion barrier properties corresponding to R_p values lower than 10^6 ohm cm^2 are considered to be insufficient.

The trends of Q_c and R_p are different for sample P_a and P_b. In Fig. 7 the trend of Q_c for the sample P_a after bending is shown. In this case also it is possible to note a higher starting value of capacitance for the samples deformed with 25% strain due to an instantaneous decrease of the water barrier properties caused by residual stresses.

For sample *P* the water barrier properties are reduced by the deformation, but not so strongly as in the case of sample *C*, as can be seen on comparing the Q_c values after 300 h of immersion. This means that the delamination process due to water presence at the interface is not yet active after 500 h and good adhesion of the coatings can be assumed. This hypothesis is confirmed by the absence of any blisters at the end of testing and by the different values of the R_p parameter. In fact the R_p values of the sample *P* approach 10^6 ohm cm^2 — considered as a threshold value of acceptability — only for 25% of strain and after 450 h immersion, which interestingly shows a better behaviour than sample *C*. The best results are shown by sample P_b for which, even in the case of bending inducing 25% of strain, the pore resistance never reaches the threshold limit of resistance (10^6 ohm cm^2). For this class of material it is possible also to assume the maximum studied deformation as acceptable.

The analysis of the chemical–physical processes occurring on the metal surface is concerned with the metal–electrolyte contact, described by the double layer capacitance Q_{dl}, and the electrochemical dissolution of the zinc, described by the charge transfer resistance R_{ct}. The time behaviours of these two parameters are reported for the standard product C in Figs 8 and 9. The Q_{dl} trend in Fig. 8 for samples C10 and C25 shows a large increase of double layer capacitance, proving the loss of adhesion due to electrolyte penetration during the first 50 h of testing — many days before any blister nucleation was visible —confirming that the large amount of water uptake shown in Fig. 5 is located at the metal–coating interface during the first two days of testing. This phenomenon, i.e. the presence of electrolyte on the metal surface,

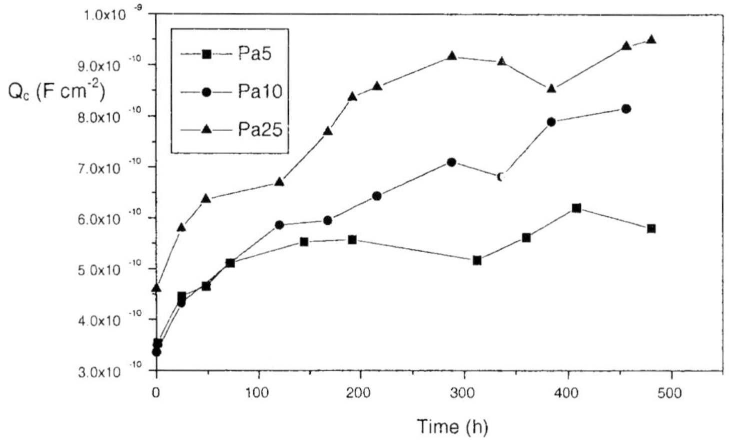

Fig. 7 *Coating capacitance* (Q_c) *vs the immersion time for the* P_a *samples after the bending test.*

has an obvious influence also on the corrosion rate. The behaviour of sample C5 is completely different from the behaviour of the other two samples as is shown in Fig. 9. The sample with 5% strain shows very high charge transfer resistance values (higher than 10^9 ohm cm^2) which means a low corrosion rate, while both materials C10 and C25 show lower R_{ct} values of below 10^6 ohm cm^2 after about 500 h testing. The samples P_a and P_b show, on the contrary, Q_{dl} values at the end of testing of the order of 10^{-8} Fcm^{-2} and R_{ct} values higher than 10^9 ohm cm^2, in agreement with the supposition that the delamination process for these materials, even in the case of maximum deformation, is negligible.

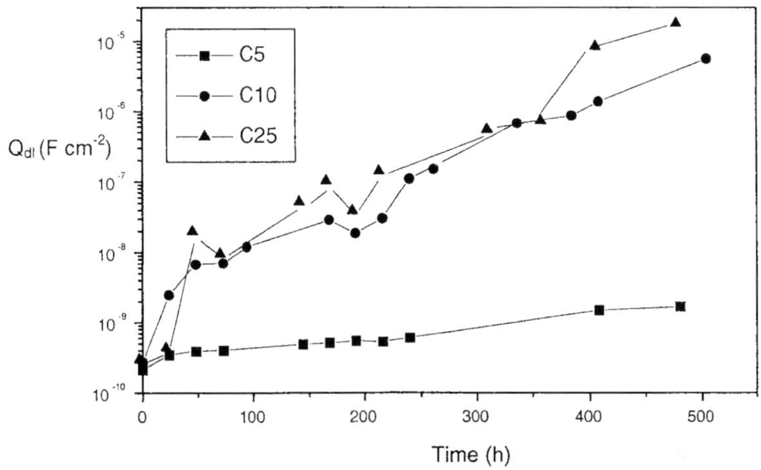

Fig. 8 *Double layer capacitance (Q_{dl}) vs the immersion time for the* C *samples after the bending test.*

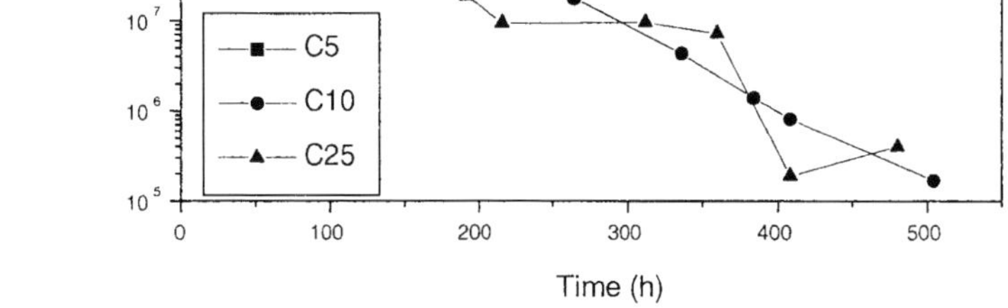

Fig. 9 *Charge transfer resistance (R_{ct}) vs the immersion time for the* C *samples after the bending test.*

This result is particularly interesting because it seems to indicate that the critical aspect in the deterioration of these materials is the metal–coating adhesion which can be reduced by deformation. The main reason for the delamination process that is active in these materials is the presence of damage at the metal–polymer interface after deformation which weakens the chemical bonds and favours the loss of adhesion. Comparing the samples under study (see Table 1), both the metal substrates (zinc or zinc–aluminium alloy) and the thickness of the primer make the class of materials *C* and *P* different.

The substrate can play an important role because it is known that the possibility of crack formation in a pure zinc coating is significantly higher than in the zinc alloys. In Figs 10 and 11 the surface morphologies of sample *C* deformed by the cupping test are shown: 2 mm (Fig. 10) and 6 mm (Fig. 11) of indentation obtained by deforming organic coated sheets and removing the polymeric coating by an organic solvent. In the micrograph of Fig. 11 the presence of cracks is evident, these are probably intergranular due to the deformation, but completely absent for sample C2 in Fig. 10. The Al–Zn alloy coating with higher strain (samples P_a6 and P_b6) shows a different morphology (Fig. 12). In this case also it is possible to note some small cracks due to the deformation but these are few and considerably smaller. In the case of the samples *C* with 10 and 25% strain the presence of cracks and voids in the metallic coating causes a local detachment of the continuous organic coating covering the metal, which thus provides a zone of nucleation for water-induced loss of adhesion (i.e. blisters). This mechanism can explain the big differences in the adhesion behaviour shown by samples *C* in comparison with samples *P*, but also the primer thickness is probably relevant. The thicker primer can assure a better covering of the metal surface, while it is possible that 2 μm of primer leaves some small zones uncovered because of the presence of pores or small pin holes which can reduce adhesion.

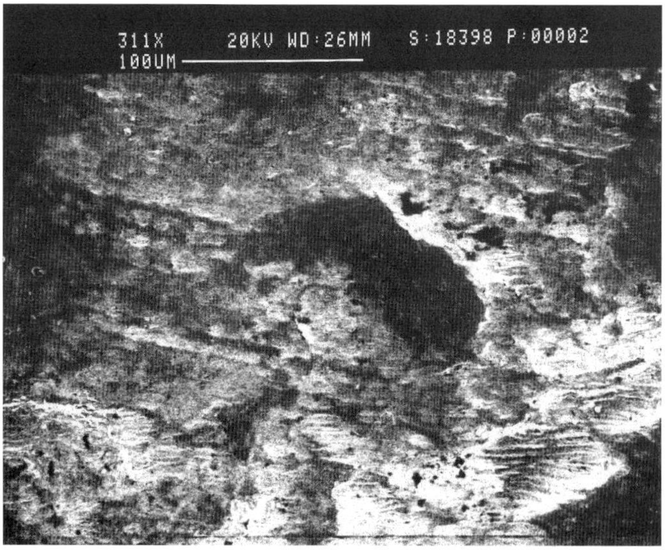

Fig. 10 SEM micrograph of the metallic surface of sample C2.

Comparing the different coil coating materials, the protective properties of the new products, samples *P*, are higher than those of sample *C* used as a standard. In order to have a more precise view, Fig. 13 reports the pore resistance results after 500 h of immersion for all the materials as a function of the depth of indentation in the cupping test. First, it is possible to note that the values of R_p as a function of the deformation for all the materials decrease drastically, which means that the protective performances of the systems under study are strongly dependent on the coating

Fig. 11 *SEM micrograph of the metallic surface of sample* C6.

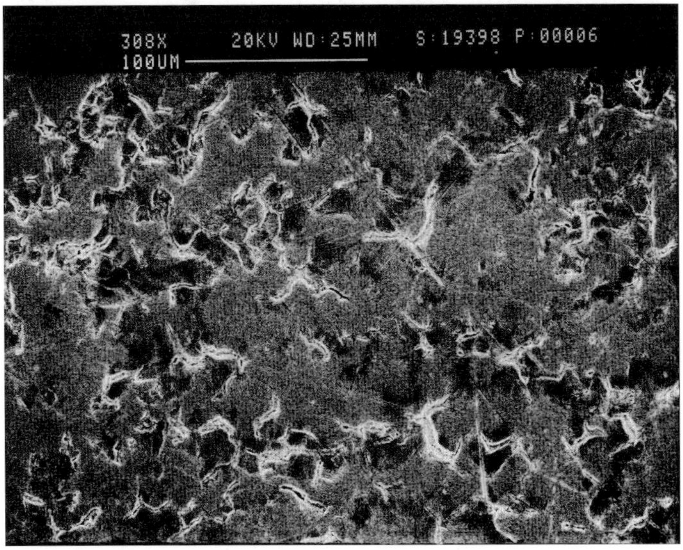

Fig. 12 *SEM micrograph of the metallic surface of sample* P_a6.

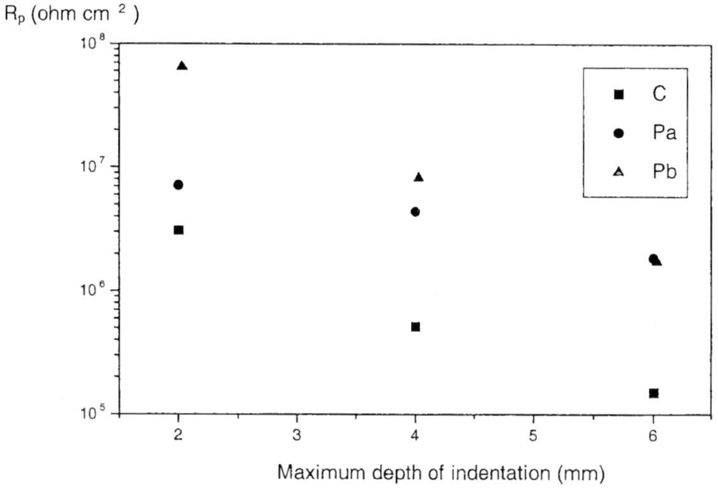

***Fig. 13** Value of pore resistance (R_p) after 500 h of testing for all the studied materials as a function of the indentation depth of the cupping test.*

deformation. Moreover, on comparing the different materials, the standard coating *C* was found to show sufficient protective properties at the minimum deformation (C2), having final values of R_p above the threshold of 10^6 ohm cm^2. This behaviour was not confirmed after the cupping tests with depth of penetration of 4 and 6 mm, where the final value of R_p was lower than 10^6 ohm cm^2. Therefore, this material can be considered suitable for maintaining good protective properties only for cases of very limited deformation, in fact, over 10% strain is too much for this kind of coating.

The sample P_a also shows a better behaviour for minimum deformation and seems to have a less sharp dependence of protective properties on the level of deformation. The threshold of R_p is never reached and this coating can be used for applications in which the deformation involves strains of about 25% (see Table 3). The best behaviour is shown by sample P_b, especially for deformation caused by 2 and 4 mm of indentation. This material can also be deformed up to 25% while maintaining adequate corrosion protection properties. This similar behaviour of samples P_a and P_b is due to the small difference between these two painting cycles; this difference relates to the top coat thickness which for sample P_b is about 40% higher than that for samples P_a. The higher thickness induces better barrier properties only for low levels of deformations (2 and 4 mm of indentation), while the behaviour is the same for 6 mm of indentation in the cupping test. In conclusion, a small increase in the top coat thickness, compatible with the coil coating production line, may cause a slight increase in the protective properties of the system for low strains, but the increase is probably not sufficient to justify the higher costs.

It is interesting to compare the results obtained on deforming the samples using the bending test and the equivalent results obtained by using the cupping test. Figures 14 and 15 show the trends of the coating capacitance Q_c and pore resistance R_p for the standard sample C after the cupping test. On examining the Q_c values it is possible to note that only in the case of the sample C2 (with depth of indentation 2 mm) is the Q_c variation small, while for the other indentation depths (4 and 6 mm) there are significant increases in Q_c. These results are consistent with the R_p data. Only the sample C2 remains always above the threshold limit (10^6 ohm cm^2); the degradation of the barrier properties of the other two materials make these unacceptable.

The results in Figs 14 and 15 confirm the characterisation obtained on samples deformed with the bending test and similar considerations can be made on comparing the materials P_a and P_b after bending and after the cupping test.

One of the aims of the research was to compare the two different ways to induce mechanical deformations in coil coated sheets. It has already been shown that finite

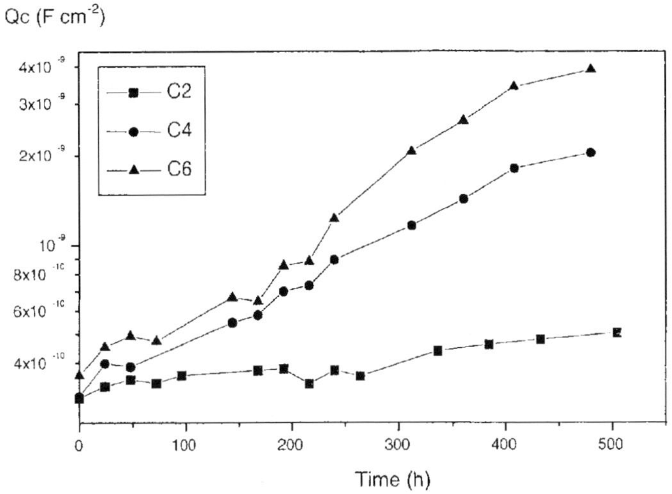

Fig. 14 Coating capacitance (Q_c) vs the immersion time for the C samples after cupping test.

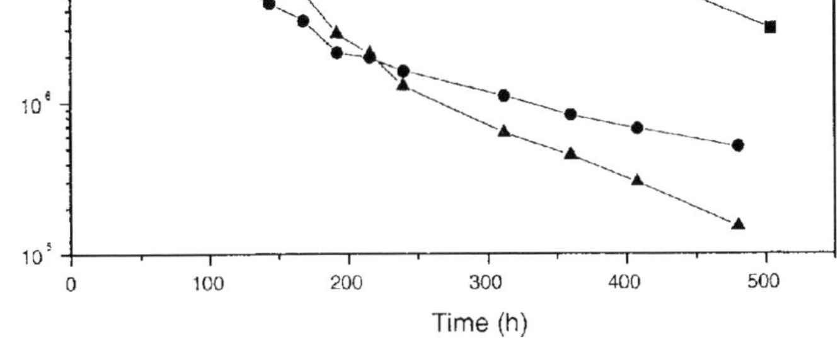

Fig. 15 Pore resistance (R_p) vs the immersion time for the C samples after cupping test.

element analysis can provide comparisons between cupping test indentation and bending strains (e.g. 2 mm ≈ 5%; 4 mm ≈ 10%; 6 mm ≈ 25%) and it is now necessary to compare the electrochemical results in depth for these two methods of deformation.

In Fig. 16 the comparison of the R_p results obtained with the two methods at the maximum and minimum deformation is reported for the sample *C*. A good agreement, similar to the result in Fig. 16, was found for all the materials and every level of deformation. This means that the substantial equivalence of the two approaches is probably a general result, applicable also for other coil coating products.

This conclusion is very important because, assuming that the EIS characterisation obtained on samples after the bending and after the cupping test are equivalent, it is sufficient to study the coil coating materials by using only one method and that can be chosen by convenience. In particular, it is important to remember that the samples deformed by the cupping test show a lot of advantages: thus, the testing area is broader than in the case of bending and it is independent of the level of deformation (depth of indentation), so allowing a better comparison of different samples, whereas the bending test produces very small samples and, in general, with a testing area which depends on the radius of the mandrel (level of deformation).

The ease of preparation of the cell for the electrochemical measurements is another aspect which makes the cupping test to be preferred over the bending test. In fact, for the latter method, as previously described, it is important to seal carefully (by using polymeric materials) the area of testing from the other parts of the sample in order to avoid localised corrosion phenomena. The preparation of the electrical contact for the working electrode is also not simple. The only disadvantage of the cupping test is that the strain is not constant over the testing area and it has to be ascertained by using numerical methods.

After the characterisation of the barrier properties of the coatings and the comparison of the different deformation methods, the behaviour of the interface between the coating and the metal substrate (corrosion, adhesion, etc.) was further investigated in the case of the cupping test.

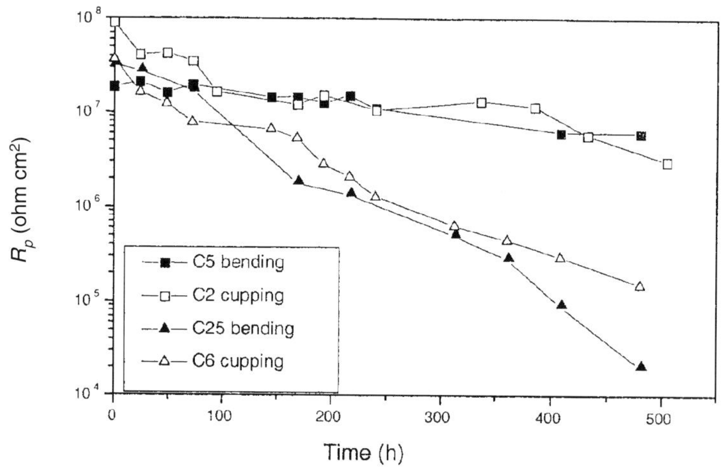

Fig. 16 *Comparison of the pore resistance (R_p) results obtained on samples after bending test and after cupping test.*

The Q_{dl} trend for the C materials is shown in Fig. 17. In this case also, although the growth of Q_{dl} is a general trend because of the increase of the wet area, it is possible to note a clear difference between the samples with 2 mm of indentation and the other two. All the materials show similar starting values of Q_{dl}, with the only difference being a slightly higher value for the C6 sample. This result means that the coating at the beginning is almost intact, without significant cracks, in agreement with the pore resistance measured at the beginning and shown in Fig. 18. Moreover, the comparison between Figs 17 and 18 leads to some interesting considerations. The increase of the wet area (increase of Q_{dl}) could be due to two different phenomena: (1) increase of metal surface area in contact with the electrolyte caused by the development of cracks

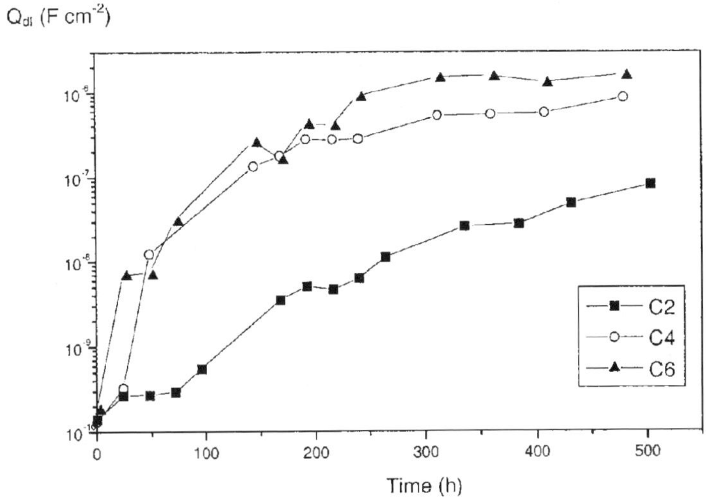

Fig. 17 *Double layer capacitance* (Q_{dl}) *vs the immersion time for the* C *samples after deformation.*

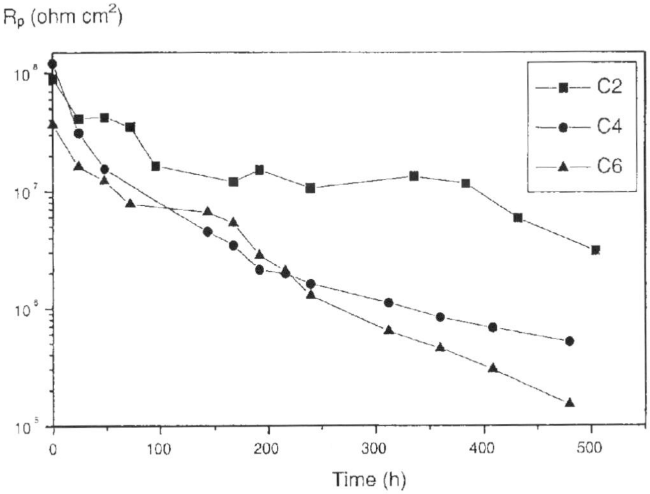

Fig. 18 *Pore resistance* (R_p) *vs the immersion time for the* C *samples after deformation.*

and pores through the coating or (2) delamination processes. One way to distinguish these two contributions is to compare the changes of Q_{dl} and R_p during the time of testing. The ratio between the final value of Q_{dl} ($Q_{dl\text{-}final}$) obtained after 500 h of immersion and the initial value Q_{dl} ($Q_{dl\text{-}initial}$), defined as the Q_{dl} variation (V_{Qdl}) is proportional to the total increase of the wet area. In a similar way the ratio $R_{p\text{-}initial}/R_{p\text{-}final}$ (R_p variation V_{Rp}, remembering that R_p is in inverse proportion to the defects area while Q_{dl} is in direct proportion) is an indication of the increase of the pore area. If the increase of the wet area is higher than the increase of the pore area ($V_{Qdl}/V_{Rp} > 1$) there is a delamination contribution which is proportional to the ratio value [17].

The results of this analysis are reported in Table 4. It is possible to note that the increase of wet area is between 6 and 50 times higher than the increase of the pore area, and the value is proportional to the depth of indentation, proving the existence of a delamination process and the influence of the level of deformation, and therefore the influence of residual stresses, on the delamination rate. The appearance of the C6 sample with many blisters at the end of testing is shown in Fig. 19. The spatial distribution of the blisters is also interesting since they are concentrated in the area of maximum strain at the top of the cup, where it can be assumed that the defect concentration is highest.

The behaviour observed for the P_a sample is different, since in spite of the better barrier properties, it shows Q_{dl} values after 100 h of immersion which are higher than those of sample C (Table 5), particularly in the case of 6 mm indentation. This behaviour can be explained by postulating lower adhesion to the substrate of this material compared to that of sample C so causing a higher delamination, especially with maximum deformation. The same variation analysis discussed for C samples is reported in Table 6. Very high delamination (V_{Qdl}/V_{Rp} of about 200–300) is confirmed. However, this hypothesis of low adhesion has yet to be confirmed with independent data.

The R_{ct} values are consistent with the Q_{dl} measurements. Both the samples with 4 and 6 mm of indentation show, from the beginning, an obvious corrosion process as suggested by the low R_{ct} values (of the order of $10^7\ \Omega\ cm^2$ after 1 day). All the samples show an increase of the charge transfer resistance values after the initial period, reaching R_{ct} values of about $10^8\ \Omega\ cm^2$ after 200–300 h. This is a very interesting result because, considering the increase of wet area indicated by the Q_{dl} values, it can be explained only by the inhibitive activity of the ion exchange pigments, as demonstrated in previous work [16–18] and as confirmed by the increase of the

Table 4. *Variation of the parameters* Q_{dl} *and* R_p *for samples* C

	C2	C4	C6
$V_{Qdl} = Q_{dl\text{-}final}/Q_{dl\text{-}initial}$	300	3000	10 000
$V_{Rp} = R_{p\text{-}initial}/Rp_{\text{-}final}$	50	200	200
V_{Qdl}/V_{Rp}	6	15	50

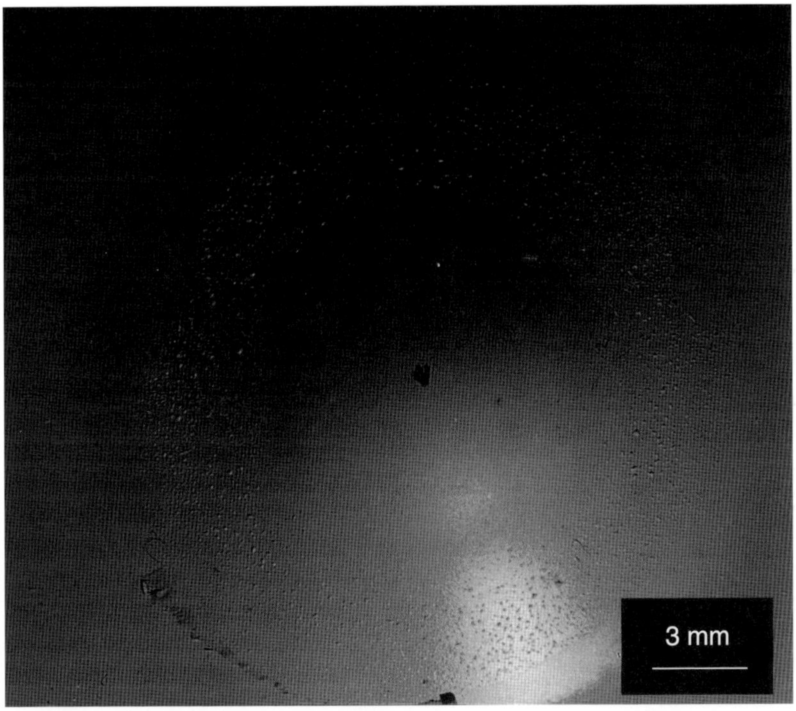

***Fig. 19** Morphology of sample C with 6 mm indentation after 500 h immersion.*

***Table 5.** Q_{dl} values after 100 h immersion*

	Q_{dl} (nF cm^{-2})		
Material	**Depth of indentation**		
	2 mm	**4 mm**	**6 mm**
C	0.55	30	50
Pa	1	50	1200

***Table 6.** Variation of the parameters Q_{dl} and R_p for P_a samples*

	P_a2	P_a4	P_a6
$V_{Qdl} = Q_{dl\text{-}final}/Q_{dl\text{-}initial}$	500	2000	10 000
$V_{Rp} = R_{p\text{-}initial}/R_{p\text{-}final}$	2.77	10	30
V_{Qdl}/V_{Rp}	180	200	333

corrosion potential from about –1000 mV Ag/AgCl (a typical value of corroding zinc in that solution) to a more noble value in the range of –700 to–800 mV.

The morphology of the P_a samples after testing, shown in Fig. 20, is completely different from the morphology of the *C* samples. The blisters, which are smaller and highly diffused, are present preferentially in the area with lower deformation. In order to understand this morphology it is important to take into account the electrochemical data previously discussed. The high value of the reactive area has been explained by a lower adhesion. It is possible that in this condition, to support the active anodic reaction (low R_{ct} values especially in the first part of the testing) a wide area of the sample will be involved in the cathodic reaction (oxygen reduction) under the paint. This reaction is concentrated where the anodic reaction is not dominant, that is over the area with the lower defect concentration thus leading, together with worse adhesion, to the development of many small blisters in the lateral zones of the sample.

The last sample to be discussed is P_b. This material is similar to sample P_a but with a different chemical composition of the polyester based primer (for reasons of confidentiality it is not possible to provide more information on the formulation) and a thicker top coat. Figure 21 shows the trend of the double layer capacitance Q_{dl} of the P_b materials versus the time of immersion. It is possible to note the absence of any relevant change in Q_{dl} for the sample deformed with 2 mm of indentation, whereas sample P_b4 shows a significant increase in Q_{dl}, pointing to an increase of the wet area only after more than 200 h immersion. There is, on the contrary, a sharp increase of Q_{dl} from the beginning for the sample with 6 mm of indentation reaching values of

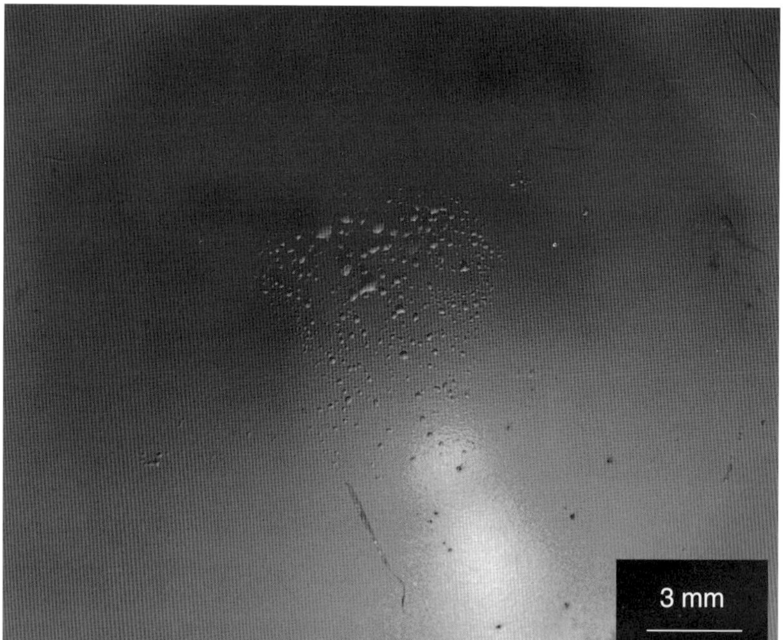

***Fig. 20** Morphology of* P_a *sample with 6mm of indentation after 500 h immersion.*

the order of 10^{-7} F cm^{-2}, which are however lower than the value from sample C (Fig. 17).

Similar differences can be seen for the R_{ct} behaviour in Fig. 22 where the good performance sample of P_b2 is confirmed.

In Table 7 the R_p and Q_{dl} ratios are reported. The good barrier properties that are retained even after deformation and the limited delamination process — similar to the behaviour of sample C — are confirmed. It is possible that the corrosion protection properties of the P_b materials, which are the best amongst the materials studied, are due to the higher thickness of the top coating, which will reduce the number of defects. This possibility is supported by the small, but significant, difference in the initial values of Q_{dl} in Figs 17 and 21 which differ by about a factor of 2, together with the absence of the adhesion problems that were evident for sample P_a. In fact, at the end of the testing the morphology of the P_b samples was different from, and better than, the appearance of the other coatings in that they had only a few small defects even in the case of maximum deformation (Fig. 23).

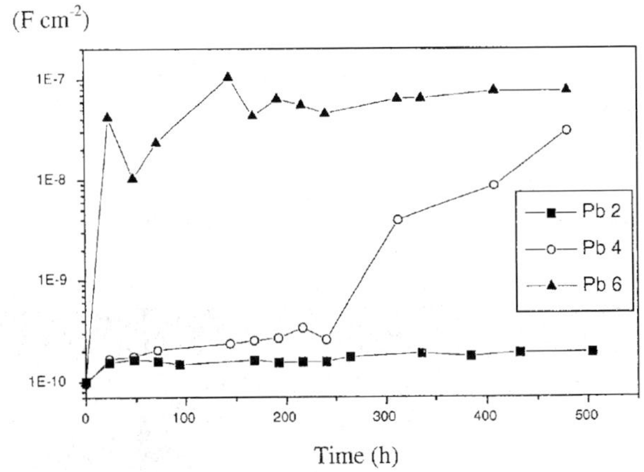

Fig. 21 *Double layer capacitance (Q_{dl}) vs the immersion time for the P_b samples after deformation.*

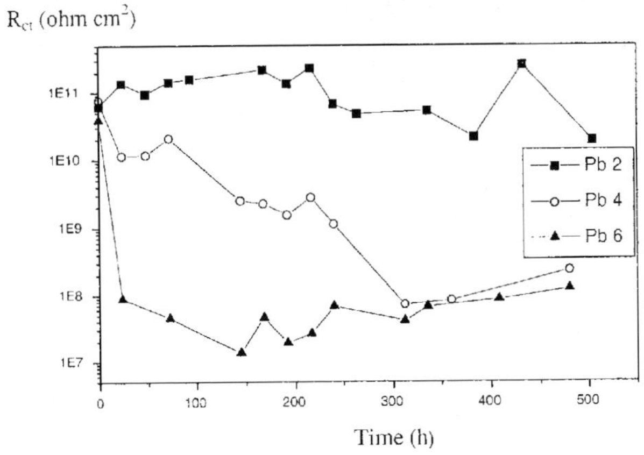

Fig. 22 *Charge transfer resistance (R_{ct}) vs the immersion time for the P_b samples after deformation.*

Table 7. Variation of the parameters Q_{dl} *and* R_p *for samples* P_b

	P_b2	P_b4	P_b6
$V_{Qdl} = Q_{dl\text{-}final}/Q_{dl\text{-}initial}$	1	90	800
$V_{Rp} = R_{p\text{-}initial}/R_{p\text{-}final}$	1	5	20
V_{Qdl}/V_{Rp}	1	18	40

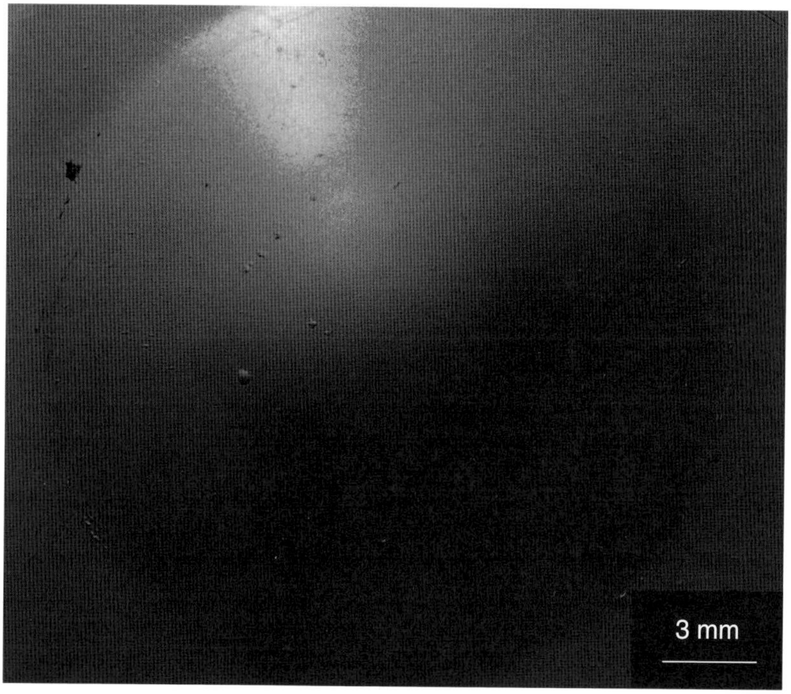

Fig. 23 Morphology of sample P_b *with 6 mm of indentation after 500 h immersion.*

The hypothesis of different adhesion properties of the coatings which follows from the interpretation of the EIS results, should be confirmed by independent data. With this aim some adhesion measurements based on immersion testing in organic solvents (NMP method) were performed on undeformed materials. Mechanical techniques for the adhesion evaluation, such as the pull-off test, were rejected after some attempts because they were unable to give acceptable results, mainly because of the difficulties in measuring the adhesion in wet conditions.

The results are shown in Table 8 where the percentages of detachment for the different materials after 1 h immersion in NMP at 60°C are reported. Both new undeformed samples and samples previously immersed in the electrochemical testing solution (0.3% Na_2SO_4) for 30 days were tested. From Table 8 it is possible to obtain confirmation that the adhesion of samples P_a is worse than that of samples C and P_b.

Table 8. *Adhesion results obtained with the NMP test*

	Detached areas (%)	
Materials	**Dry conditions**	**Wet conditions**
C	2–3%	10–15%
P_a	>90%	100%*
P_b	10–15%	10–15%

* Complete detachment after about 15 min.

Samples C and P_b show similar performance but with slightly better dry adhesion values for sample C and with sample P_b less sensitive to humid conditions — being the only material with similar adhesion values in dry and wet conditions.

4. Conclusions

The results of this work can be summarised in the following points:

- By combining EIS measurements and standardised methods for deformation of coated sheets it is possible to characterise the protective properties of coil coating products that are usually deformed in the final application. It is possible to measure differences in protective properties that are otherwise not evident simply by optical observations or immersion tests.
- On comparing samples deformed by the bending test and by the cupping test with similar strain (calculated for the latter by numerical methods) it was found that the two methods are equivalent. The simplest method for the electrochemical characterisation can therefore be chosen. By using the experimental approach for working electrode preparation described in this paper and taking into account the advantages that a larger area involves, this method is the cupping test.
- Both the double layer capacitance Q_{dl} and charge transfer resistance R_{ct} parameters can give information on the degradation of the metal surface which is useful to add to the adhesion information. Moreover, the analysis and the comparison of the Q_{dl} and R_p variations is able to distinguish coating degradation due to the development of defects through the coating and to the delamination phenomena thus giving adhesion information.
- The NMP test, although qualitative and empirical, is an interesting alternative to the traditional testing methods for the adhesion characterisation of organic coatings when the standard methods (pull-off, peel, etc.) are very difficult to

implement, as in the case of wet adhesion measurements. The data obtained are in good agreement with the EIS results.

- Among the materials studied, the best performances after deformation were found for the specially developed coatings referred to as P_b because of better barrier properties caused by the higher thickness of the top coat and better adhesion than given by the similar P_a coating. For sample *C* — standard industrial product — the poor barrier properties after deformation together with the thinner primer containing the inhibitive pigments induce faster degradation as also observed visually at the end of testing.

- The poor adhesion of sample P_a caused a higher initial under-paint corrosion rate for this kind of material although later inhibitive actions by the ion exchange pigments led to a different blister distribution compared to that with sample P_b and *C*.

5. Acknowledgements

The authors are grateful to Boher S.r.l. for preparation of samples. The authors acknowledge also L. Benedetti and E. Siena for their assistance in the electrochemical measurements, C. Bressanini for the SEM observations and V. Fontanari for assistance in the finite elements analysis.

References

1. E. Gaske, *Coil Coatings*. FSCT, Philadelphia, 1987.
2. X. Hess, *Paint Film Defects — Their Cause and Cure*. Chapman and Hall, London, 1979.
3. A. Bittner, *J. Coatings Technol.*, 1989, **61**, 11.
4. G. Adrian and A. Bittner, *J. CoatingsTechnol.*, 1986, **58**, 59.
5. R. D. Armstrong and S.Zhou, *Corros. Sci.*, 1988, **28**, 1177.
6. B. P. F. Goldie, *J. Oil Colour Chem. Assoc.*, 1988, **9**, 257.
7. B. Boukamp, *Solid State Ionics*, 1986, **18 & 19**, 136.
8. K. L. Mittal (Ed.), *Adhesion Measurements of Film and Coatings*. VSP, Utrecht, 1995.
9. W. J. van Ooij, R. A. Edwards, A. Sabata and J. Zappia, *J. Adhes. Sci. Technol.*, 1999, **7**, 897.
10. F. Deflorian, L. Fedrizzi and P. L. Bonora, *Advances in Corrosion Protection by Organic Coatings — III*, Ed. Electrochemical Society, Pennington, 1998, pp.45–56.
11. C. Desluis, M. Duprat and C. Tournillon, *Corros. Sci.*, 1983, **29**, 13.
12. C. Cachet and R. Wiart, *J. Electroanal.Chem.*, 1983, **129**, 103.
13. L. Fedrizzi, F. Deflorian, G. Boni, P. L.Bonora and E. Pasini, *Prog. Org.Coatings*, 1996, **29**, 89.
14. F. Deflorian, L. Fedrizzi and P. L. Bonora, *Corros. Sci.*, 1996, **38**, 1697.
15. E. P. M. van Westing, G. M. Ferrari and J. H. W. de Wit, *Corros. Sci.*, 1994, **36**, 979.
16. F. Deflorian, L. Fedrizzi and P. L. Bonora, *Advances in Corrosion Protection by Organic Coatings — II*, Ed. Electrochemical Society, Pennington, 1994, p. 15.
17. F. Deflorian and L. Fedrizzi, *J. Adhes. Sci. Technol.*, 1999, **13**, 629.
18. T. E. Fletcher, *Proc. 11th Int. Corros. Congr.*, Florence, 2, (1990), 265–273. Published Associazione Italiana di Metallurgia, I–20121, Milano Italy.

6
A Scanning Kelvin Probe Study of the Delamination Process at the Carbon Steel–Polymer Interface

A. NAZAROV and D. THIERRY
Swedish Corrosion Institute, Roslagsvägen 101, hus 25, 104 05 Stockholm, Sweden

ABSTRACT
The Scanning Kelvin probe (SKP) was used for *in situ* determination of the Volta potential distribution at a metal/polymer interface. It was shown that the measurements can give information on the nature of the active chemical groups in the polymer forming the adhesive bonds with the metal as well as on the kinetics of the disbonding during water penetration at the metal/organic coating interface. Monitoring the change in Volta potential during wet–dry atmospheric transitions can be useful for the evaluation of the influence of surface pretreatment on the adhesion of the organic coating and on the rate of water diffusion through the organic coating. The results are compared with those obtained by Electrochemical Impedance Spectroscopy.

1. Introduction

1.1. Volta Potential and Theory of the Kelvin Probe

The scanning Kelvin probe (SKP) is a non-contact, non-destructive device for the determination of the work function of conducting and semi-conducting materials. The theory of creation of the Volta potential by a vibrating capacitor has been given in much detail elsewhere [1–3]. The Volta potential (V_{el}) of the electrode is proportional to the electronic work function (Φ_{el}). For a metal covered by a thin electrolyte film formed during atmospheric weathering conditions, the Volta potential depends on the change in potential at all the involved interfaces according to:

$$V_{el} = \Phi_{el}/e = \Phi_m/e^- + \Psi_m + \beta_{ion} + \beta_{dip} + \theta_{as} \tag{1}$$

where Φ_m is the work function of the metal, e^- is the charge on the electron, Ψ_m is the potential change inside the metal (or oxide film) due to the influence of adsorbed species from the electrolyte, β_{ion} and β_{dip} are the ionic and dipole contributions to the potential difference across the solution side of the metal–solution interface, and θ_{as} is the dipole potential at the solution–air interface [4]. The adsorption of surface active species will also influence the electronic work function of a metal (and thereby the Volta potential) according to:

$$\Delta\Phi_m = 4\pi\mu N_s \tag{2}$$

where N_s is the surface coverage and μ is the effective dipole moment of the individual metal–adsorbate complex. Hence, according to eqn (2), the variation in Volta potential during adsorption at the metal–electrolyte interface can be used to obtain valuable information on the degree of coverage of the adsorbate, and on the sign of surface dipoles formed by adsorbed molecules. Using the vibrating capacitor, the Volta potential difference at the interface water–quaternary ammonium salt monolayer–oil was measured [5]. A potential difference near 400 mV was observed at maximum coverage (0.7–0.8 nm^2/molecule) by the ammonium salt. A decrease in the coverage of ions (2 nm^2 per molecule) led to a decrease in the Volta potential difference to 190 mV.

In a previous investigation by the present authors, self-assembled monolayers of sodium oleate $C_{17}H_{33}COONa$ and dodecylamine $C_{12}H_{25}NH_2$ were deposited on carbon steel [6]. The adsorption of oleate increased the Volta potential by 235 mV whereas that of dodecylamine decreased the Volta potential by 210 mV compared to the metal without adsorbate. This indicates that the adsorption of negatively charged acid groups and the subsequent dipole formation (negative at the adsorbed carboxyl of the oleate anion and positive at the metal oxide) increased the electronic work function of the metal. Amine adsorption leads to a dipole formation with an opposite sign (e.g. positive at the ammonium cation and negative at the metal oxide groups at the top of the metal oxide) and consequently decreases the electronic work function of the metal. Hence from these results and in agreement with eqn (2), it seems that the nature of the metal–organic monolayer interaction significantly influences the work function of the metal.

The SKP technique has also been applied to investigate the kinetics of delamination from defects for polymer coated metals [7,8]. It has been shown that the technique was able to locate the sites of the anodic and cathodic reactions below highly resistive organic coatings.

1.2. Metal Surface–Polymer Coating Interactions

The protective properties of polymer coatings depend greatly on the character and strength of the adhesive bonds at the metal/organic coating interface. Many coatings show excellent adhesion to metals in dry conditions but lose their adhesion after water penetration to the interface (wet adhesion) [9]. Bolger [10] found that adhesion of functional polymers depends on acid-base interactions across the interface that originate from the isoelectric point of the surface (IEPS) and pK_a of the polar groups in the polymer.

Metal oxide covered by hydroxyl groups can form a variety of electrostatic bonds with polar organic compounds:

Organic acid $$-MOH + HXR \rightarrow -MOH_2^{+}\,{}^{-}XR \qquad (3)$$

Organic base $$-MOH + XR \rightarrow -MO^{-}\,{}^{+}HXR \qquad (4)$$

where X is an oxygen, sulfur or nitrogen atom. The role of Lewis acid-base interactions was pointed out for iron covered by polyacrylic acid (eqn 5) [11].

$$-COOH + HO–Fe \rightarrow -COO^- \, ^+Fe + H_2O \quad (5)$$

The alkoxysilanes $R^*–Si(OMe)_3$ are known as adhesion promoters of organic polymers to the surfaces of different inorganic substrates. The vapour phase adsorption of methoxysilanes was studied by SKP [6] and it was found that a monolayer of octyltrimethoxysilane increased the Volta potential of iron by 250–360 mV. Hence, it is possible that the adsorption increased the electronic work function as a result of the formation of dipoles with a negative charge on the adsorbate ($^-$O–Si) and a positive charge on the surface of the metal oxide. This indicates the formation of ferrosiloxane bonds = $Fe^{+-}O–Si\,(O)_2–R^*$ at the interface (eqn 3). The potential difference for a grafted siloxane monolayer did not change in humid air (96–100%RH) indicating that the interfacial bonds were stable to hydrolysis.

Scanning Kelvin probe measurements (SKP) can also be obtained for metals covered by highly resistive polymer coatings in dry and wet air conditions. It was shown [6] that the metal (Zn, Al)-polymer (alkyd and epoxy resins) interface contains dipoles formed by the ion groups of the polymer and the metal oxide. This acid-base interface interaction (eqns 3,4) was observed and measured by potential scanning across the coated metal/metal boundary. The formation of dipoles as described in eqn (3) increased the Volta potential whereas that described in eqn (4) decreased the Volta potential relative to non-coated metal. During the ingress of water to the interface and the hydrolysis of adhesion bonds, a thin electrolyte layer separates the metal and the polymer. This leads to the formation of two interfaces (e.g. metal/electrolyte and electrolyte/polymer) and changes the interfacial potential difference. Hence, a decrease in the wet adhesion properties at the metal/polymer interface can be monitored using the scanning Kelvin probe [6]. The main aim of the present paper was to investigate further the feasibility of using the scanning Kelvin probe to study the metal/polymer interface during atmospheric weathering conditions.

2. Experimental

2.1. Samples

Carbon steel samples were used as substrates. The samples were polished, washed in methanol and water and finally dried. An alkyd resin containing TiO_2 as pigment (e.g. Beckers Syntem spray) and an epoxy coating (e.g. Ciba HY 991 with triethylenetetramine cross-linker) were used in this study. The siloxanes were synthesised from silanes (APS γ-aminopropyl (trimethoxysilane) and OcS octyl (trimethoxysilane) by ABCR-Huls-Petrarch) with dissolution in water (1% wt, pH 10.6). The deposition of siloxane APS and OcS on the metal surface was carried out from a silane–water solution with consecutive drying of the surface.

2.2. Scanning Kelvin Probe

Volta potentials were measured using a commercial Scanning Kelvin probe (UBM Messtechnik). The equipment can be used for measuring the change in Volta potential vs time at a single point of the surface or for measuring the surface distribution in Volta potential on a part of the sample.

The reference electrode (probe) was a Ni–Cr alloy needle with a tip diameter of 50 μm. A surface contour measurement was made prior to each potential measurement. This allows the separation between the tip (Ni–Cr alloy) and the specimen to be kept constant during potential measurements. The lateral resolution of the probe was 50–100 μm. The vibration amplitude was 20 μm and the frequency was 2 kHz. The probe potential was calibrated relative to the $Cu/CuSO_4$ electrode and recalculated on the standard hydrogen electrode scale. The samples were exposed to air or argon at a controlled relative humidity.

2.3. Electrochemical Impedance Spectroscopy

Electrochemical Impedance Spectroscopy (EIS) measurements were performed using a two-electrode configuration. The electrochemical sensor used to perform EIS measurements under atmospheric weathering condition has been described in detail elsewhere [12]. Basically, it consists of a gold or titanium grid deposited on the top of the organic coating [12]. The sensors with Ti grids were exposed for at least 24 h at 96% RH prior to the EIS measurements.

3. Results and Discussion

3.1. Influence of the Nature of Adhesion Bonds at the Metal–Polymer Interface on the Volta Potential

According to the theory of the creation of Volta potential several factors will contribute to the Volta potential at the metal/polymer interface such as:

(1) the formation of ordered dipole layers (groups with oriented dipole moments) in the polymer close to the metal surface,

(2) the separation of charges at the metal/polymer interface,

(3) the dielectric properties of the polymer itself, and

(4) the influence of the formation of oriented dipoles at the polymer/air interface.

The presence of this last contribution has been observed for ion-exchange membranes [3,13].

Figure 1 shows the Volta potential distribution over a carbon steel sample coated with an alkyd resin, paraffin or epoxy resin. A part of the sample was non-coated (right-hand side in Fig. 1). Alkyd increased the Volta potential of the steel by 300 mV (Fig.1a). The value of the Volta potential measured at low RH (30%) was independent of the coating thickness (in the range of 10 to 110 μm). The alkyd was a polymerised polybasic organic acid (phthalic acid) and polyhydric alcohol (glycerin). Both carboxylate and ester groups can participate in the metal/polymer binding. The potential at the steel/alkyd interface was shifted in the same direction as by monolayers of oleate or octylsiloxane. On this basis it is possible to suggest that the metal–alkyd adhesion bonds include interaction of ionic dipoles with positive charges on the passive film and negative charges on carboxylate groups of the polymer.

Paraffin was melted and deposited on carbon steel. The Volta potential was measured in air at 30%RH across a coated (left-hand side) and non-coated (right-hand side) steel surface. A small increase in Volta potential in the range of 50–70 mV (10 μm film thickness) was observed at coated locations (see Fig. 1b). This may be due to a low concentration of dipole groups at the interface (carboxylate or ester groups that can be formed due to the oxidation of paraffin). Increase in paraffin film thickness to 110 μm gives a potential difference of steel-coated steel near 90 mV. This shows that it is possible to perform SKP measurements even for high resistive coatings with very low concentration of ionic groups.

As opposed to alkyd or paraffin, the epoxy resin decreased the Volta potential of carbon steel by 300–600 mV (depending on the composition of the resin). The measurements were carried out by scanning across the boundary of coated and non-coated steel at 30%RH. This is shown in Fig.1(c). Again no large Volta potential dependence on coating thickness (30–190 μm) was observed. Increasing thickness of the polymer led to an increase in the potential of only 20–50 mV, which can be related to a change in the distance between probe tip and metal/polymer interface. The active groups in the epoxy resin can be amino groups fixed in the polymer matrix. The interfacial interaction (eqn 6) includes the ionic bonds: negatively charged oxide and positively charged protonated amino groups.

$$=\text{Fe–OH} + \text{HN (P)–R} \rightarrow =\text{Fe–O}^{-}\ {}^{+}\text{H}_2\text{N(P)–R} \tag{6}$$

The formation of the surface compounds (eqn 6) during vapour adsorption of amines was observed by X-ray photoelectron spectroscopy (XPS) [14]. This kind of interface dipole can give a contribution in the measured potential change. To demonstrate further this hypothesis, the triethylenetetramine cross-linker was adsorbed from a saturated vapour phase for 12 h on to a carbon steel surface. The surface was then kept in the air for 24 h. Under these conditions nearly one monolayer of adsorbate can be deposited. The adsorption of the cross-linker decreased the Volta potential of steel by 250 mV. The results are similar to that obtained for a dodecylamine monolayer [6]. The maximum potential difference of 400 mV was observed at the highest degree of coverage of the water surface by ammonium salt [5]. The observed potential decrease (600 mV) in presence of the epoxy resin was approximately twice that observed for a monolayer of dodecylamine or triethylenetetramine. This can be due to additional contributions to the Volta potential from the polymer itself (oriented dipoles inside the structure of the polymer) or to the polymer/air interface. To study this first possibility the potential distribution across a sandwich cell, i.e. carbon steel–epoxy resin–carbon steel was measured. This is shown in Fig. 2. The right-hand part of the figure corresponds to non-covered steel, the middle part is epoxy covered steel (*ca.* 100 μm thick) and at the left-hand side to an assembly with a second steel plate pasted on the top of the epoxy resin. The potentials of the two steel surfaces were at similar values. The polymer inside the two Me/epoxy/Me interfaces has

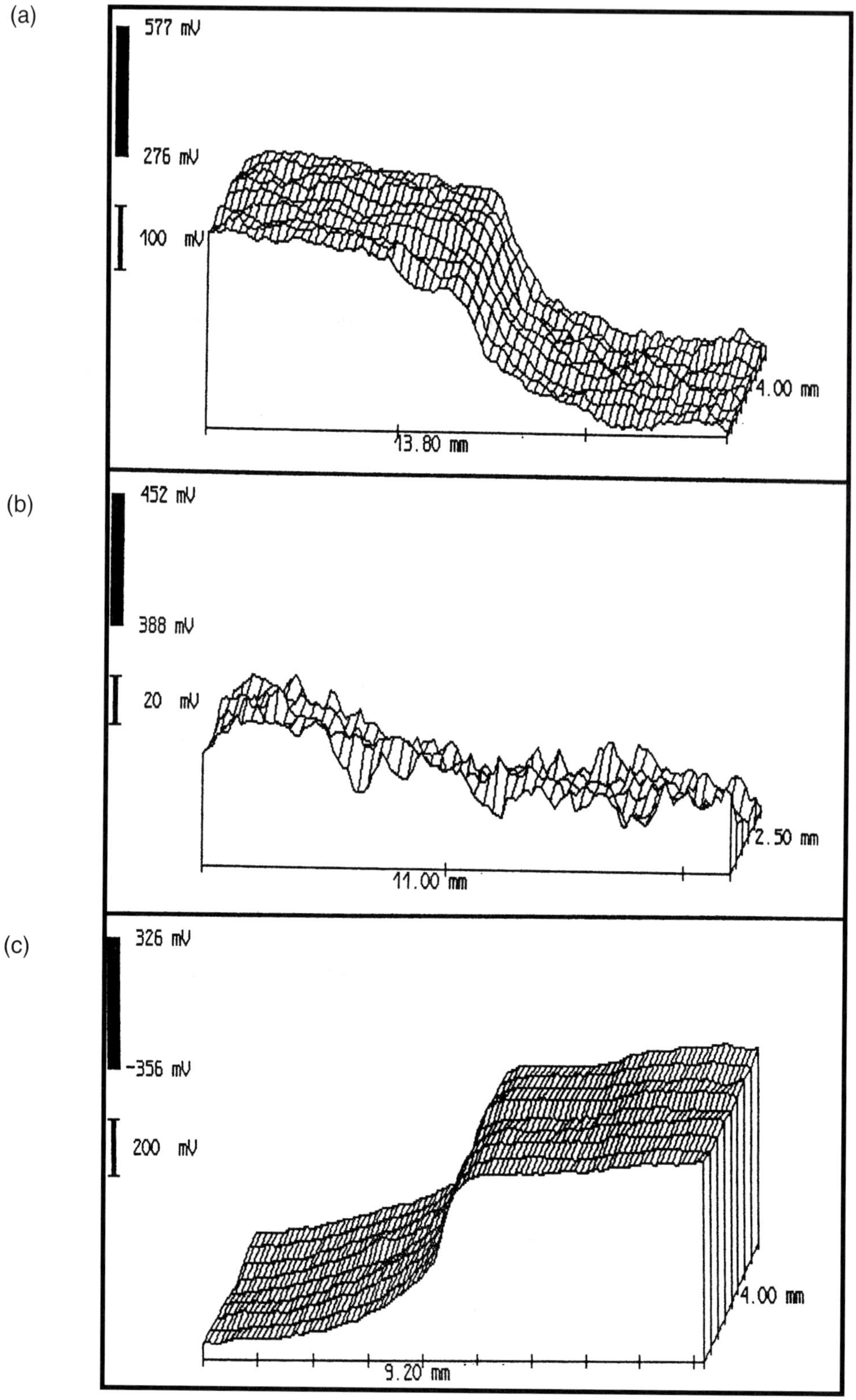
(a)
577 mV
276 mV
100 mV
13.80 mm
4.00 mm
(b)
452 mV
388 mV
20 mV
11.00 mm
2.50 mm
(c)
326 mV
-356 mV
200 mV
9.20 mm
4.00 mm

dipoles in opposite directions to the electrical moments and the potential changes at the interfaces were consequently compensated. These data show that the dielectric properties of the epoxy resin do not introduce a large contribution to the measured value of the Volta potential at the epoxy/steel interface.

For an evaluation of the contribution of the epoxy/air interface to the measured Volta potentials a coating composed by two layers (alkyd (40 μm) and epoxy (80 μm)) was deposited. This provides the formation of steel/alkyd, alkyd/epoxy and epoxy/air interfaces. The coverage by epoxy decreased the Volta potential from 580 to 300 mV compared to the alkyd coated steel surface. We can suppose that the alkyd/epoxy interface makes a small contribution to the Volta potential since the epoxy was deposited after polymerisation of the alkyd resin thus leading to a decreased amount of the acid groups in the alkyd that are able to interact with the amino groups of the epoxy resin. The observed potential drop by 280 mV can be related to the epoxy–air dipole layer. However, it should be noted that the exact determination of

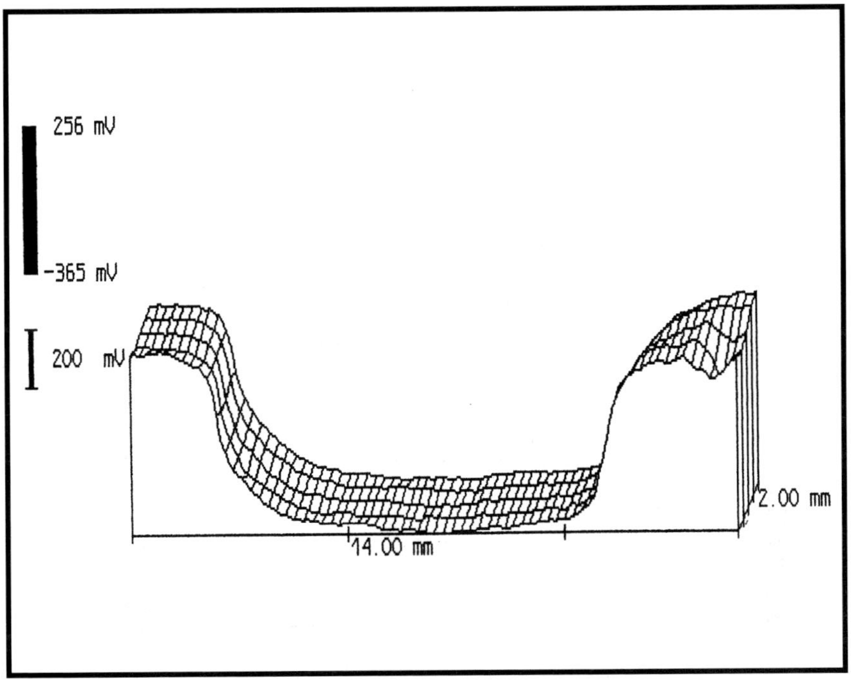

Fig. 2 (above) *The Volta potential distribution across a sandwich sample consisting of carbon steel–epoxy resin–carbon steel at 30% RH. The upper thick bar indicates the extent of the change in Volta potential; the lower thin bar relates directly to the Figure.*

Fig. 1 (opposite) *The Volta potential distribution across a coated steel (left-hand side)–steel (right hand-side) boundary at 30%RH. (a) alkyd resin, (b) paraffin and (c) epoxy resin. The potentials are given relative to SHE. The upper thick bar indicates the extent of the change in Volta potential; the lower thin bar relates directly to the Figure.*

the Galvani potential difference for this interface is a complicated task. The structure of the air/epoxy interface can include the location of positively charged amino-groups on the top of the polymer and counter-ions (for instance hydroxide ion) deeper inside the polymer [3,13]. This orientation of the structured dipoles will decrease the overall potential of the system.

Hence, from these results it seems that the decrease in the Volta potential for steel coated with an epoxy resin may be explained by dipole formation at the steel/epoxy and at the epoxy/air interfaces.

To investigate further the influence of ion-exchange groups in polymers on the Volta potential, two siloxanes (aminopropylsiloxane (APS) and octylsiloxane (OcS)) were synthesised on the steel surface by deposition of the droplets of their respective water solutions of silanes. After drying, the thickness of both polymers was in the range of 10–15 μm. The siloxanes were cross-linked by Si–O–Si bonds and contained different organic radicals. The APS contained amino-groups that can carry positive charges whereas OcS had uncharged organic radicals. The OcS siloxane increased the potential by 300–500 mV. The main positive contribution to the Volta potential was given by ferrosiloxane bonds [6]. Figure 3 (a,b) shows the geometrical profile and the potential profile over an island of APS siloxane. Aminosiloxane decreased the potential of steel by 350 mV. The difference in the potentials can reflect a different orientation of the dipoles at the siloxane–air interfaces. As shown in Fig. 3(c)(*overleaf*), aminosiloxane has fixed proton-containing amino-groups above and negative counter ions (for example Si–O–) in the polymer matrix. This dipole orientation decreased the overall potential at the APS/steel interface.

3.2. Water Disbonding at the Metal–Polymer Interface

The influence on the Volta potentials of the diffusion of water through the organic coating was investigated. It has been shown in previous investigations that the potential of carbon steel in the passive state is in the range of 200–450 mV [15]. An increase in RH increased the Volta potential whereas drying decreased the potential by 50–60 mV. As described earlier, the alkyd coating in dry conditions (30%RH) increased the Volta potential at the steel/alkyd interface by 300–350 mV. During prolonged exposure at 96% RH the potential of the painted steel decreased to values close to that of the non-covered metal. This effect can be related to hydrolysis of the metal–polymer bonds and formation of a thin water layer at the interface. Figure 4 shows the variation of the Volta potential as a function of the exposure time at 96% RH for different surface treatment before the painting operation. The time corresponding to the potential fall was largely dependent on the surface preparation. For non-degreased surfaces (Fig. 4, curve 1), the Volta potential reached a minimum after 3 h and the paint started to delaminate (tested by knife cutting) during further exposure in air at 96% RH. The time corresponding to the potential drop increased after cleaning the metal surface with ethanol (Fig. 4, curve 3) and the most stable interface was found after surface alkali degreasing and rinsing in water (curve 2). In this case the potential reached a minimum after two days of exposure in humid air. This indicates that the penetration of water into the organic coating neutralised the dipoles (see eqn 7).

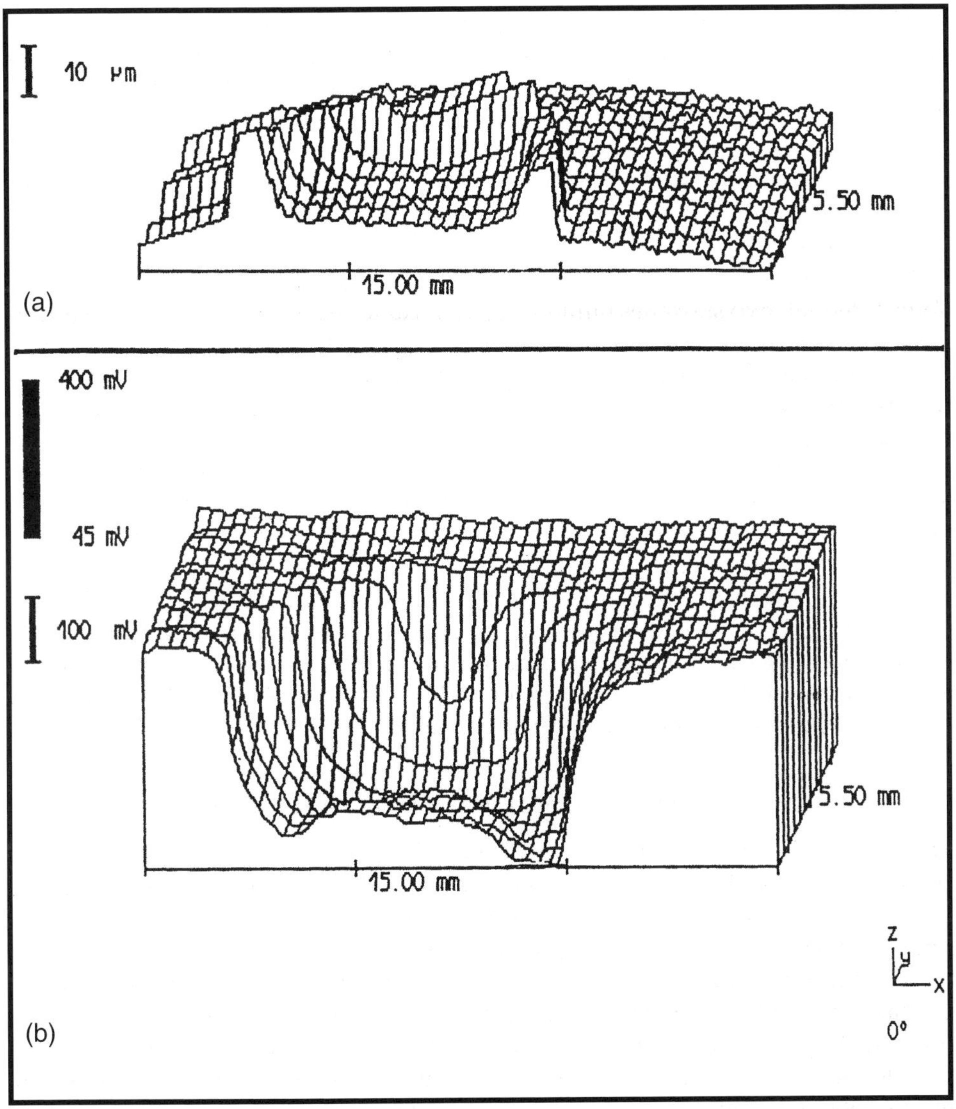

Fig. 3 (a,b) *Surface contour (a) and Volta potential distribution (b) over an APS siloxane island deposited on carbon steel. The measurements were performed in air at 30% RH. (c) the proposed structure of the siloxane–air interface.*

$$(R\text{–}COO^{-}\ {}^{+}Fe\text{=}) + n\ H_2O \rightarrow R\text{–}COOH\ \ H_2O + H_2O\ HO\text{–}Fe\text{=} \quad (7)$$

Drying of the samples led to an ennobling of the potential, which indicates a reformation of interfacial dipoles and of interfacial bonds. The time necessary to recover the potential during drying was in the range of 18–20 h, which should be compared to the much lower time necessary to start the hydrolysis process. It should be noted that each wet/dry transition decreased the maximum potential obtained during dry conditions. This indicates that the level of adhesion in dry conditions

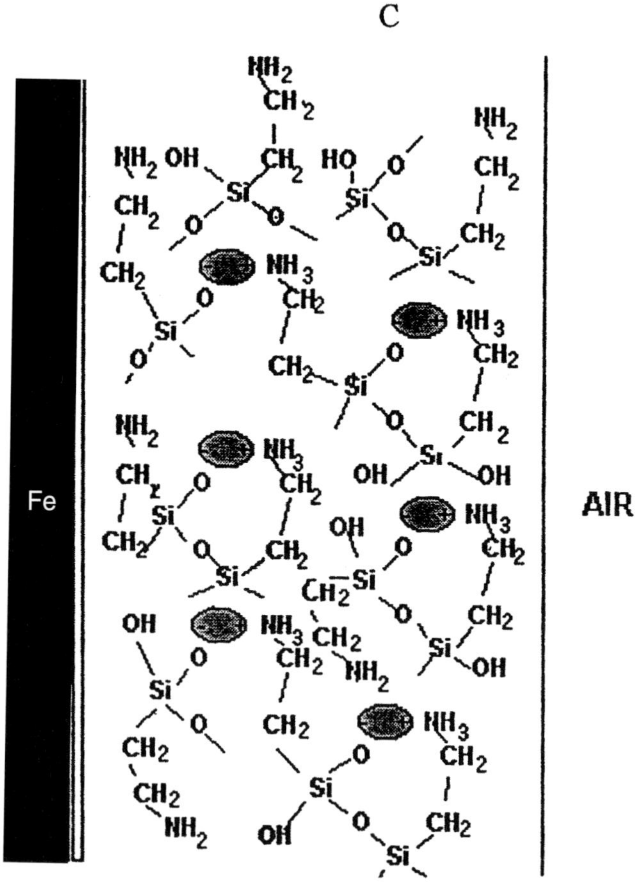

Fig. 3 (c) The proposed structure of the siloxane–air interface.

decreased as a function of the number of wet/dry transitions. A prolonged exposure to humid air (6 days) led to a change in the potential in the opposite direction. At this stage wetting led to a potential ennoblement whereas drying led to a potential decrease by about 60 mV. The potential transitions were fast and correspond to the water adsorption–desorption process at the steel surface. No influence of the coating on the potential of steel was observed due to weakening of the coating adhesion.

The change in Volta potential during wet–dry transitions was studied for the epoxy/steel interface (Fig. 5). As already demonstrated, the epoxy resin decreased the Volta potential of the steel at 30%RH by about 600 mV. Exposure to wet air at 96%RH led to a fast potential ennoblement from –350 mV to –50 mV (Fig. 5a). This potential was stable during four hours at 96% RH. Extended drying decreased the potential to negative values (–220 mV). Hence, as for the alkyd/steel, the hydrolysis of interfacial bonds changed the potential towards the value of the non-coated metal. However, the potential obtained in the dry state was more noble than the initial potential and this effect may be related to a decrease in the coverage of the interface by active groups participating in the adhesion bonding. The epoxy–steel interface was exposed to humid air (96%RH) for 4 days. The dry–wet transition increased the

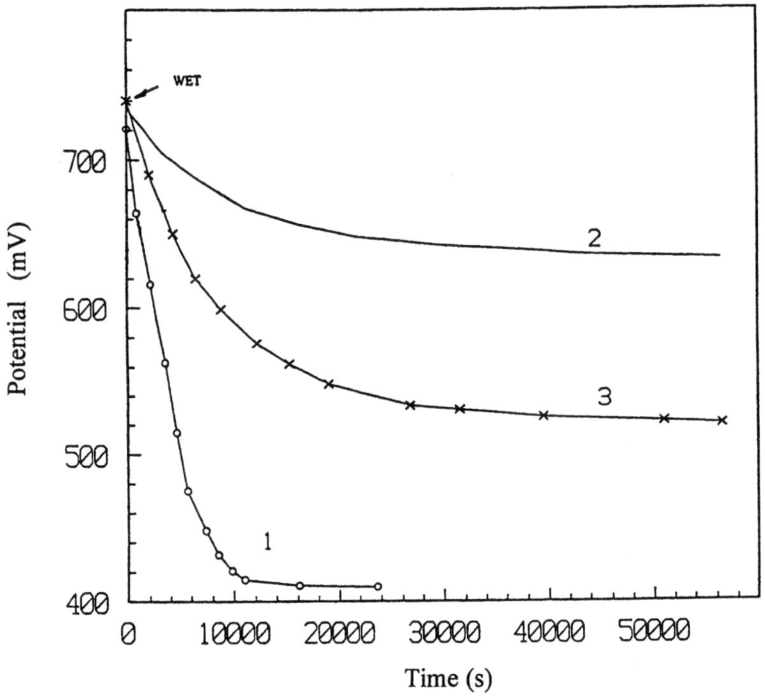

Fig. 4 Volta potential vs time of exposure for the alkyd/steel interface as a function of the surface pretreatment. 1: no degreasing; 2: alkali degreasing; 3: degreasing by ethanol. Curves 2 and 3 correspond to the beginning of the exposure.

potential from –220 mV to 80 mV (Fig. 5b). A higher potential in the wet state may also be related to lower adhesion properties. This last potential (e.g. 80 mV) was stable during next 5 days exposure at 96%. The potential difference during one wet–dry cycle decreased from 430 mV at the beginning of exposure to 100 mV after 15 days of exposure in humid air (Fig. 5c). The last potential transient is close to the potential change observed during wet–dry transitions for non-coated passive steel surface (50–70 mV) [6]. At this stage of the exposure, the adhesion at the metal/polymer interface was low and the polymer could rapidly be delaminated. However, a negative difference (near 270 mV) exists between the potential of the free steel (350 mV) and that of the epoxy coated steel (80 mV) with a hydrolysed interface. This may be related to the contribution from the epoxy/air interface, which is not subjected to hydrolysis. The potential transitions (Fig. 5c) during wet–dry air humidity cycling were the same after two months of the exposure in 96%RH air.

Electrochemical impedance spectroscopy studies were conducted for the epoxy coated steel surface during exposure at 96%RH. As inferred from the experimental section the measurements were performed *in situ*. As only a single time constant was visible from the EIS spectra, the coating capacitance (C_c) and the coating resistance (R_c) were determined using a simple RC circuit. Figure 6 shows the time dependence of C_c and R_c during exposure to air at 96% RH. After 15 days of exposure the resistance decreased by three order of magnitude whereas the coating capacitance increased by a factor of 3. This corresponds to the time at which SKP measurements performed

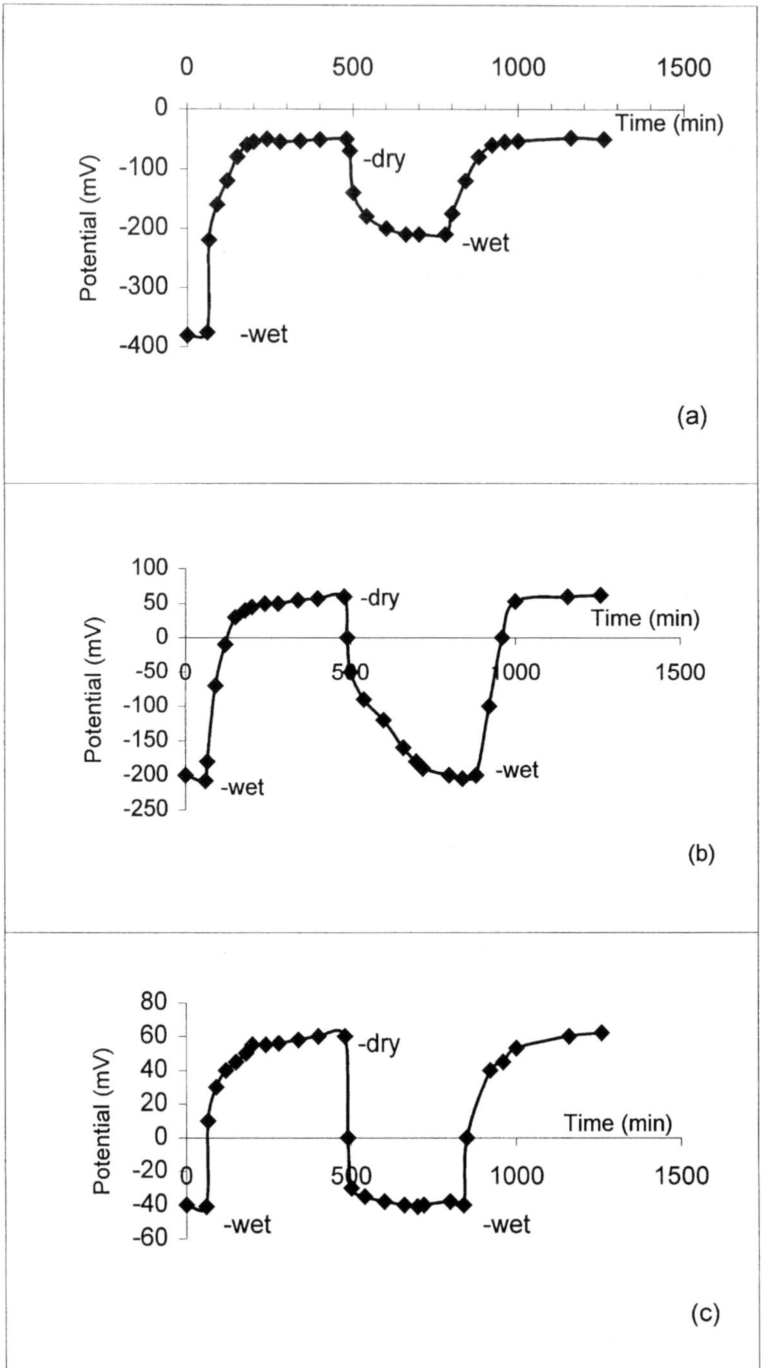

***Fig. 5** The influence of the wet–dry (30–96% RH) transitions on the Volta potential of the carbon steel epoxy interface; (a) at the start of exposure; (b) after 5 days of exposure, and (c) after 15 days of exposure in humid air (96% RH).*

during wet/dry transitions indicate evidences of the water adsorption–desorption process at the metal surface below the organic coating. The potentials in wet conditions reached the level obtained for non-coated metal.

4. Conclusions

- The scanning Kelvin probe is a unique in situ tool for studying the charge distribution in the metal–polymer interface.

- The measured Volta potential includes contributions from potential changes at the metal/polymer and polymer/air phase boundaries. The structures of interface dipoles and nature of the chemical groups of the polymer responsible for adhesion interactions with metals can be evaluated. The presence of amino-groups (epoxy resins) leads to formation of dipoles with positive charge on polymer boundary and negative on the metal oxide and decreases the Volta potential in comparison with coating-free metal. The acid groups (alkyd resins) increase the Volta potential due to formation of the interface dipoles with negative charge on the polymer and positive on the metal oxide.

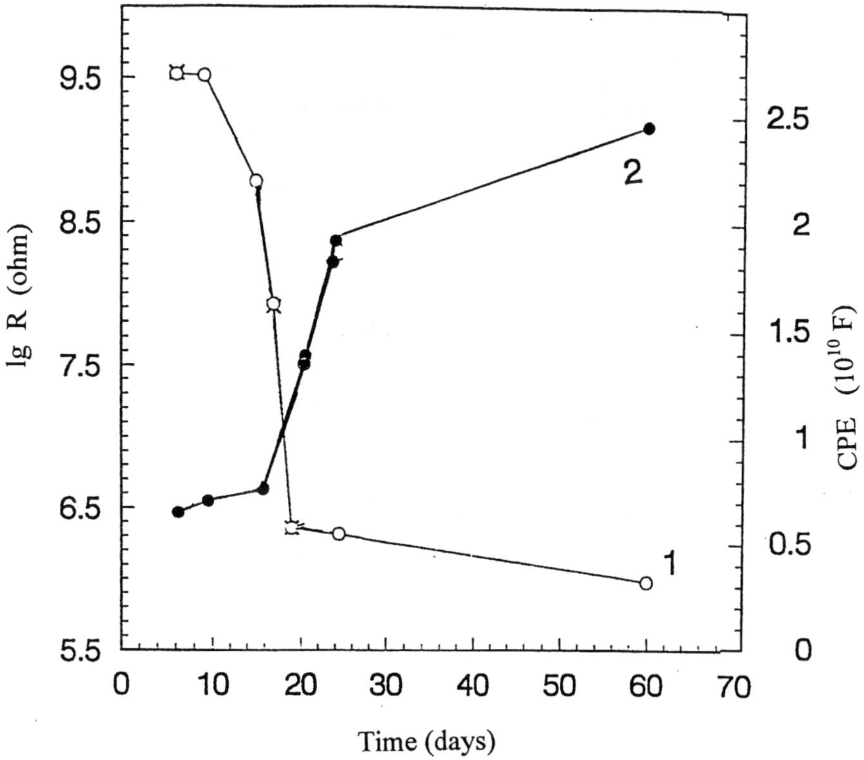

***Fig. 6** The coating resistance (1) and coating capicatance (2) determined by EIS vs time of exposure in humid air (96% RH). The CPE exponents were in the range 0.8–0.85.*

- Water ingress to the metal–polymer interface neutralised the interface dipoles and changed the Volta potentials to values closer to the coating-free metals. From measurement of the Volta potential during exposure of the interfaces in humid air the kinetics of water penetration and loss of coating adhesion can be determined. On this basis the quality of the metal surface preparation can be evaluated.

References

1. S. Yee, R. A. Oriani and M. Stratmann, *J. Electrochem. Soc.*, 1991, **138**, 55–64.
2. M. Stratmann and H. Streckel, *Corros. Sci.*, 1990, **30**, 681–697.
3. Z. Samec, B.W. Johnson, M. Cappadonia, M. Jauch and K. Doblhofer, *Sensors and Actuators B*, 1993, **13–14**, 741–743.
4. S. Trasatti, in *Modern Electrochemistry*, 13, B. E. Conway and J. O'M. Bockris (Eds). Plenum Press, New York, 1979, pp.81–93.
5. J. T. Daves and E. K. Rideal, *Interfacial Phenomena*, Academic Press, NY & London, 1963.
6. A. Nazarov and D. Thierry, *Proc. 3rd Int. Conf. ACT '98*, paper No. 45, Katowice, Poland, 1998.
7. M. Stratmann, M. Wolpers, H. Streckel and R. Feser, *Ber. Bunsenges. Chem.*, 1991, **95**, 95–102.
8. M. Stratmann, R. Feser and A. Leng, *Electrochim. Acta*, 1994, **39**, 1207–1214.
9. H. Leidheiser and W. Funke, *J. Oil Colour Chem. Assoc.*, 1987, **5**, 121–127.
10. J. C. Bolger, in *Adhesion Aspects of Polymeric Coatings*, M. L. Mittal (Ed.) Plenum Press, NY, 1983, pp. 4–12.
11. H. Yamabe, *Progr. in Org. Coatings*, 1996, **28**, 911–917.
12. T. C. Simpson *et al.*, *J. Electrochem. Soc.*, 1989, **136**, 2761–2769.
13. M. Cappadonia and K. Doblhofer, *Electrochim. Acta*, 1989, **34**, 1815–1819.
14. M. J. Incarvia and S. Contarini, *J. Electrochem. Soc.*, 1989, **136**, 2493–2499.
15. A. Nazarov and D. Thierry, in *Proc. EUROCORR' 98*, Utrecht, the Netherlands, 1998.

7
Anti-Corrosion Properties of Polyaniline Coatings

M. KRALJIĆ, Z. MANDIĆ* and LJ. DUIĆ
Faculty of Chemical Engineering and Technology, University of Zagreb, Zagreb, Croatia
*PLIVA, Research Institute, Zagreb, Croatia

ABSTRACT

Polyaniline (PANI) coatings were electrosynthesised on steel samples (13% and 5% Cr) using sulfuric and phosphoric acids as supporting electrolytes. The protective properties of the coatings in the supporting electrolytes were investigated by monitoring open-circuit potentials vs time, and by applying EIS (electrochemical impedance spectroscopy). Polyaniline films have been found to provide corrosion protection. Thicker PANI films have capacitive behaviour at low frequencies with small resistance at high frequencies. Similar results were obtained on Pt. Thinner films exhibit much higher resistance probably because a significant amount of PANI is transformed from emeraldine (EM) to leucoemeraldine (LE). The protective film deposited in a phosphate solution seemed to have better properties than the film deposited in a sulfate solution and therefore this film was also tested in 0.1 molL^{-1} HCl. In the chloride-containing solution, the time of protection was significantly shorter.

1. Introduction

Polyaniline (PANI) is a conductive polymer that is considered for many applications because of its specific properties; one of the applications is corrosion protection. There are numerous reports on the corrosion protection of PANI where PANI was synthesised either chemically or electrochemically. Chemically synthesised polymer is deposited from a dispersion [1] or solution [2]. Since the electrochemical synthesis takes place directly on the metal surface it is expected to have better adherence than in the case of chemically synthesised PANI.

Surface metal oxides stabilise many metals in aggressive environments i.e. the dissolution rates are very low. The stability of such films depends on the properties of the film and on the solution composition. In a very aggressive medium the oxide film is damaged and the potential of the metal is shifted toward negative values. DeBerry [3] reported that in acid solution PANI stabilises metal oxide films within the passive region for a long period of time. When electrochemically synthesised PANI is used for protection, oxide is formed during the synthesis and is stabilised by PANI. In the case of chemically synthesised PANI it appears that the polymer promotes the formation of oxide [1].

Polyaniline has been used for protection of stainless steel [3], iron [1,4], mild steel [5] and copper [1]. To protect iron and mild steel some unconventional acids were used for the supporting electrolyte like oxalic [4,5] and *p*-toluenesulfonic [6] acids because electrodeposition of conductive polymers on such substrates is not easy. In the case of stainless steel electrodeposition should be easier because oxide is formed on the metal and the dissolution rate is reduced.

The aim of this study was to electrosynthesise an adherent polymer coating on stainless steel. The conditions of the synthesis were determined as well as the protective properties. Two different coatings, from sulfuric and phosphoric acid respectively, were compared and samples of two different steels (13% and 5% Cr) were used.

2. Experimental

The measurements were carried out in a standard one-compartment three-electrode cell. The working electrode was a steel sample or Pt-electrode. The synthesis was carried out on steel samples and on the Pt electrode. The counter electrode was Pt-foil and the reference electrode Ag/AgCl. The samples used in this study were stainless steel AISI 420 (13% Cr) and low-alloy steel (5% Cr) mounted in a Teflon holder with exposed electrode area of 0.95 cm^2. Steel samples were polished using 600 grit emery paper, washed with double-distilled water and finally with ethanol. The voltammetric experiments were carried out using a potentiostat (Wenking, LB75L), a function generator (Wenking, VSG 83) and PS 486 computer for storing and evaluating the results.

Polyaniline was electrochemically synthesised from 0.5 $molL^{-1}$ aniline in 1.5 $molL^{-1}$ H_2SO_4 and from 0.5 $molL^{-1}$ aniline in 3 $molL^{-1}$ H_3PO_4 by cycling the working electrode potential between –400 and 1000 mV for stainless steel and between +300 and 1000 mV for low-alloy steel at a scan rate of 50 mVs^{-1}. The number of cycles varied depending on the supporting electrolyte that was being used. Polyaniline films of different thickness were synthesised and the thickness was estimated from the amount of charge, *Q*, necessary to switch from the leucoemeraldine (LE), the reduced form of polyaniline, to the emeraldine (EM), the oxidised form of polyaniline. The conversion takes place according to the following reaction:

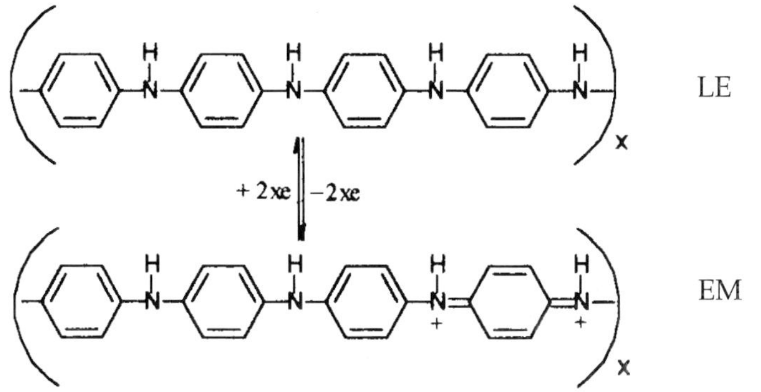

(1)

The amount of charge, *Q*, necessary to switch from LE to EM is represented by the charge under the current peak A in Fig. 1. Table 1 presents the thickness and the charge of PANI films. Coatings were examined in 1 $molL^{-1}$ H_2SO_4 and 3 $molL^{-1}$ H_3PO_4.

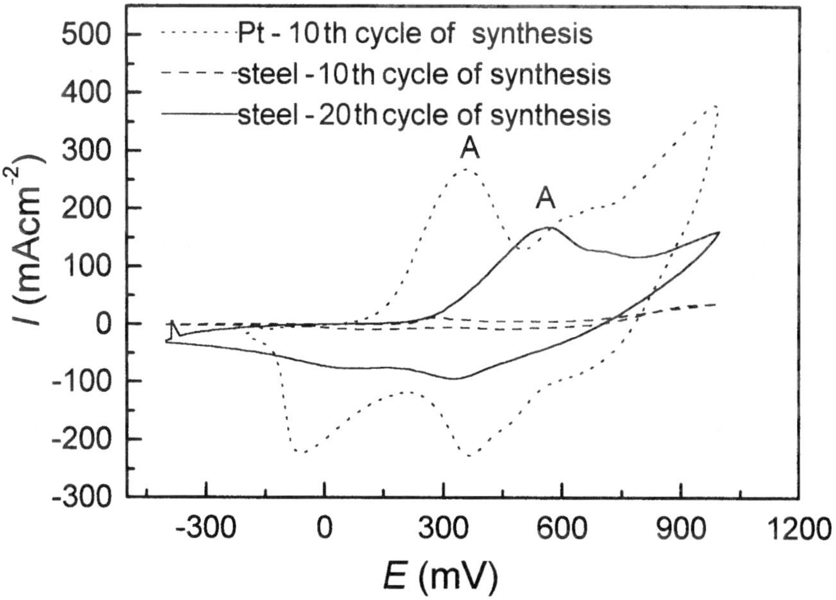

Fig.1 *A comparison of voltammograms for PANI synthesis on stainless steel and on Pt electrode.* c_{an} = *0.5 molL*$^{-1}$, c_{el} = *1.5 molL*$^{-1}$ H_2SO_4, v = *50 mV s*$^{-1}$.

Table 1. *Charge Q and the calculated thickness of PANI films*

Designation of PANI electrode synthesised from H_2SO_4	Designation of PANI electrode synthesised from H_3PO_4	Charge Q (mC cm^{-2})	Thickness d (nm)
PANI1SO_4	PANI1PO_4	16	80
PANI2SO_4	PANI2PO_4	73	350
PANI3SO_4	PANI3PO_4	479	2300
PANI4SO_4	PANI4PO_4	871	4500

EIS measurements were carried out using a potentiostat (EG &G Princeton Applied Research, Model 283), Frequency Response Detector (EG&G Princeton Applied Research, Model 1025) and Model 398 Electrochemical impedance software. The impedance was measured at the potential of EM (+530 mV) and, also, at the steel corrosion potential. The superimposed sinusoidal voltage signal of 5 mV amplitude was applied. Data were collected in the frequency range 10^5–10^{-2} Hz, taking five points per decade. The electrodes were kept at each selected potential during 60 s before the measurement. The impedance data were analysed by Boukamp's fitting program (Version 5,71).

3. Results and Discussion

The synthesis of PANI was carried out on the stainless steel electrode, and for comparison also on the Pt electrode. Figure 1 illustrates voltammograms showing that synthesis on steel is a slower process and that the current peak presenting the LE–EM transformation develops at higher potentials as compared to the voltammogram from the Pt electrode. This is in agreement with previous work on non-noble metals [7]. Figure 2 illustrates PANI synthesis on stainless steel but in phosphoric acid solution. It is evident that polyaniline phosphate ($PANIPO_4$) layer growth is a slower process than polyaniline sulfate ($PANISO_4$).

3.1. Open Circuit Potential Monitoring

After the synthesis the electrode with PANI layer was transferred to the supporting electrolyte solution and was kept at 500 mV to ensure the conversion of LE to EM (schematic eqn 1). After that the electrode was left at the open circuit potential (OCP) which was monitored. Figure 3 illustrates the potential–time dependence for the $PANISO_4$ electrode. It was established that the stability of the OCPdepends on the thickness of the layer. Thus, the OCPof $PANI3SO_4$ layer during 30 days of monitoring stayed in the passive region. In the case of the bare stainless steel electrode, the electrode potential after the passivation reached a corrosion potential value of –470 mV almost instantly. The OCP of stainless steel covered by a $PANIPO_4$ layer (Fig. 4)

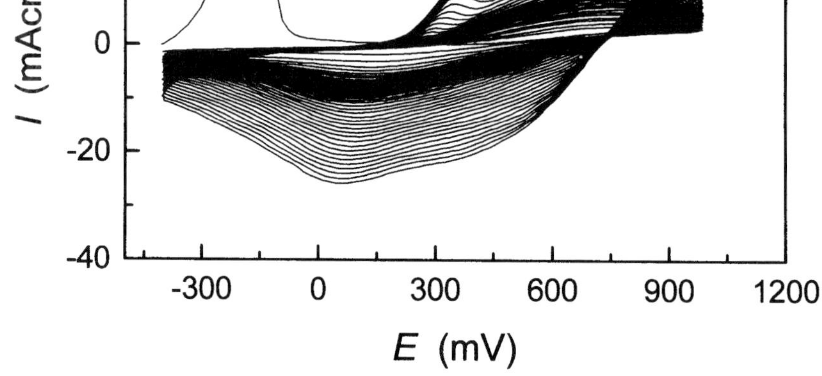

Fig. 2 The illustration of PANI film growth on stainless steel electrode. $c_{an} = 0.5\ molL^{-1}$, $c_{el} = 3\ molL^{-1}\ H_3PO_4$, $\nu = 50\ mV\ s^{-1}$, *40 potential cycles of synthesis.*

remained in the passive region during the time of monitoring, irrespective of the layer thickness but thinner layers did result in somewhat lower values of OCP.

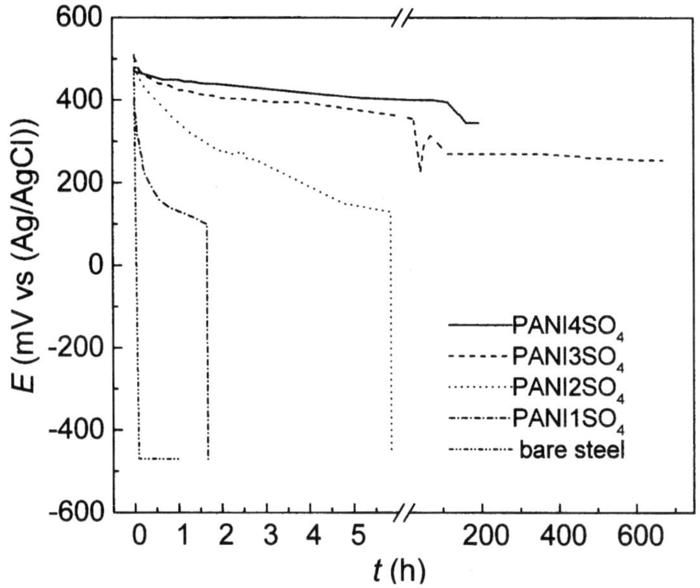

Fig. 3 *The OCP– time dependence for the bare stainless steel, and for the stainless steel-PANISO$_4$ electrode in 1.0 molL^{-1} H_2SO_4.*

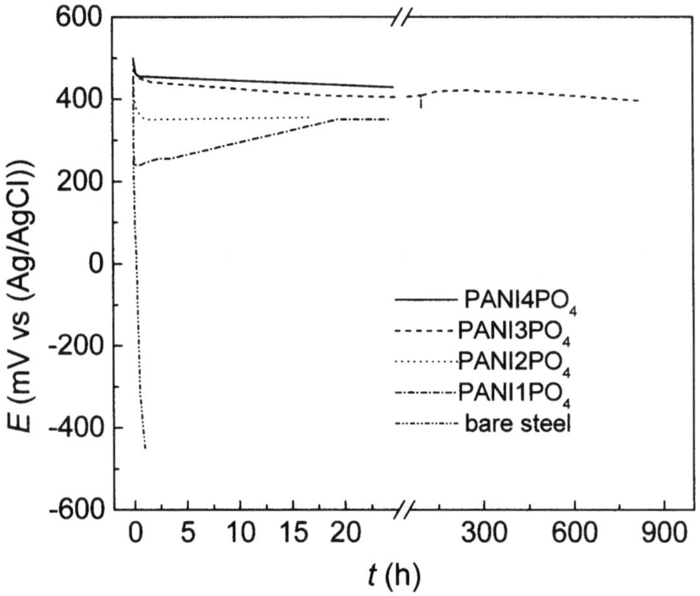

Fig. 4 *The OCP–time dependence for the bare stainless steel and for the stainless steel-PANIPO$_4$ electrode in 3 molL^{-1} H_3PO_4.*

To compare the efficiency of the protection of $PANISO_4$ and $PANIPO_4$ tests in sulfuric acid solution were carried out for both layers. Thus, $PANIPO_4$ was formed in phosphoric acid solution, and the OCP monitored for some time, the electrode was then transferred into sulfuric acid solution and the monitoring continued (Fig. 5). There was no change in the *ocp* value which means that $PANIPO_4$ protects stainless steel equally well in phosphoric and in sulfuric acid solution. It is obvious that $PANIPO_4$ offers better protection than $PANISO_4$ for which the OCP stability in sulphuric acid solution lasted for only two hours (Fig. 3).

Since a $PANIPO_4$ layer exhibits better protection it was also tested in hydrochloric acid solution. As is shown in Fig. 6 the OCP remained only for a short period of time at the passivation potential region before approaching the corrosion potential region. The time the potential remained at the OCP of the passive region was somewhat longer for thicker layers, presumably due to the longer distance that Cl^- ion has to travel to reach the stainless steel substrate. Therefore a PANI layer does not offer good corrosion protection in a chloride-containing medium.

The $PANIPO_4$ layer was also tested for protection of low alloy steel. Figure 7 illustrates the OCP–time dependence. As one can see the period of protection is considerably shorter than in the case of stainless steel, and it depends on the thickness of the layer.

The tests carried out in sulfuric acid solutions show that PANI stabilises the OCP in the passive region, what can be explained through the stabilisation of the oxide layer formed at the stainless steel surface during the electrochemical synthesis of PANI, thus protecting the substrate. Since the OCP of protected steel depends on the

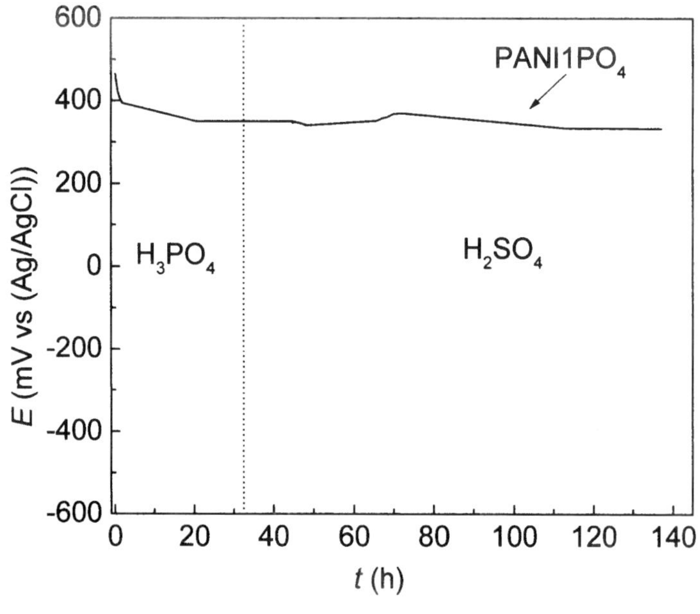

Fig. 5 The OCP–time dependence for the stainless steel-$PANIPO_4$ electrode in two different supporting electrolyte solutions.

PANI layer thickness, and in the case of thinner layers after a period of time reaches the corrosion potential region, one may assume that EM has been converted to LE

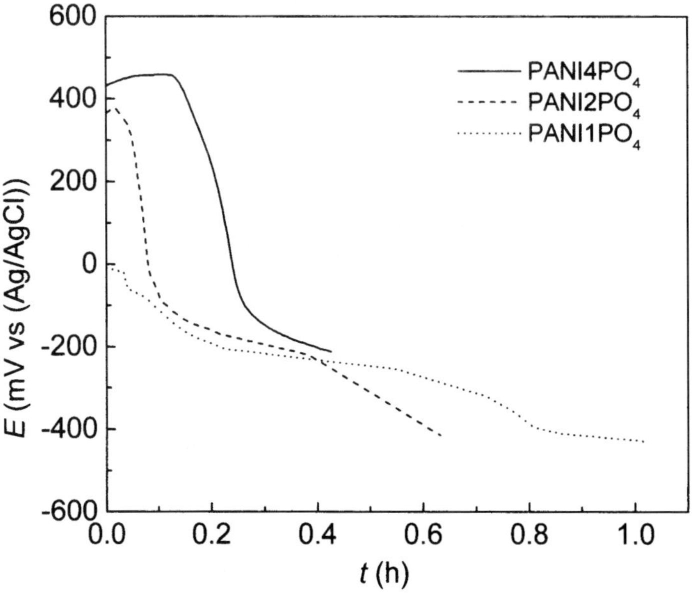

Fig. 6 The OCP–time dependence for the stainless steel-PANIPO$_4$ electrode in 0.1 molL^{-1} HCl.

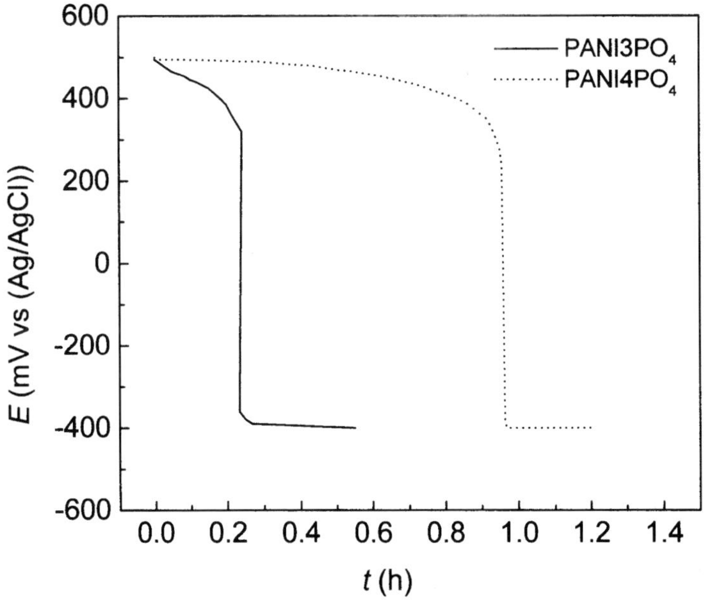

Fig. 7 *The OCP–time dependence for the low alloy steel-PANIPO$_4$ electrode in 3 molL^{-1} H$_3$PO$_4$.*

and the protection is terminated. The role of PANI layer can be assumed by the following mechanism [8]:

$$2Fe + 3EM^{2+} + 3H_2O \rightarrow Fe_2O_3 + 3LE + 6H^+ \quad (2)$$

$$^1/_2\,O_2 + LE + H_2O \rightarrow EM^{2+} + 2OH^- \quad (3)$$

$$EM^{2+} + 2e^- \rightleftarrows LE \quad (4)$$

That mechanism assumes the re-oxidation of LE to EM by oxygen from the solution (i.e. from air), and the time of protection by PANI will depend on the relative rates of reactions (2) and (3), as well as on the available amount of EM.

3.2. Electrochemical Impedance Spectroscopy Investigations

To enable the characterisation of the protection mechanism of PANI on stainless steel, EIS measurements were carried out on the bare stainless steel electrode, on the Pt-PANI electrode, and on the stainless steel-PANI electrode. The EIS measurements were carried out at the potential of the passive region (EM potential region) of +530 mV. Figure 8 shows the Nyquist diagram for the bare stainless steel tested in sulfuric acid and phosphoric acid solutions. A semicircle at high frequencies is registered in both cases, from which charge transfer resistance (R_{ct}) is evaluated. Although the measurements in phosphoric acid solution were not very reproducible, it was possible to evaluate R_{ct}, which is significantly higher for the phosphoric acid medium than

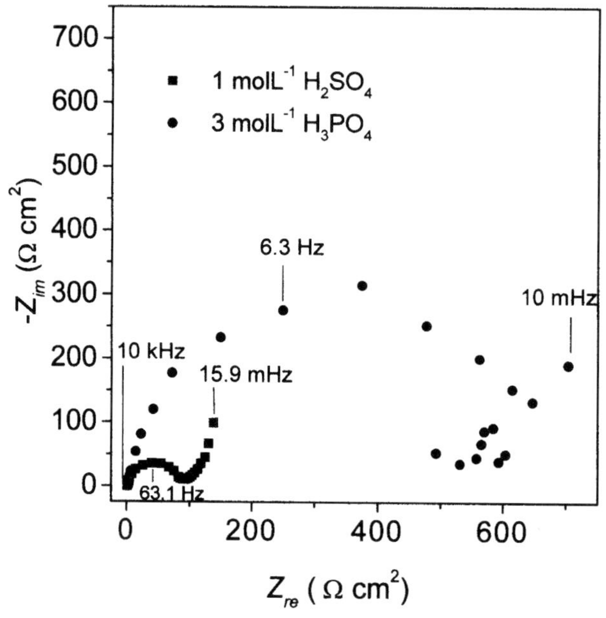

Fig. 8 Nyquist diagram for the bare stainless steel electrode in two different supporting electrolyte solutions.

for the sulfuric acid solution (Tables 4 and 5). The Nyquist diagram for H_2SO_4 solution at lower frequencies exhibits the behaviour attributed to Warburg diffusion impedance and to capacitive behaviour at very low frequencies. These phenomena are in accordance with the theory of a passive layer formation [9].

The behaviour of Pt-PANI electrode was examined at the same potential (Fig. 9), at which the Nyquist diagram represents pure capacitive behaviour, which is in accordance with expectations since at that potential PANI is in a conductive state [10]. These measurements are important for the understanding of the results obtained for the stainless steel-PANI electrode. The Nyquist diagram for stainless steel-PANI3 electrode (Fig. 10) exhibits pronounced capacitive behaviour accompanied by a very small value of R_{ct} at high frequencies, and it is close to the results obtained on the Pt-PANI electrode. Such behaviour suggests that there is no charge transfer process associated with the steel substrate, and the obtained values of R_{ct} are associated with the EM to LE transition. These results confirm the protective property of PANI under the experimental conditions.

The Nyquist diagram for the thinner PANI1 layer (Fig. 11) on stainless steel at 530 mV also gives a semicircle at high frequencies, while at lower frequencies there is a diffusion component and a capacitive segment. A semicircle that results in higher value of R_{ct} than in the case of PANI3 layer should be attributed to the increased ratio of LE/EM. The existence of LE at the potentials of EM was proved to be possible by Malinauskas *et al.* [11], in the presence of a substance being oxidised by EM. In this case that substance is the stainless steel substrate. From the experimental results and from the fitted values a model of the equivalent circuit is assumed for the stainless steel-PANI system, and the values are given in Tables 2 and 3.

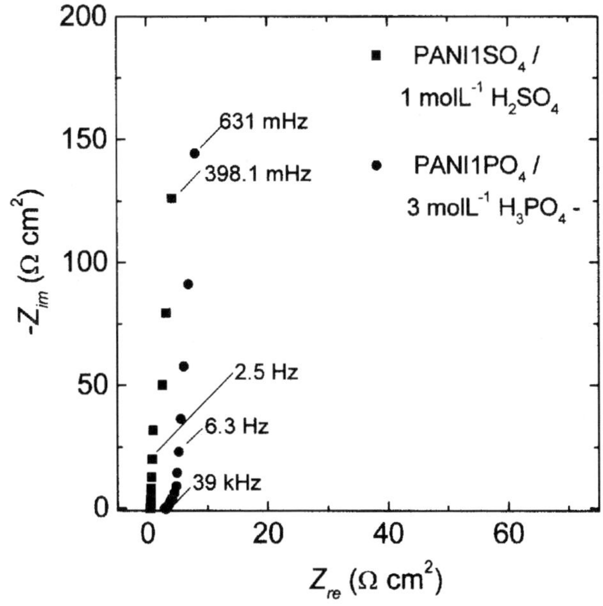

Fig. 9 *Nyquist diagram for the Pt-PANISO$_4$ and PANIPO$_4$ electrodes in corresponding supporting electrolyte solution.*

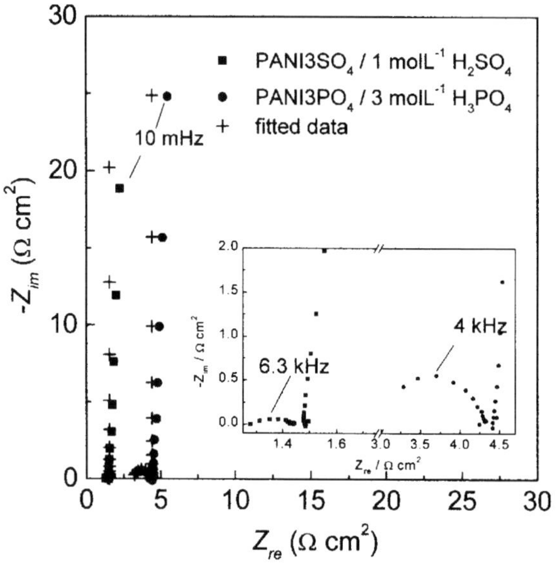

Fig. 10 Nyquist diagram for stainless steel-PANI3 electrodes in two different supporting electrolyte solutions at +530 mV.

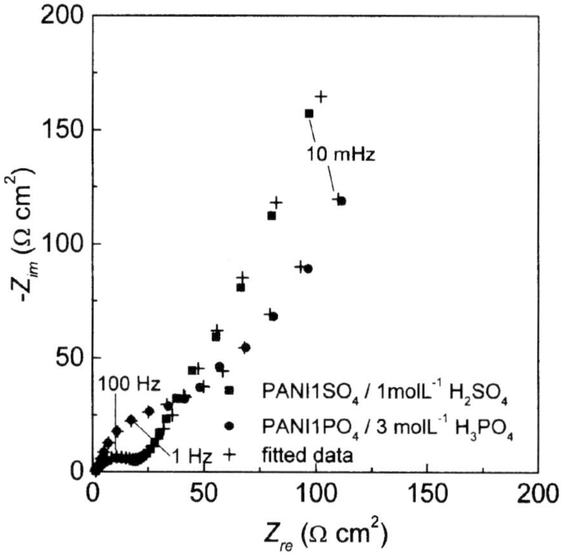

Fig. 11 Nyquist diagram for stainless steel PANI1 electrodes in two different supporting electrolyte solutions.

***Table 2.** The values obtained from the equivalent circuit for the stainless steel-PANI3 electrode*

	R_S (Ωcm^2)	$C_{dl} \times 10^4$ (Fcm^2)	R_{ct} (Ωcm^2)	C_L (Fcm^2)
PANI3SO$_4$	1.38	3.16	0.16	0.64
PANI3PO$_4$	3.17	0.38	1.23	0.79

The equivalent circuit for the thicker PANI layer (Fig. 12a) is represented by the solution resistance (R_s), the charge transfer resistance (R_{ct}) in a conversion reaction

***Table 3.** The values obtained from the equivalent circuit for the stainless steel-PANI1 electrode*

	R_s (Ωcm^2)	$Q_1 \times 10^3$ ($\Omega^{-1}s^n\ cm^{-2}$)	n_1	$R_{ct} \times 10^{-1}$ (Ωcm^2)	$Z_w \times 10^2$ (Ωcm^2)	Q_2 ($\Omega^{-1}s^n\ cm^{-2}$)	n_2
$PANI1SO_4$	1.32	0.43	0.74	1.82	3.77	0.67	0.85
$PANI1PO_4$	3.13	4.40	1.00	3.58	2.84	2.62	1.00

n_1, n_2 indicate the capacitor-like behaviour of the constant phase elements Q_1 and Q_2 through the relationship: $c = Q^n$.

of EM to LE, the double layer capacitance at PANI/solution boundary (C_{dl}), and the capacitance which has a tendency to approach the frequency independent value (C_{L}– low frequency capacitance).

The equivalent circuit of the thinner PANI layer (Fig. 12b) is represented by R_s, R_{ct}, the constant phase elements (Q_1, Q_2) and the Warburg diffusion impedance (Z_w).

The EIS measurement was carried out at the value of the steel corrosion potential (at –470 mV). At that potential PANI is in the LE form and it is not expected to protect steel. The Nyquist diagrams (Figs 13 and 14) show that the stainless steel-PANI electrode exhibits a higher value of R_{ct} than the bare steel electrode and that this increases with increase of the layer thickness. This behaviour shows that although LE has no protective effect on steel, it imposes a certain barrier to the corrosion process.

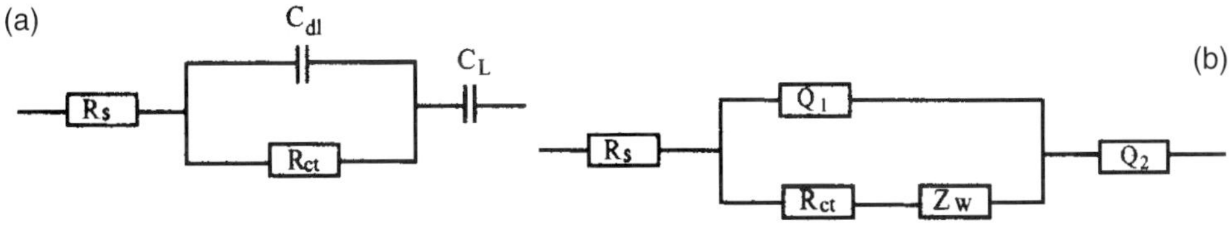

***Fig. 12** The equivalent circuit for the stainless steel–PANI electrodes for (a) PANI3, and (b) PANI1 layers; R_s: solution resistance, R_{ct}: charge transfer resistance, C_{dl}: double layer capacitance, C_L: low frequency capacitance, Q_1, Q_2: constant phase elements, Z_w: Warburg diffusion impedance.*

***Table 4.** R_{ct} values of different electrodes for the sulphuric acid solution*

***E* (mV)**	R_{ct} (Ωcm^2)				
	$Pt-PANISO_4$		**Stainless steel-$PANISO_4$**		**Bare stainless steel**
	1	**3**	**1**	**3**	
+530	0.63	0.36	18	0.16	87
–200	727	1395			
–470			21	43	7

Table 5. R_{ct} *values of different electrodes for the phosphoric acid solution*

E (mV)	R_{ct} (Ωcm²)				
	Pt-PANIPO$_4$ 1	3	**Stainless steel-PANIPO$_4$** 1	3	**Bare stainless steel**
+530	1.59	1.30	35.80	1.23	638
–200	8770	12 500			
–470			185	2560	35

However, after a certain period of time the layer is damaged due to hydrogen evolution, and the electrode assumes the behaviour of a bare steel electrode.

4. Conclusions

It has been shown that it is possible to synthesise polyaniline (PANI) layers on stainless steel, as well as on low alloy steel electrodes. Such PANI layers, when in the emeraldine (EM) state, protect stainless steel from corrosion in sulfuric and phosphoric acid solutions by stabilising the oxide layer formed on the steel surface during PANI polymerisation — thus preventing the metal dissolution process. Comparing PANISO$_4$ and PANIPO$_4$ layers, PANIPO$_4$ proved to offer better protection,

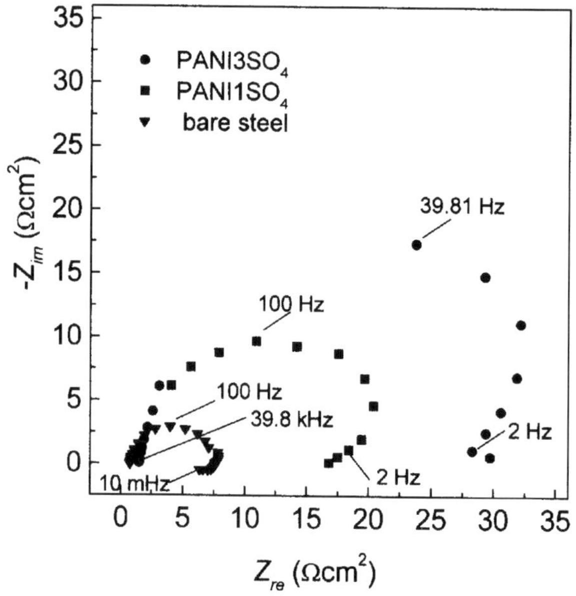

Fig. 13 *Nyquist diagram for the bare stainless steel and PANISO$_4$ covered electrodes at the corrosion potential (–470 mV).*

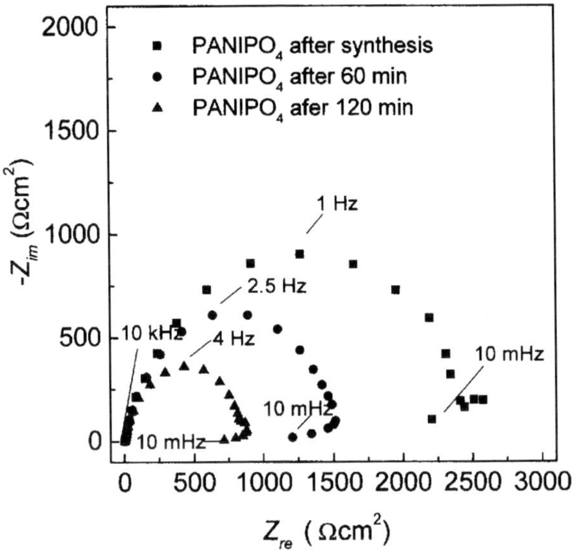

***Fig. 14** Nyquist diagram for stainless steel-PANIPO$_4$ electrode in 3 molL^{-1} H_3PO_4 at the corrosion potential (–440 mV).*

i.e. the open circuit potential stabilised in the passive region for a longer period of time under the experimental conditions. It was also shown that thicker layers offer protection for a longer time. For low alloy steel, however, the time of protection was much shorter.

Electrochemical impedance spectroscopy measurements show that a PANI layer in the EM state prevents the dissolution of steel, and the capacitive behaviour of PANI prevails in the passive potential region. At the values of the corrosion potential the leucoemeraldine (LE) state of PANI does not prevent, but only slightly hinders, the metal dissolution process.

References

1. B. Wessling, *Adv. Mater.*, 1994, **6**, (3), 226–228.
2. Y. Wei, J. Wang, X. Jia, J. M. Yeh and P. Spellane, *Polymer*, 1995, **36**, (23), 4535–4537.
3. D. W. DeBerry, *J. Electrochem. Soc.*, 1985, **132**, (5), 1022–1026.
4. D. Sazou and C. Georgolios, *J. Electroanal. Chem.*, 1997, **429**, 81–93.
5. J. L. Camalet, J. C. Lacroix, S. Aeiyach, K. Chane–Ching and P. C. Lacaze, *J. Electroanal. Chem.*, 1996, **416**, 179–182.
6. J. L. Camalet, J. C. Lacroix, S. Aeiyach and P. C. Lacaze, *J. Electroanal. Chem.*, 1998, **445**, 117–124.
7. V. V. Abalyaeva and O. N. Efimov, *Polymers for Advanced Technologies*, 1997, **8**, 517–524.
8. B. Wessling, *Synth. Met.*, 1997, **85**, 1313–1318.
9. J. R. Macdonald, *Impedance Spectroscopy — Emphasizing Solid Materials and Systema*, John Wiley & Sons, NY, USA, 1987.
10. P. Fiordiponti and G. Pistoia, *Electrochim. Acta*, 1998, **34**, (2), 215–221.
11. A. Malinauskas and R. Holze, *Electrochim. Acta*, 1998, **43**, (18), 2563–2575.

8

Evaluation of Polymer Coating Properties with Electrochemical Impedance Spectroscopy (EIS), Electrochemical Noise Analysis (ENA) and Artificial Neural Networks (ANN)

F. MANSFELD and C. C. LEE

Corrosion and Environmental Effects Laboratory (CEEL), Department of Materials Science and Engineering, University of Southern California, Los Angeles, CA 90089-0241, USA

ABSTRACT

Electrochemical impedance spectroscopy (EIS) and electrochemical noise (ECN) data have been collected for steel samples with 12 different polymer coating systems exposed to natural sea water at two remote marine test sites and to artificial sea water in laboratory studies. The impedance spectra have been analysed using suitable equivalent circuits (EC) and the delaminated area A_d has been determined from the breakpoint frequency, f_b, and/or pore resistance, R_{po}, as a function of exposure time. These data were normalised using visual observation data obtained after 90 days according to ASTM D 610. For the remaining exposure time excellent agreement between the impedance data and visual observation was obtained in most cases. The ECN data have been analysed in the time and the frequency domains. The dependence of the noise resistance, R_n, on the bandwidth, Δf, of the ECN measurement has been discussed. Potential and current power spectral density plots have been prepared from which spectral noise plots have been calculated. Excellent agreement between spectral noise plots and impedance spectra has been observed provided instrumental artifacts and the low current noise levels for protective coatings did not affect the ECN measurements. Two artificial neural networks (ANN) have been trained using theoretical impedance spectra for 'good', 'intermediate' and 'poor' coating quality. The trained ANNs have been tested using experimental impedance spectra for different polymer coating systems on steel collected during exposure in natural sea water for up to one year. In general, excellent agreement between the predictions of coating quality made by experienced operators based on general features of the impedance spectra and parameters such as f_b and R_{po} and the classification results obtained from the ANNs was obtained. Coated samples with intentional defects have been exposed to natural and artificial sea water for 30 days. One set was exposed at the open-circuit potential, while another set was connected to a zinc electrode. It was found that bacteria colonised only those samples exposed to natural sea water which were not cathodically protected suggesting that identification of microbially induced corrosion (MIC) cannot be based solely on spatial relationships between large accumulations of bacterial cells and iron corrosion products.

1. Introduction

In a project devoted to the evaluation of the effects of microorganisms on corrosion protection by polymer coatings twelve coating systems applied to steel have been exposed to natural sea water at Port Hueneme, CA and Key West, FL and to artificial sea water as control electrolyte [1–4]. Of special interest were the effects of different primers, midcoats and topcoats on polymer coating degradation caused by microorganisms. Degradation of the protective properties of these coatings has been monitored using electrochemical impedance spectroscopy, electrochemical noise analysis and visual observation. A special monitoring approach has been developed for collection of EIS and ENA data from remote test sites. ENA has been performed in the time and the frequency domains. Potential and current power spectral density (PSD) plots were transformed into spectral noise plots for comparison with impedance spectra in the Bode plot format. From analysis in the time domain, the noise resistance, R_n, which has been suggested to be useful in the evaluation of coating properties [5], can be determined. The dependence of R_n on the bandwidth Δf of the ECN measurements and the relationship of R_n to fit parameters obtained from EIS were of interest in order to ascertain the application and reliability of ECN measurements in evaluation of the protective properties of polymer coatings especially with respect to corrosion monitoring. To determine the effects of potential on colonisation by microorganisms two sets of coated steel samples with intentional defects were exposed to natural sea water at Key West and to artificial sea water. One set was exposed at the open-circuit potential, while the second set was connected to a zinc electrode to provide cathodic protection.

Electrochemical impedance spectroscopy has become a very powerful tool for evaluation of the protective properties of polymer coatings and their degradation during exposure to corrosive environments. However, the need of the correct choice of an equivalent circuit, and the use of sophisticated software for the analysis of impedance spectra for polymer coated metals limits the use of the EIS technique to experienced operators. In addition, there is a need for instantaneous information of coating quality during on-site inspection of large structures. In an extension of previous evaluations of the feasibility of using artificial neural networks (ANN) for the classification of EIS data for polymer coated steel [6] a commercial software package [7] was used to establish the most efficient ANN for evaluation of experimental impedance data for polymer coated steel [8]. A training set was established for a number of theoretical impedance spectra which were classified as 'good', 'intermediate' and 'poor' based on the previous quantitative analysis of the experimental EIS data. A testing set was established using monthly data for six different polymer coating systems on steel obtained during the first year of exposure at Key West. Comparisons were made between the predictions based on quantitative analysis of the EIS data by fitting to the appropriate EC and the classification results obtained with the use of the trained ANN [8].

2. Experimental Approach

2.1. Materials and Methods

2.1.1. Materials

In the CR series of coatings systems (Table 1), the nature of the primer, midcoat or topcoat was varied to allow comparisons of the effects of coating type on corrosion protection in general and attack by microorganisms in particular. Coating systems CR 1 and 2 had the same primer, but different mid- and topcoats; CR 5 and 6 had the same primer and midcoat, but different topcoats, while for CR 6 and 7 the midcoat and topcoat were the same, but the primer was different. The total coating thickness ranged from about 125 μm for CR 2 to 275 μm for CR 5 (Table 1a). The coating series JJ (Table 1b) was designed to allow evaluation of the effects of different primers and an additional polyurethane topcoat over an epoxy midcoat on coating performance.

2.1.2. Methods

Impedance spectra were collected once a week for more than one year at the two sites. The traditional approach using a frequency response analyser and a potentiostat was applied in laboratory studies. For collection of electrochemical impedance spectra at remote test sites, a different approach had to be applied in which the potentiostat was eliminated and a multiplexer was used to connect the recording system to the different coating systems (Fig. 1) [1–4]. A PC and the equipment shown in Fig. 1(a) were placed at the marine test site. Communication between a computer at University of Southern California (USC) and a computer at the test site was established as shown in Fig. 1(b). Data obtained at the two remote test sites were transmitted by commercial telephone line to the computer at USC. Electrochemical potential and current fluctuations were collected simultaneously for two coated samples of the same type (Fig. 1). Relative large samples (15.2 × 10.2 cm) were used in order to lower the impedance of the coating systems. The exposed area was about 250 cm^2 for each sample. Visual observation was performed once a month according to ASTM D 610 at Port Hueneme and in the laboratory. Electrochemical impedance and ECN data were analysed using appropriate models and software developed in this laboratory. Samples with artificial coating defects were exposed at Key West and in artificial sea water for 30 days [4,9,10]. Four 0.32 cm dia.holes (intentional defects) (two per side) were drilled through the coating to the metal surface to create defined holidays. Some panels were cathodically protected by attaching a zinc anode; others were allowed to corrode freely. In natural sea water ECN data were collected only for samples not cathodically protected, while in laboratory tests ECN and EIS data were obtained for both types of samples. Field data were compared with ECN data for artificial sea water collected in the laboratory [4,9,10].

Classification of polymer coating quality using an ANN was based on impedance data obtained between 10 kHz and 0.1 Hz [8]. Training of the ANN was performed using phase angle Φ ($-90° < \Phi < 0°$) data in the Bode plot format or the logarithm of the impedance modulus /Z/ (/Z/ in ohm.cm^2). Three types of training data were constructed which were categorised as 'good', 'intermediate' and 'poor'. The program was tested using experimental data which had been collected during exposure to

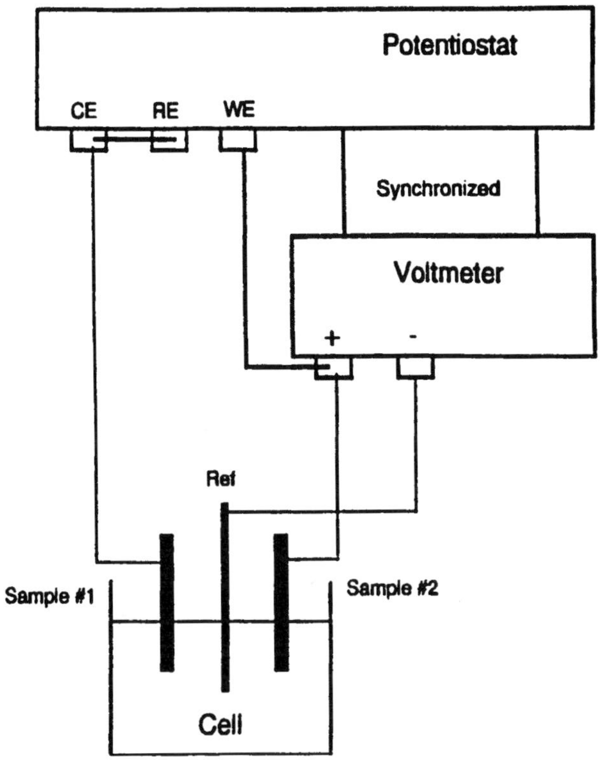

Fig. 1 (Opposite and below) Experimental approach for collection of electrochemical impedance and noise data from a remote test site.

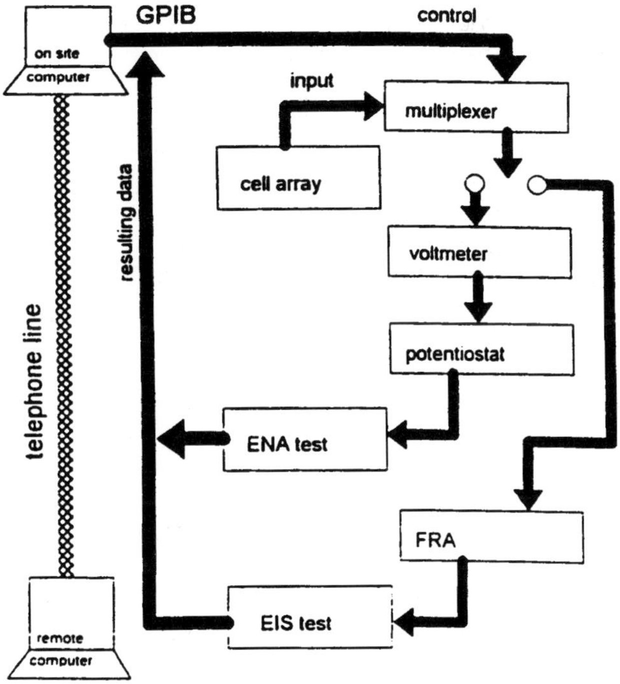

natural sea water at Key West. A training set consisted of three different groups of data files in the Bode plot format for 'good', 'intermediate' and 'poor' quality coatings. Each data file was created using the EC which is most commonly applied for simulation and analysis of impedance data for polymer coated metals (Fig. 2a). In this EC R_s is the ohmic resistance in the electrolyte, $C_c = \varepsilon\varepsilon_0/d$ is the capacitance for a coating with a dielectric constant ε and thickness d, R_{po} is the pore resistance of the coating, while R_p and C_{dl} (the polarisation resistance and double layer capacitance) describe the corrosion reaction at the coating/metal interface. Since the contributions by R_p and C_{dl} do not occur for the 'good' and 'poor' coatings in the frequency range chosen for the analysis by the ANN, the one-time-constant model (OTCM) shown in Fig. 2(b) was used for these two cases. The particular values of the parameters in the EC of Fig. 2 were chosen based on the properties of polymer coated steel exposed to natural sea water shown in Table 1, which describes the composition of the primer, midcoat and topcoat. The range of the C_c values used for the training set of the 'good' coating was based on $50 < d$ (μm) < 250 and $\varepsilon = 4$. For the 'intermediate' and 'poor' coating training data ε was varied between 4 and 80 to take into account the changes due to water uptake. Fifteen impedance spectra were constructed for each group of training data by random choice of the parameters in Table 2 within the indicated range. Figure 3 shows the three groups of training data as log /Z/ – log f plots (Fig. 3a) and Φ – log f plots (Fig. 3b). For the 'good' coatings the impedance spectra are

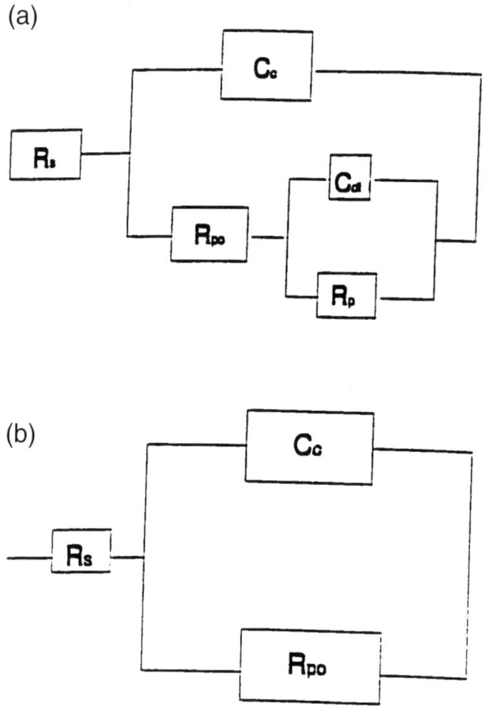

Fig. 2 Equivalent circuits for simulation and analysis of impedance data for polymer coated metals; (a) the coating model; (b) the one-time-constant-model (OTCM).

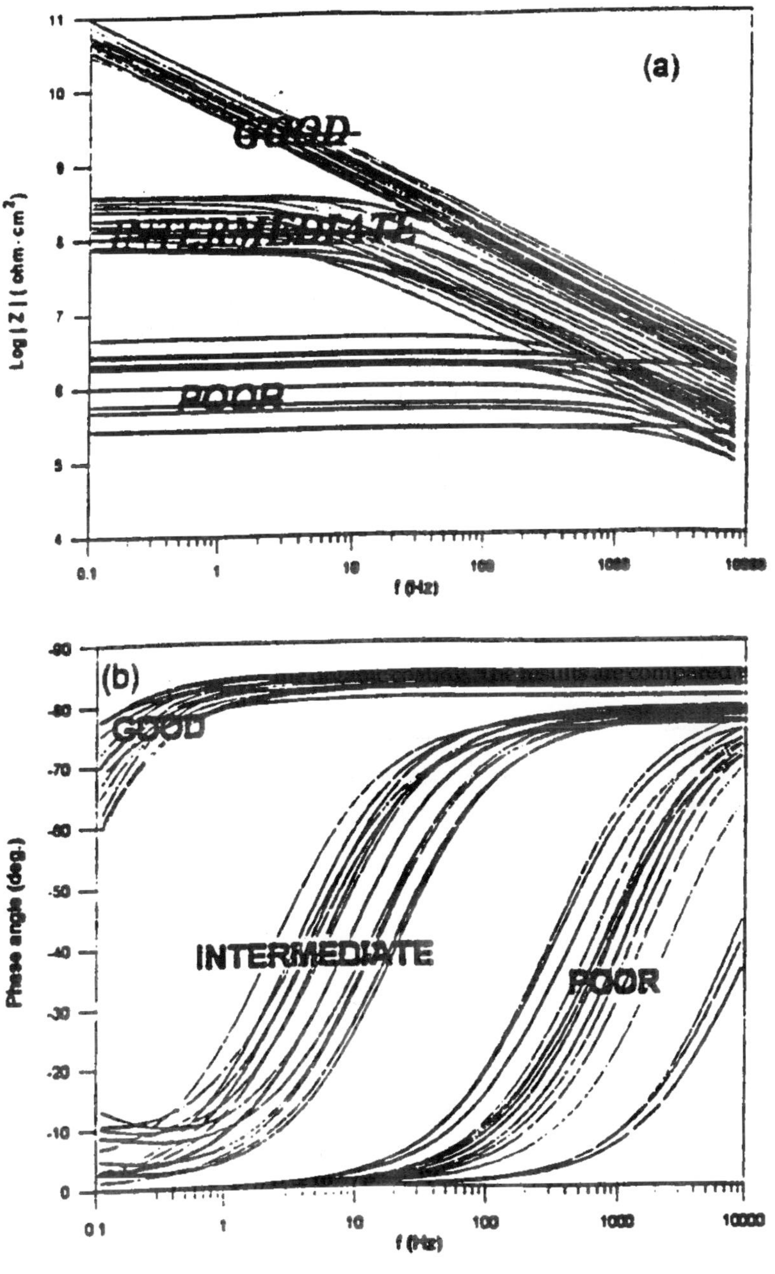

***Fig.* 3** *Theoretical impedance spectra used as training sets for "good", "intermediate" and "poor" coating quality.*

capacitive in the entire frequency range chosen for the analysis, while for the 'poor' coating they are resistive below 1 kHz. For the 'intermediate' coatings the spectra show the contribution by C_c at high frequencies and that by R_{po} at low frequencies (Fig. 3). The entire training set consisted of 15 data files for each type of coating

Table 1. *Composition of coating systems*

(a) CR-series

Sample name	CR1	CR2	CR5	CR6	CR7	CR9
Substrate	Hot rolled steel	Hot rolled steel	Hot rolled steel	Hot rolled steel	Hot rolled steel	Hot rolled steel
Primer	Alkyd high solid[1]	Alkyd high solid[1]	Zn-rich primer[4]	Zn-rich primer[4]	Epoxy polyamide[5]	Epoxy polyamide[5]
Midcoat	Low VOC alkyd[2]	Silicone alkyd[3]	Epoxy polyamide[5]	Epoxy polyamide[5]	Epoxy polyamide[8]	Epoxy polyamide[8]
Topcoat	Low VOC alkyd[2]	Silicone alkyd[3]	Poly-urethane[6]	Latex[7]	Latex[7]	Epoxy polyamide[9]

(b) JJ-series

Sample name	JJ1	JJ2	JJ3	JJ4	JJ5	JJ6	JJ7
Substrate	Cold rolled steel	Cold rolled steel	Cold rolled steel	Cold rolled steel	Cold rolled steel	Cold rolled steel	Cold rolled steel
Primer	Zinc plate[a]	Zinc plate[a]	Phosphate coat[d]	Phosphate coat[d]	IVD-Al[e]	IVD-Al[e]	Epoxy polyamide[f]
Midcoat	Epoxy[b] polyamide	Epoxy[b] polyamide	Epoxy[b] polyamide	Epoxy[b] polyamide	Epoxy[b] polyamide	Epoxy[b] polyamide	Epoxy polyamide[g]
Topcoat		Poly-urethane[c]		Poly-urethane[c]		Poly-urethane[c]	Epoxy polyamide[h]

NOTE:

1. TT-P-645B.
2. TT-E-489H.
3. TT-E-49.
4. SSPC-20, type 2.
5. MIL-P-24441, Formula 150.
6. MIL-C-85285.
7. MIL-P-28578.
8. MIL-P-24441, Formula 151.
9. MIL-P-24441, Formula 152.

(a) Zinc plate per QQ-Z-325 Rev. C (Type II, Class 2).

(b) 2 coats of epoxy polyamide primer MIL-P-23377 Rev. F (Type I class C) per MIL-F-18264 REV.D AMD.I.

(c) 1 coat of MIL-C-85285 REV.B (AS) AMD.2 polyurethane per MIL-F-18264 REV.D AMD.1 colour #36375 lusterless gray of FED-STD-595.

(d) Phosphate coat DOD-P-16232 REV.F (Type Z class 3); hydrogen embrittlement relieved for 8 h at 210–225°C.

(e) IVD aluminium per MIL-C-83488-C (Notice 1, Type II Class I).

(f) MIL-P-24441/20, Rev. B Formula 150, type III.

(g) MIL-P-24441/21, Rev. B Formula 151, type III.

(h) MIL-P-24441/22, Rev. B Formula 152, type III.

quality for a total of 45 data files. The output was designated as (1, 0, 0) for a coating of 'good' quality, while outputs of (0, 1, 0) and (0, 0, 1) were set for 'intermediate' and 'poor' coating quality, respectively.
The commercial software package NeuralWorks Professional II/Plus [7] was used to create two back-propagation networks. There were 51 processing elements (PE) and 3 PEs for inputs and outputs, respectively. Based on the results of Hack and Matteson [6], who found that a 3 hidden layer design gave the best training results, 26 PEs, 13 PEs and 7 PEs were chosen for the 1st, 2nd and 3rd hidden layers, respectively. The training process has been discussed in detail elsewhere [8]. Coating quality was evaluated by these two trained ANNs based on the frequency dependence of Φ or log $|Z|$ according to the training standards shown in Fig. 3. The test data files were selected from the impedance spectra obtained for the first year of immersion in natural sea water at Key West by choosing one measurement/month and normalising the data by the immersed area resulting in 10 different test data sets, i.e. one for Φ and one for log $|Z|$ for each of the five different coating systems. For less protective coatings, which were removed before the end of one year, fewer than 10 test files were available for one test data set.

In order to facilitate the comparison of the predicted and determined results the coating quality index (CQI) has been introduced which has values of 1 for 'good' coating quality and 0 for 'poor' coating quality [8]. The coating quality index is calculated from the expected or determined results by multiplying the components of the results by a certain factor and summing of the three elements. For example, the expected result (0, 1, 0) becomes $CQI_{in} = 1 \times 0 + 0.5 \times 1 + 0 \times 0 = 0.5$, while the output value (0.057, 0.964, –0.024) becomes $CQI_{out} = 0.539$. Plots of CQI_{in} and CQI_{out} as a function of exposure time allow easy comparison of the agreement between predicted coating quality and the classification given by the ANN. The results of the classification of polymer coating quality using ANNs will be discussed for the five different coating systems.

3. Experimental Results and Discussion

The type of information obtained with EIS and ENA during exposure to artificial and natural sea water will be discussed for coating systems CR 5, 6, 7 and 9 (Table 1). Figure 4 shows impedance spectra obtained in natural sea water at Port Hueneme for CR 5, 6 ,7 and 9 after 1, 4 and 7 months. Spectra obtained in artificial sea water for the corresponding time periods had similar features. It can be assumed that the continued decrease of the impedance was due to an increase of the delaminated area at which corrosion occurred and an increasing influence of mass transport phenomena as pores developed in the coating and very small rust spots were observed. For CR 5 visual observation after one year indicated a total area of rust spots between 0.03 and 0.1% of the exposed area. Both samples of CR 6 and one sample of CR 7 were removed when the total rust area had reached 0.3% after 7 and 9 months exposure at Port Hueneme, respectively. Removed samples were subjected to observation with an environmental scanning electron microscope (ESEM) to determine the spatial relationships between corrosion damage and location of bacteria [4]. For the remaining

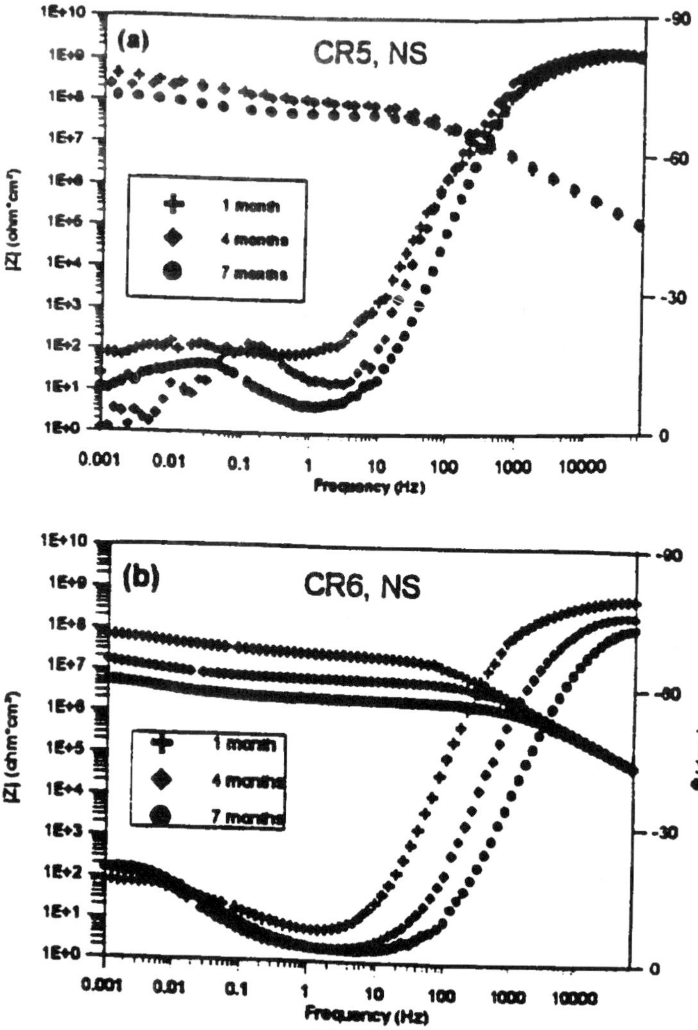

Fig. 4 (a,b) Impedance spectra for CR5 and CR6 exposed to natural sea water for 1, 4 and 7 months at Port Hueneme, CA.

CR 7 sample less than 0.03% of rusted area was observed. No loss of corrosion protection was indicated in EIS or ECN data for CR 9 samples. A comparison of the time dependence of the pore resistance R_{po}, which was assumed to be inversely proportional to the delaminated area A_d, is given in Fig. 5(a) for CR 5, 6 and 7 exposed at Port Hueneme. The lowest values of R_{po} were observed for CR 6 which performed the poorest according to visual observation. The increase of R_{po} for CR 5 after about 80 days seems to be due to plugging of pores in the coating by corrosion products and perhaps also bacteria. Observation with ESEM indicated that bacteria were concentrated in pores of the coating [4]. The breakpoint frequency f_b increased continuously for CR 6, but levelled off at values below 100 Hz for CR 5 and CR 7 (Fig. 5b). For CR 9 f_b remained below 1 Hz at all times. The delaminated or corroding

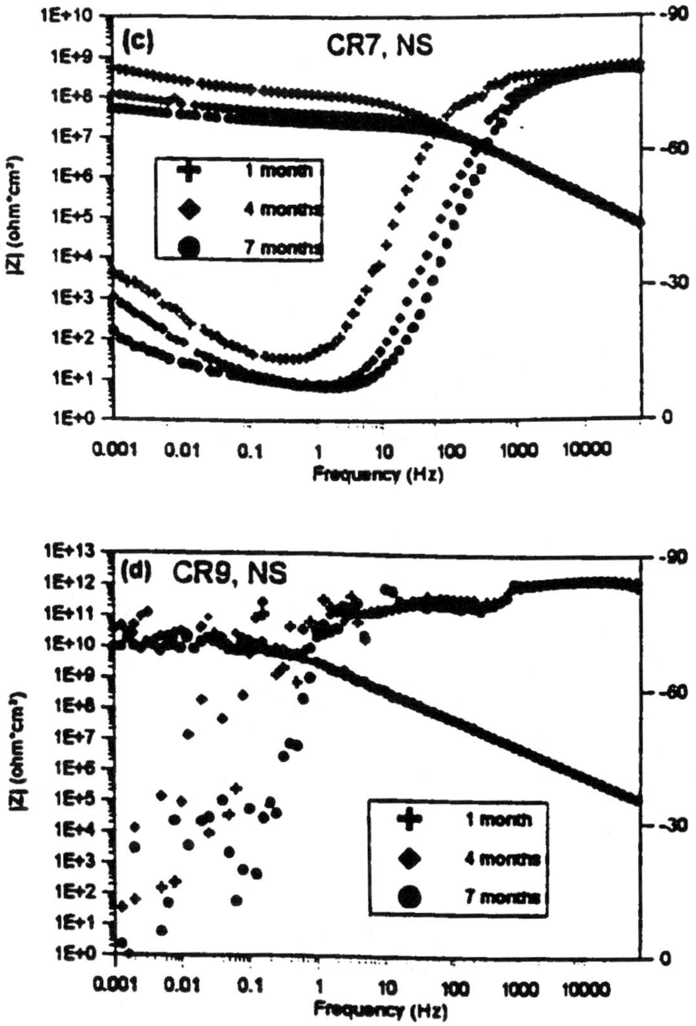

Fig. 4 (c,d) Impedance spectra for CR7 and CR9 exposed to natural sea water for 1, 4 and 7 months at Port Hueneme, CA.

area A_d can be estimated from f_b which for coating systems with impedance spectra such as those shown in Fig. 2 is defined as:

$$f_b = (2\pi R_{po} C_c)^{-1} = A_d(2\pi R^o_{po} C_c)^{-1} = \Delta(2\pi R^o_{po} C^o_c)^{-1} \qquad (1),$$

where R^o_{po} is the specific pore resistance in ohm.cm^2 and C^o_c is the specific coating capacitance. The area fraction $\Delta = A_d/A$, where A is the total exposed area, can be estimated based on eqn (1) provided R^o_{po} and C^o_c known. Figure 6 gives a comparison of Δ values calculated based on eqn (1) and results of visual observation for CR 6 exposed to natural sea water at Port Hueneme and Key West, and to artificial sea water. The value of R^o_{po} was fixed by assuming that the Δ value obtained by visual

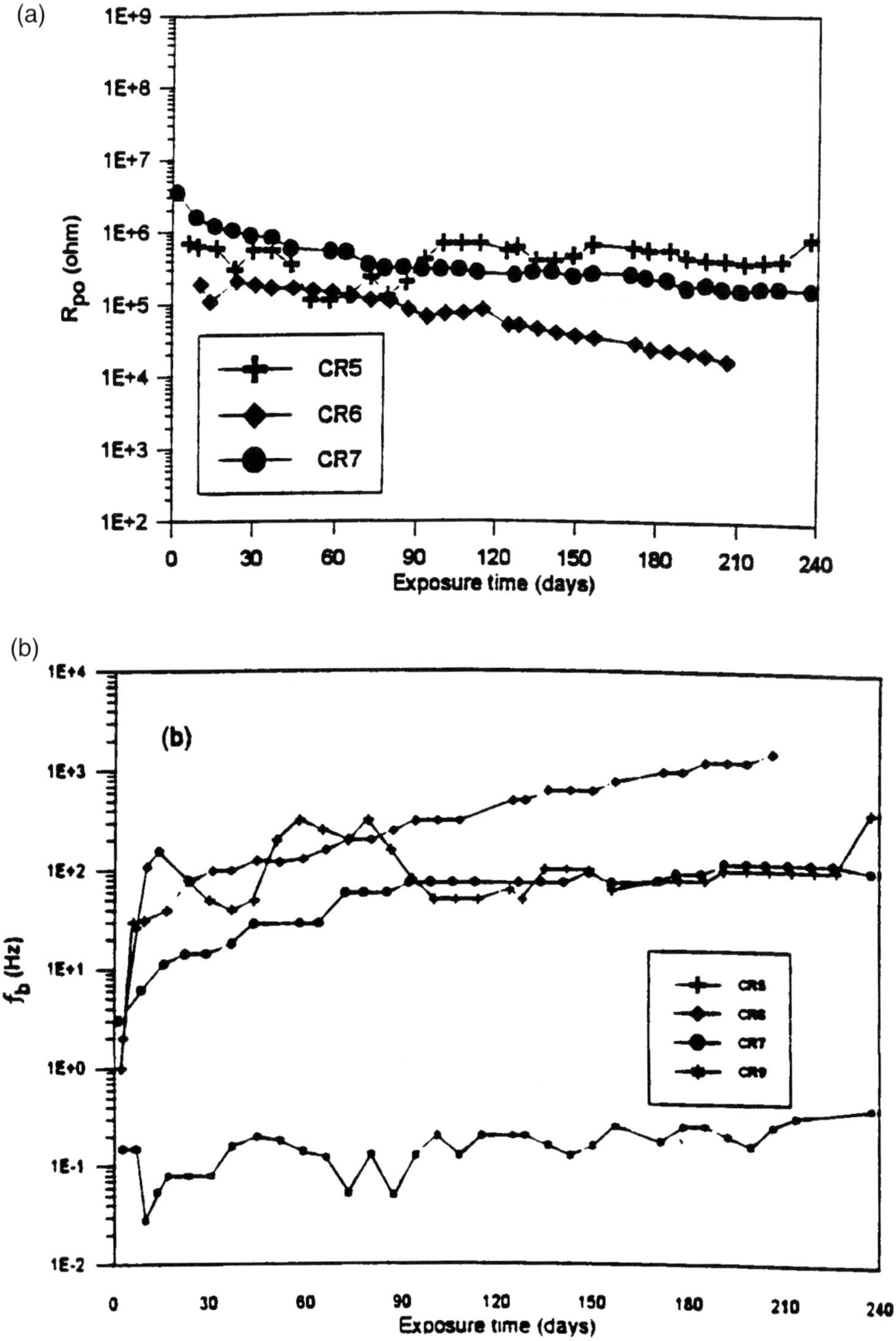

Fig. 5 *Time dependence of pore resistance* R_{po} *(Fig. 3a) and breakpoint frequency* f_b *(Fig. 3b) for CR5, CR6, CR7 and CR9 exposed at Port Hueneme, CA.*

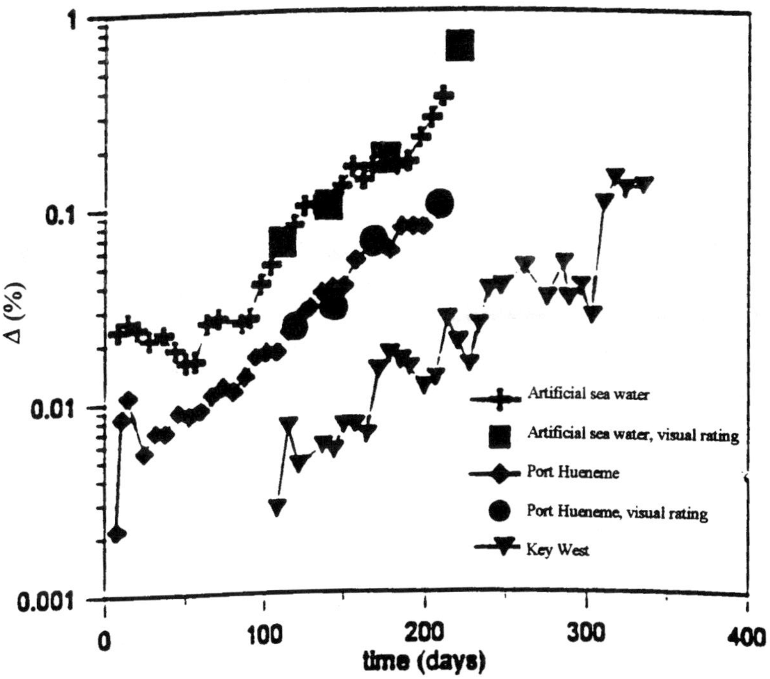

Fig. 6 *Time dependence of delaminated area fraction* Δ *obtained from* f_b *and by visual observation according to ASTM D 610 for CR 6 exposed to artificial sea water and to natural sea water at Port Hueneme and Key West.*

observation after about 110 days and that calculated from f_b were the same. Excellent agreement between the Δ values obtained by the two methods was observed over a four month period (Fig. 6).

The impedance spectra for CR 6 have been analysed using the equivalent circuit in Fig. 7. The coating capacitance, C_c, represented here by a generalised constant phase element (CPE), is shorted by R_p which is in series with an open boundary finite length diffusion (OFLD) element which models the diffusion impedance when the diffusion layer has finite dimensions (Fig. 7). The mass transport impedance is given by:

$$Z_{OFLD} = \{\tanh(B(j\omega)^{1/2})\}/Y_o(j\omega)^{1/2}, \qquad (2a),$$

where $B = 1/(D)^{1/2}$ is the characteristic diffusion parameter, D is the diffusion coefficient and $Y_o = (\sigma(2)^{1/2})^{-1}$. For infinite values of the diffusion length l this model becomes the Randles circuit, in which the Warburg impedance Z_w is in series with R_p and is given by:

$$Z_w = \sigma(1 - j)\omega^{-1/2}, \qquad (2b),$$

where σ is the Warburg coefficient. Excellent agreement between the experimental spectra and the fit to the model was observed (Fig. 7). As A_d increased, σ decreased and an increase of Δ with time similar to that shown in Fig. 6 was determined.

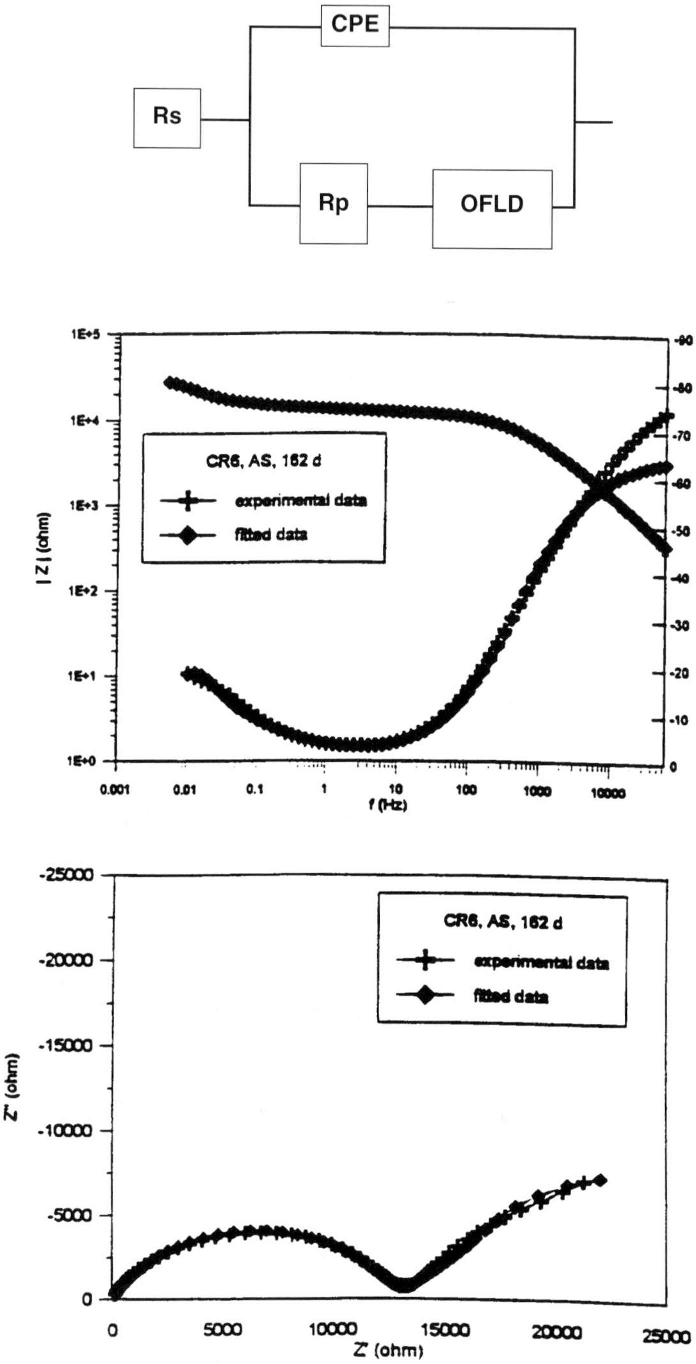

Fig. 7 Open boundary finite length diffusion (OFLD) model for impedance spectra with charge transfer and mass transport contributions and fit results compared with the experimental data shown as (a) Bode plot and (b) Nyquist plot for CR 6 exposed to AS for 162 days.

The ECN data obtained in the time domain have been transformed into potential and current power spectral density (PSD) plots. Spectral noise plots can be obtained from PSD plots as $R_{sn}(f) = |V_{PSD}/I_{PSD}|^{1/2}$, where V_{PSD} and I_{PSD} are the potential and current power levels, respectively. A comparison of spectral noise plots and impedance spectra can be performed in the Bode format. As discussed elsewhere [4,11], spectral noise plots obtained at the two sites were lower than the corresponding impedance spectra especially in the early stages of exposure. These discrepancies have been explained by the low levels of current noise for protective coatings and limited current measurement capabilities of the equipment used at these marine test sites [4,11]. More recently, noise data have been obtained with instrumentation having better current resolution (AutoAC DSP, ACM Instruments, Cart-in-Carmel, UK). Experimental potential and current noise data are shown in Fig. 8(a) and (b) for CR 9 exposed to artificial sea water for 780 days. The corresponding PSD plots are shown in Fig. 8(c) and (d). The current PSD data at higher frequencies are close to the instrument limit of about 10^{-25} A^2/Hz due to the very protective nature of the coating on CR 9 (Fig. 8d). Despite this problem, excellent agreement between the spectral noise plot and the impedance spectrum collected on the same day was observed (Fig. 8e). This result demonstrates that the impedance of the system can be measured either by applying an external signal or by analysing the naturally occurring potential and current fluctuations. Figure 8 also lists the potential of the coupled electrodes E_{coup} and the coupling current I_{coup} as well as the standard deviations σV and σI used to calculate R_n.

The effect of the bandwidth $\Delta f = f_{max} - f_{min}$ of the noise measurement on the computed value of R_n has been determined by stepwise decreasing the maximum frequency f_{max}, while maintaining the minimum frequency f_{min} which is determined by the measurement time which was 1024 s in the present investigations. The results obtained for coating system CR 9 in Fig.9, in which $R_n(f)$ is plotted as a function of $f_{max} = 0.5 f_s$, where f_s is the sampling frequency, indicated that R_n increased with decreasing Δf. Spectral noise or "noise impedance" plots $R_{sn}(f)$ are presented in Fig. 9 for $f_s = 32$ Hz and 2 Hz. Also shown in Fig. 9 is the impedance spectrum obtained for CR 9 exposed to artificial sea water for 801 days. Good agreement was observed again between the spectral noise ($\log R_{sn}(f) - \log f$) plots and the Bode ($\log |Z| - \log f$) plot. Since $R_n(f)$ depends on Δf it cannot be related to a particular coating property in a simple manner.

Inspection of spectral noise plots and impedance spectra in Figs 8 and 9 demonstrates the limited frequency range in which spectral noise plots can be obtained. The high-frequency limit could be extended by increasing the data collection rate beyond the presently used 2 points/s, however instrumentation noise limits this possibility. The low-frequency limit could be decreased by increased time for data collection (1024 s at present), however this possibility is not very attractive.

The impedance spectra for CR2 suggested that the coating quality changed from 'intermediate' to 'poor' after about five months. The classification results obtained with the use of the trained ANNs based on $\log |Z|$ and Φ agreed with this prediction. After 149 days, f_b increased sharply and continued to increase as coating damage and corrosion of the steel surface increased. Pore resistance, R_{po}, showed a more gradual decrease with time changing by about a factor of 100 between five and seven months. The predicted and determined values of CQI agreed completely during the

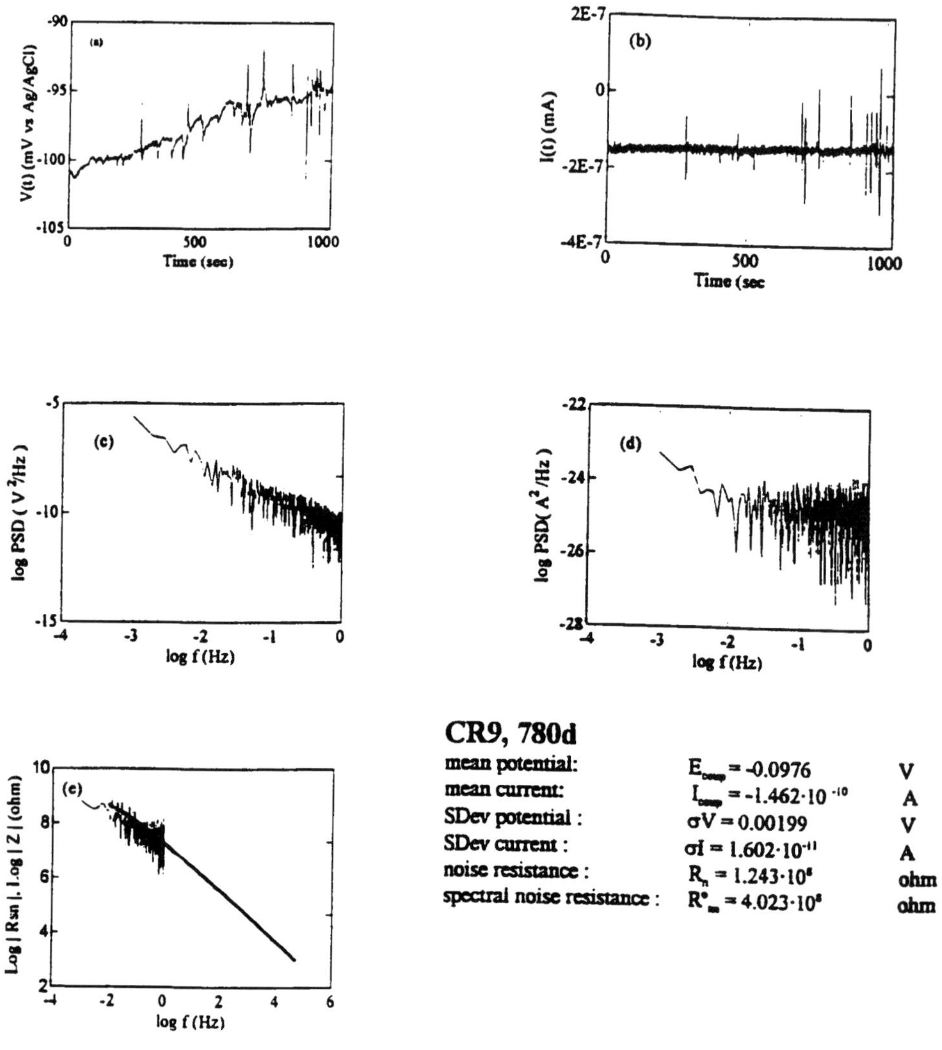

Fig. 8 Record and analysis of ECN data for CR 9 exposed to artificial sea water for 780 days.

entire exposure period (Fig. 10a). A drop from CQI = 0.5 ('intermediate') to 0 ('poor') occurred between 149 and 189 days.

For CR5 coating quality was classified as 'intermediate' for the first 157 days in both cases in agreement with the predicted results, while 'poor' coating quality was indicated for the data exceeding 247 days of exposure. The predicted and determined values of CQI agreed completely during the first 157 days (Fig. 10b). For the impedance spectra obtained after exposure for 190 days, excellent agreement was observed for the log |Z| data, however the predicted value of CQI for the phase angle data was 0.25, while the determined value was 0 and remained at this value for the remaining test period. Starting with the exposure period of 247 days, the CQIs based on log |Z| remained at zero (Fig. 10b).

The predicted and determined values of CQI for CR6 based on log |Z| agreed

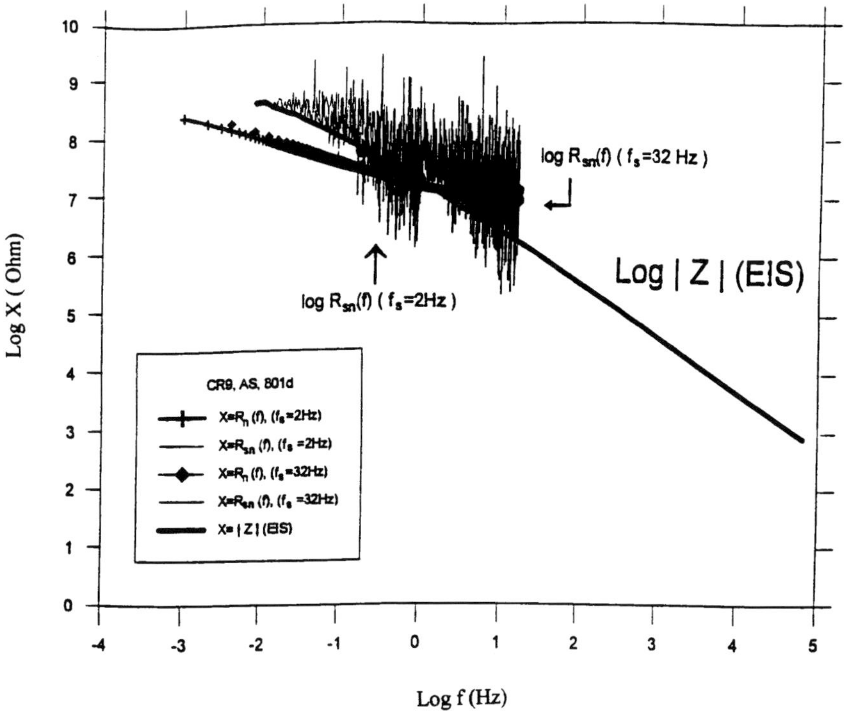

Fig. 9 *Comparison of impedance spectrum (log |Z|), spectral noise plots (log* $R_{sn}(f)$*) and noise resistance* $R_n(f)$ *for CR 9 exposed to artificial sea water for 801 days.*

completely during the first 190 days (Fig. 10c) similar to the results for CR5 (Fig. 10b). For the impedance spectra obtained after exposure for 220 days total agreement was obtained for the CQI values based on the phase angle. After 190 days the predicted value of CQI based on the phase angle was 0.25, while the determined value was 0 and remained at this value for the remaining test period. A decreasing trend of the determined CQI values based on Φ was observed after about 150 days. This trend also started to occur for the CQI values based on log |Z| after about 190 days. A fairly large disagreement between the predicted and determined CQI values based on log |Z| was observed at the end of the exposure period (Fig. 10c).

The predicted and the classification results for CR7 based on Φ and log |Z| are shown in Fig. 10(d). After exposure for 88 days excellent agreement between these two sets of results was observed. An asymmetry problem [12] was observed in the impedance spectra for the first 59 days. Fairly low f_b values of less than 30 Hz typical for 'intermediate' coating quality were observed for the entire exposure time. Based on the R_{po} data, coating performance between 'good' and 'intermediate' was predicted for the first 59 days of exposure and 'intermediate' coating performance for the remaining time. The impedance data for 269 days placed CR7 at the lower end of the 'intermediate' training data set. The predicted results and the classification results agreed well except for the data showing asymmetric behavior. These results are reflected in the CQI values in Fig. 10(d) which had values close to 0.5 starting with an exposure period of 95 days.

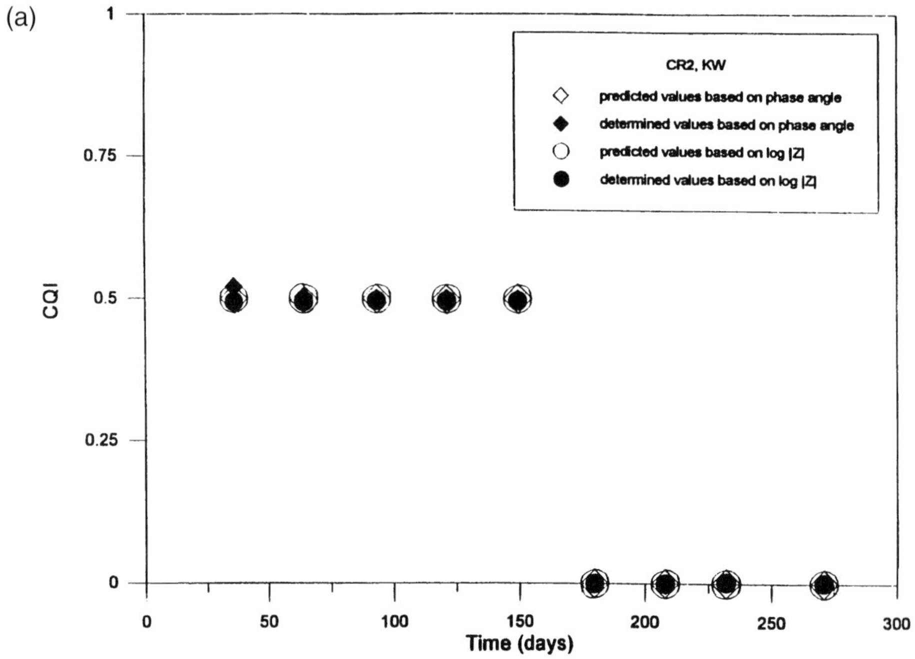

Fig. 10(a) *Time dependence of the coating index CQ1 for CQ2.*

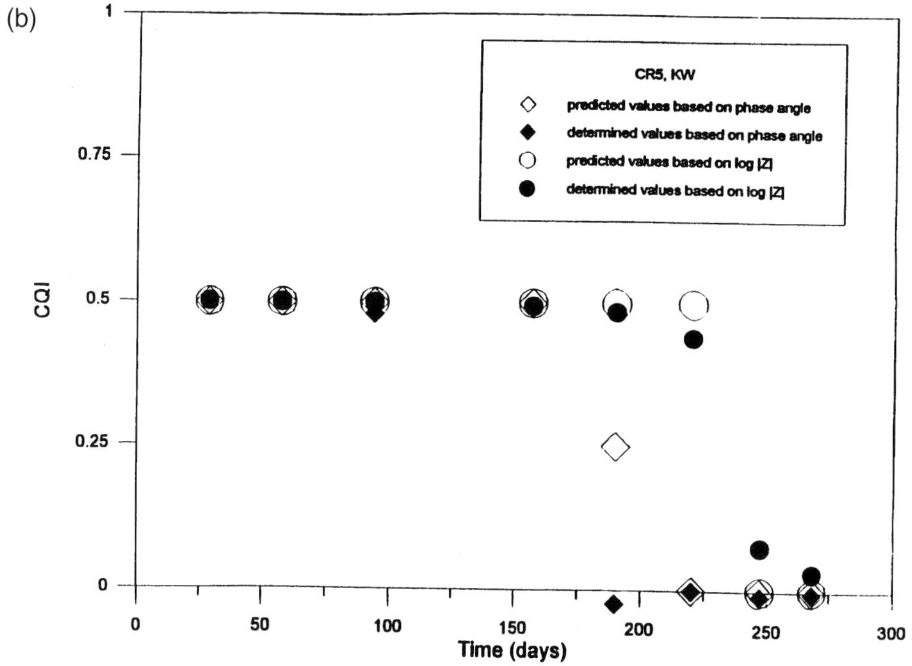

Fig. 10(b) *Time dependence of the coating quality index CQI for CR5.*

(c)

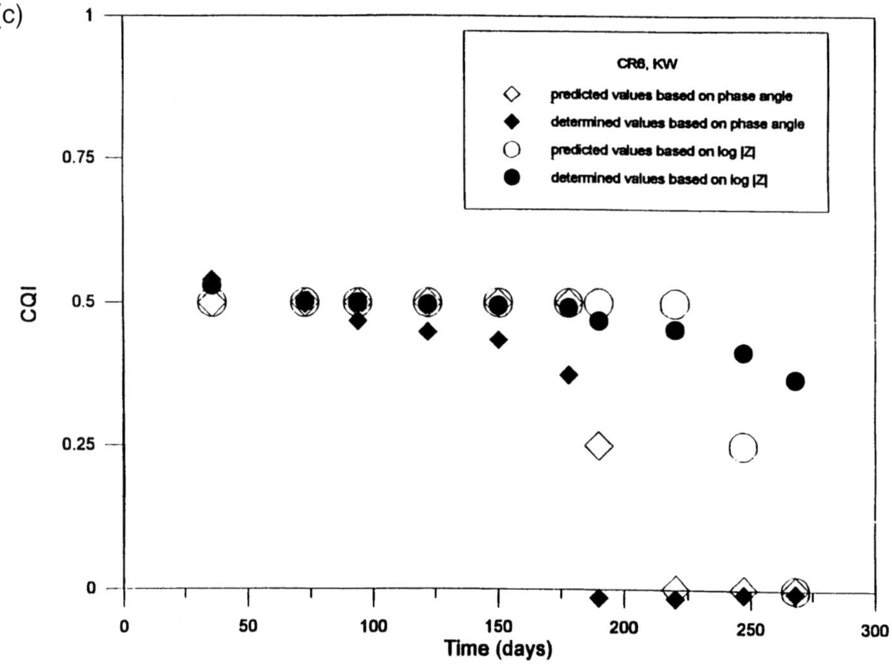

Fig. 10(c) *Time dependence of the coating quality index CQI for CR 6.*

(d)

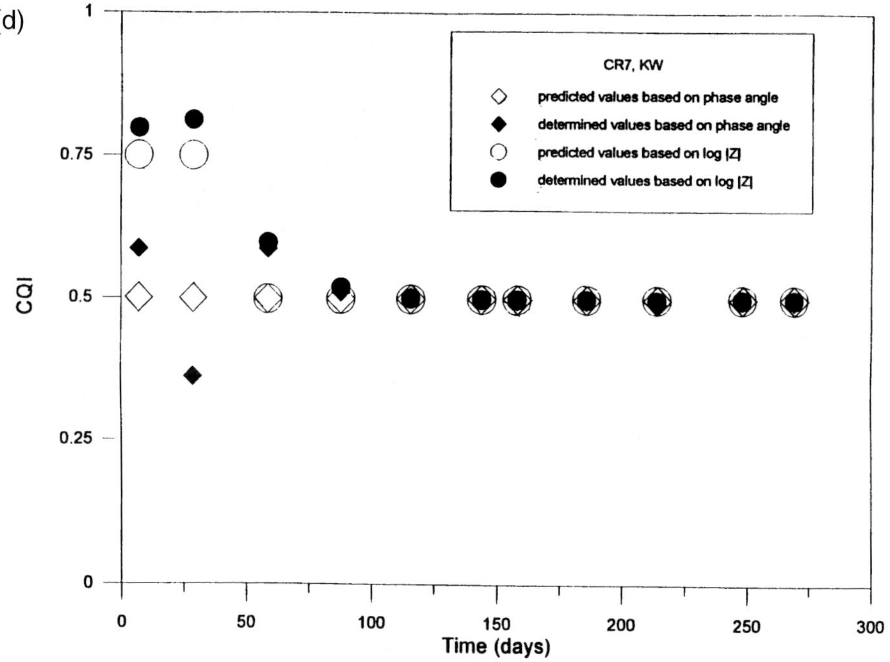

Fig. 10(d) *Time dependence of the coating quality index CQI for CR 7.*

For coating system CR9, for which no indications of degradation were observed in tests up to three years [1–3,8], excellent agreement was observed between the predicted and the classification results for the Φ and the log |Z| data sets resulting in a 'good' performance score. The CQI values were all close to unity [8].

Samples of the JJ-series with artificial coating defects were exposed at Key West and in artificial sea water for 30 days [4,9,10]. Coupled electrode potential values, E_{coup}, for JJ1 and JJ2 with a zinc plate primer were close to –1000 mV vs Ag /AgCl (Fig. 11), i.e. similar to E_{corr} for Zn in sea water; the E_{coup}-values for all other JJ samples were close to – 700 mV which is similar to E_{corr} for iron in sea water. While the zinc layer on steel in JJ1 and JJ2 provided cathodic protection to exposed steel in the defects, the same effect was not observed for JJ5 and JJ6 with an ion vapor deposited (IVD) Al primer (Fig. 11). Visual observation showed that the holes were covered with rust except for the JJ1 and JJ2 samples where calcareous deposits seemed to have formed. More detailed information was obtained by examination in the ESEM which showed that bacteria colonised only those defects which were not cathodically protected and were filled with rust particles [4,9,10]. Due to the cathodic protection provided by the metallic zinc primer for JJ1 and JJ2 only calcareous deposits, but no bacteria were detected in the intentional defects for samples exposed at Key West which were not connected to the zinc electrode [4,12]. For cathodically protected samples, R_n increased with time during immersion in artificial sea water (Fig. 12). The value of E_{coup} for all coupled samples was about –1000 mV vs Ag/AgCl. The increase of R_n with time was most likely due to formation of calcareous deposits in the defects which increased R_{po}. Impedance spectra obtained during immersion in

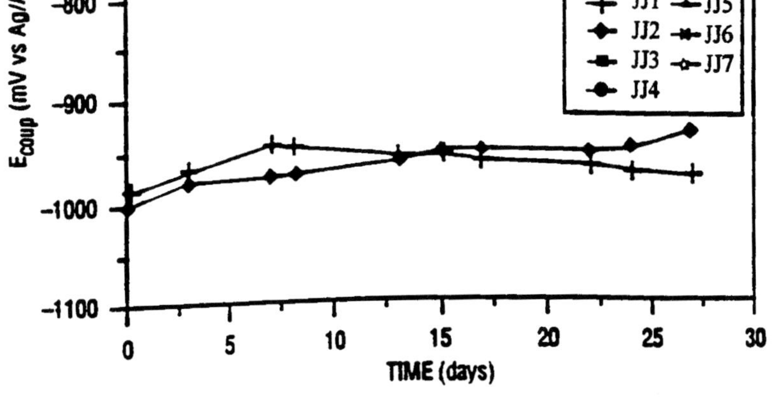

***Fig. 11** Time dependence of* E_{coup} *for JJ-series samples with intentional defects exposed at Key West.*

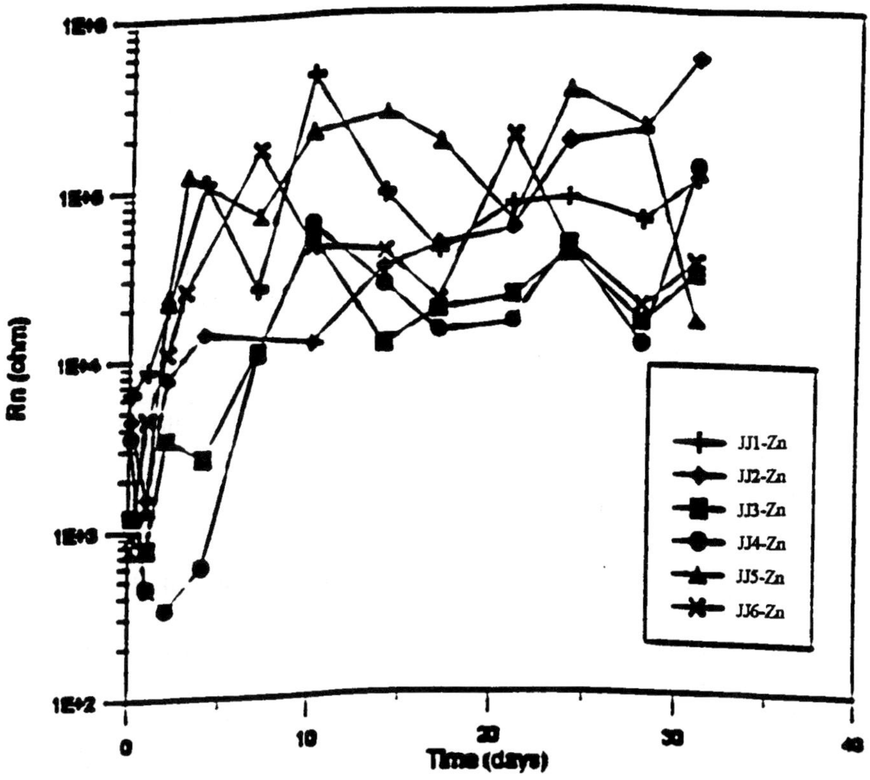

***Fig. 12** Time dependence of* R_n *values for cathodically protected JJ series samples.*

artificial sea water confirmed the assumption that R_{po} increased as defects were filled with calcareous deposits. As shown in Fig. 13 the spectra for JJ5 changed from a one time-constant mechanism at early exposure times to spectra which contained C_c at the highest frequencies and R_{po} at the lower frequencies where the ECN data were recorded. For the system under investigation R_n was therefore related to R_{po}.

4. Summary and Conclusions

Data from EIS and ECN have demonstrated significant effects of coating properties on the corrosion resistance of polymer coated steel exposed to sea water. The delaminated area fraction Δ could be determined from the breakpoint frequency f_b and its time dependence could be followed after normalisation with visual observation data. For certain coating systems a significant influence of mass transport on the frequency dependence of the impedance was observed. In these cases the open boundary finite length diffusion model could be used for analysis of the EIS data. The ECN data have been analysed in the time and frequency domains. The noise resistance R_n has been determined from the ECN data in the time domain, while spectral noise ($R_{sn}(f)$) plots were determined from PSD plots. In the absence of instrumental artifacts very good agreement between $R_{sn}(f)$ plots and impedance

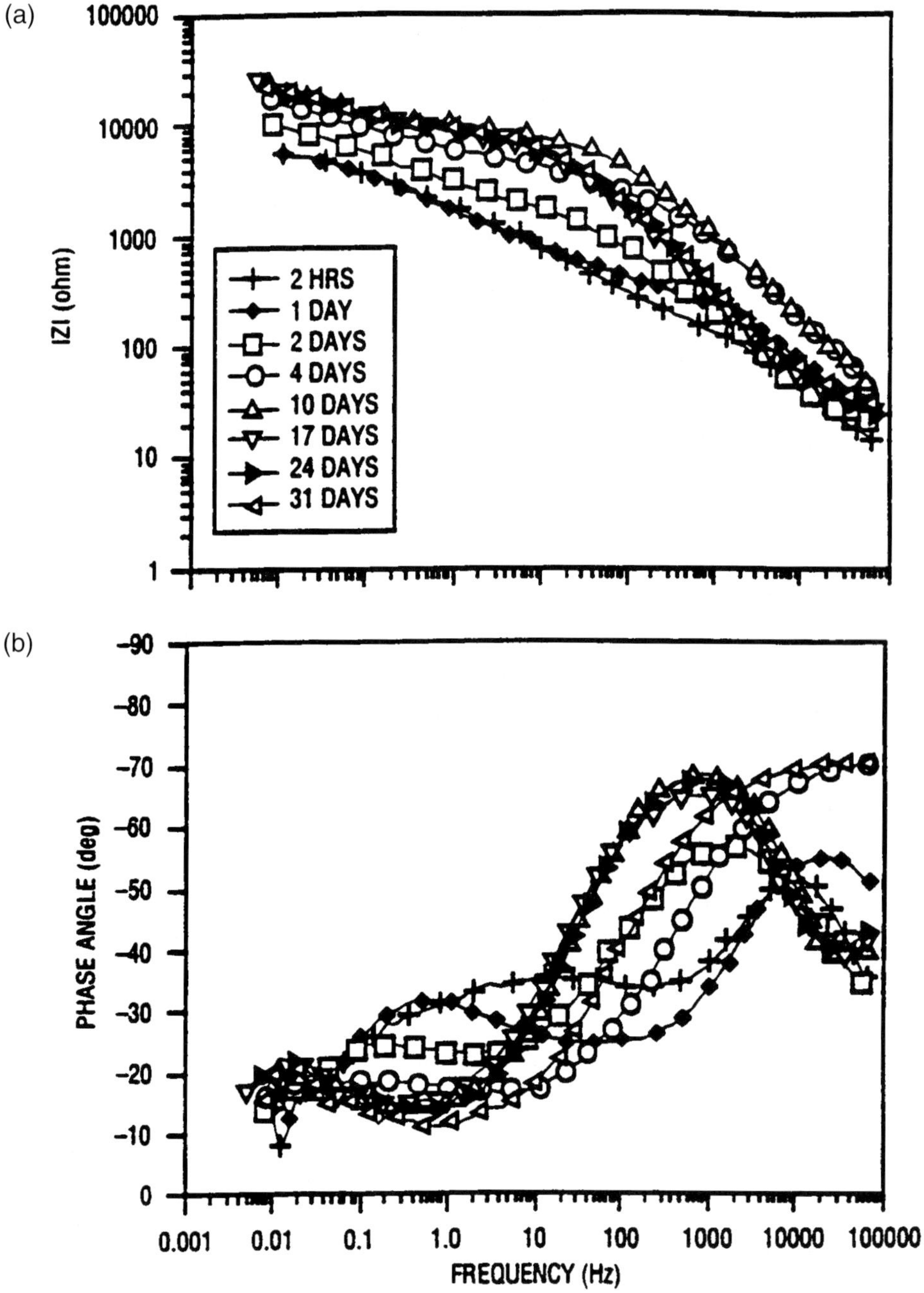

***Fig. 13** Bode plots for cathodically protected JJ5 with intentional defects exposed in artificial sea water for 31 days.*

(Bode) plots was observed. The value of the R_n was found to depend on the bandwidth Δf of the ECN measurement for cases in which the impedance was capacitive within Δf. In such case R_n does not have a simple relationship to coating properties.

Two ANNs — one for classification based on $\Phi - \log f$ data and one for classification based on log $|Z|$ – $\log f$ data — have been trained using three sets of theoretical impedance spectra for 'good', 'intermediate' and 'poor' coating quality. The trained ANNs have been tested using experimental impedance spectra for five different polymer coating systems on steel collected during exposure to natural sea water at a remote marine test site for exposure periods up to one year. In general, excellent agreement between the predictions of coating quality made by experienced operators based on general features of the impedance spectra and parameters such as f_b and R_{po} on the one hand and the classification results obtained from the ANNs on the other hand was obtained. Evaluation of the results of these analyses was made easier by introduction of the coating quality index (CQI) which has values between 0 and 1. In the present investigation the ANNs have been trained to evaluate coating quality for a number of different types of coating systems. In most applications it can be expected that only one particular coating type is of interest. In this case the training set could be re-defined based on the known impedance characteristics of this system producing even more reliable results.

For samples with artificial defects observation with an ESEM has demonstrated that bacteria did not colonise such defects for samples which were cathodically protected either by connection to an external zinc electrode or by the presence of an metallic zinc primer. This result suggests that identification of microbially influenced corrosion cannot be based solely on spatial relationships between large accumulations of bacterial cells and iron corrosion products [13].

5. Acknowledgements

This project has been supported by the Office of Naval Research (Dr A. J. Sedriks) under Contract No. N00014-94-1-0026. The authors acknowledge helpful discussions with H. P. Hack and M. A. Matteson concerning the training procedure for the ANNs in a project which has been supported under Contract No. N0016796M1416 by the Carderock Division, Naval Surface Warfare Center (R. DeNale).

References

1. H. Xiao, L. T. Han, C. C. Lee and F. Mansfeld, *Corrosion*, 1997, **53**, 412.
2. F. Mansfeld, L. T. Han, C. C. Lee, C. Chen, G. Zhang and H. Xiao, *Corros. Sci.*, 1997, **39**, 255.
3. F. Mansfeld, H. Xiao, L. T. Han and C. C. Lee, *Prog. Org. Coat.*, 1997, **30**, 89
4. F. Mansfeld, C. C. Lee, G. Zhang, L. T. Han, B. J. Little, P. A. Wagner, R. Ray and J. Jones-Meehan, "The Impact of Microbiologically Influenced Corrosion on Protective Polymer Coatings", Final Report , Contract No. 00014-94-1-0026, April 1998.
5. D. J. Mills, G. B. Bierwagen, B. Skerry and D. Tallman, *12th Int. Corros. Congr.*, NACE, TX, 1993, p. 182.
6. H. P. Hack and M. A. Matteson, "Feasibility of Using Artificial Neural Networks with Electrochemical Impedance Spectroscopy Data from Coated Steel", Technical Report, CARDIVNSWC-TR-61-95/15, Dec. 1995.
7. NeuralWare Professional II/Plus, NeuralWare, Inc., Pittsburgh, PA 15275.
8. F. Mansfeld and C. C. Lee, *Corros. Sci.*, 1999, **41**, 439.

9. B. J. Little, P. A. Wagner, R. I. Ray, J. Jones-Meehan, C. C. Lee and F. Mansfeld, 193rd Meeting of The Electrochem. Soc., Paper No. 125, May 1998, The Electrochemical Society, USA, 1998.
10. B. J. Little, R. I. Ray, P. A. Wagner, J. Jones-Meehan, C. C. Lee and F. Mansfeld, *Biofouling*, 1999, **13**, 310.
11. F. Mansfeld, L. T. Han, C. C. Lee and G. Zhang, *Proc. Corrosion '97 Research Topical Symp.*, p. 217, NACE, 1997.
12. F. Mansfeld, C. Chen, C. C. Lee and H. Xiao, *Corros. Sci.*, 1996, **38**, 497.
13.B. J. Little, R. I. Ray, P. A. Wagner, J. J. Jones-Meehan, C. C. Lee and F. Mansfeld, Marine Bacteria and Localized Corrosion on Polymer Coated Steel: Cause and Effect, *Corrosion '99*, Paper No. 183, NACE International, Houston, TX, 1999.

9
Galvanic and Barrier Protection in Zinc Rich Paints

C. M. ABREU, M. IZQUIERDO, P. MERINO, X. R. NÓVOA and C. PÉREZ
Universidade de Vigo, E.T.S.E.I.M., Lagoas-Marcosende, 9, E-36280 - Vigo, Spain

ABSTRACT

The behaviour of zinc-rich epoxy paints has been studied in 3% NaCl solution. The study focuses on the influence of film thickness and zinc particle size on coating performance. Electrochemical Impedance Spectroscopy was the main technique employed to carry out the experimental work. The impedance spectra are interpreted by using a fluidised bed model that takes into account the contact impedance between zinc particles. A new parameter called Total Film Resistance is defined as the parallel combination between contact impedance and electrolyte resistance. This new parameter can be used as a design tool because it allows the end of the cathodic protection period to be clearly identified. The film thickness as well as the zinc particle size and distribution in the binder define the duration of the cathodic period and the performance of the resulting barrier properties.

1. Introduction

Zinc-rich Paints (ZRP) play a very important role as protective coatings for avoiding steel corrosion. Their protection mechanism is based on the galvanic couple existing between the zinc particles in the coating and the steel substrate and consequently, ZRP protect steel from corrosion even when small damages or holes exist in the paint film. This feature is a significant advantage of ZRP when compared with traditional barrier coatings. The electronic conductivity between the zinc particles inside the coating and between the particles and the steel substrate is the main characteristic of these paints. For this electrical conductivity to be guaranteed, a high pigment volume concentration (PVC) is necessary in paint formulations [1] and so the films become highly porous. Thus, contrary to barrier coatings, ZRP allow easy electrolyte uptake, i.e. they show lower resistance to ionic movement. The electronic conductivity and the film porosity have a major influence on anticorrosive properties of ZRP, because both ionic and electronic conductivities are necessary for the galvanic couple to function.

Two different mechanisms are well established [2,3] for ZRP protection. Cathodic protection operates in the first stage when zinc protects the steel substrate while the second stage corresponds to a barrier effect resulting from the sealing of coating pores by zinc corrosion products. The galvanic period is of major importance for the coating performance although its duration is still the subject of discussion despite a general agreement existing about its dependence on several aspects such as PVC, type of binder, grain size of the zinc particles, zinc-to-steel area ratio, film thickness and type of electrolyte.

Electrochemical Impedance Spectroscopy (EIS) in the range 10 kHz–1 mHz has been extensively applied to study the behaviour of ZRP. The analysis and interpretation of the two capacitive loops usually obtained for these paints in immersion conditions in 3% NaCl have been performed in terms of Randles equivalent circuits or transmission line models. In some cases the high frequency time constant is attributed to the charge transference process, and the oxygen diffusion is assigned to the low frequency time constant [4–8]. Other studies correlate the low frequency time constant with the charge transference for zinc dissolution and the high frequency one with the dielectric properties of the polymeric matrix [9–13].

Extensive work by the authors [14–16] has established a model for the steel/ZRP/NaCl system based on fluidised bed electrochemical reactors [17–18]. The model assumes:

(a) A porous nature of the electrode (zinc particles in the film) that results in a heterogeneous distribution of the current, i.e. a transmission line (TL) type behaviour.

(b) The existence of some contact impedance between zinc particles which results initially from the natural oxide covering the zinc particles and later, during exposure, from the zinc corrosion products that tend to separate zinc particles from each other.

(c) The galvanic system operates under an oxygen limiting current so that the overall impedance approaches the zinc impedance.

According to these assumptions, the experimental impedance spectra can be modelled using the equivalent circuit depicted in Fig. 1.

The impedance of the transmission line in Fig. 1(a), Z_{TL}, is given in eqn (1). In (1) L is the length of the TL (taken as the film thickness). $\gamma = (Z_m + R_e)/Z_1$, Z_m is the contact impedance between zinc particles along the transmission line and Z_1 the interfacial impedance per unit of TL length for zinc dissolution. R_e is the electrolyte resistance inside the film per unit of TL length.

$$Z_{TL} = \frac{Z_m R_m}{Z_m + R_m} L + \frac{(Z_m^2 + R_e^2)\cosh(L\sqrt{\gamma}) + 2Z_m R_e}{(Z_m + R_e)\sqrt{\gamma}\,\sinh(L\sqrt{\gamma})} \tag{1}$$

$$Z_m = \frac{R_m}{1 + (j\omega R_m C_m)^{\alpha_m}} \tag{2}$$

The contact impedance is given in eqn (2) where R_m is the contact resistance, C_m the corresponding dielectric capacitance and α_m accounts for the dispersion of the $R_m C_m$ time constant along the transmission line. $j^2 = -1$ and $\omega = 2\pi f$. The interfacial

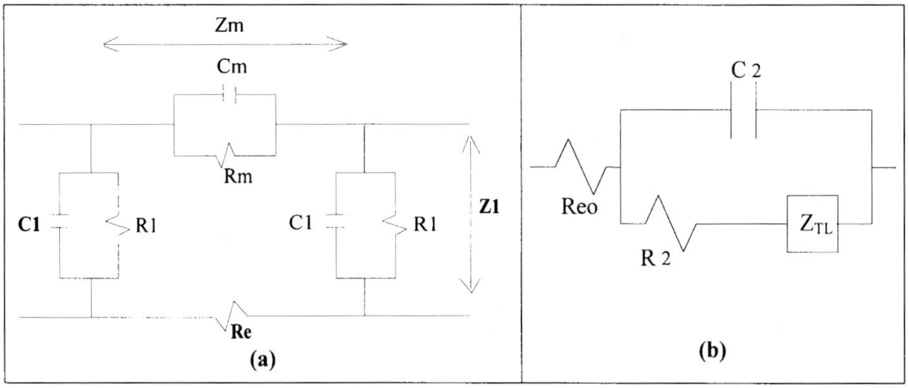

Fig. 1*(a) Scheme of the transmission line used to model the behaviour of a ZRP/steel system.* Z_m *represents the contact impedance between zinc particles,* Z_1 *the interfacial impedance and* R_e *the electrolyte resistance throughout the binder. (b) Equivalent circuit used to model the behaviour of the ZRP/steel system by the end of ZRP lifetime.* R_{eo} *corresponds to the solution resistance outside the film,* R_2C_2 *to the oxygen reduction at the outer most part of the film and* Z_{TL} *the circuit in Fig.1(a).*

impedance, Z_1 in eqn (3), is obtained by considering the double layer capacitance C_1, and the faradaic resistance, R_1, associated with the zinc dissolution; α_1 accounts for the distribution of the R_1C_1 time constant.

$$Z_1 = \frac{R_1}{1+(j\omega R_1 C_1)^{\alpha 1}} \tag{3}$$

This equivalent circuit model explains the two time constants appearing in the impedance spectra. The high frequency loop is related to the contact impedance between zinc particles, and the low frequency loop corresponds to the zinc dissolution process [15].

According to thermodynamics, the cathodic protection of iron is attained when the electrode potential is lower than –0.86 V vs SCE (saturated calomel electrode). This value corresponds to the commonly accepted criterion [19] of a maximum Fe^{2+} concentration of 10^{-6} M. When the ZRP electrode potential exceeds the iron cathodic protection limit, a third capacitive loop appears in the impedance spectrum. This time constant was related [14,15] to the cathodic reaction starting to develop at the outer part of the paint (paint/electrolyte interface), when the ohmic drop throughout the film makes the galvanic effect between this part of the film and the steel substrate vanish.

The equivalent circuit depicted in Fig. 1(b) describes this new situation. Equation (4) gives the related impedance, Z.

$$Z = R_{e_0} + \frac{R_2}{1+(j\omega R_2 C_2)^{\alpha_2} + \dfrac{R_2}{R_2 + Z_{TL}}} \tag{4}$$

where R_{e_0} represents the solution resistance outside the film, and R_2C_2 is associated with the oxygen reduction on zinc particles at the outermost part of the film.

2. Experimental

Commercial ZRP based on epoxy resin were applied by roll bar to carbon steel plates previously cleaned and degreased. The zinc content in the ZRP was 62 volume % of the dry film. The thickness applied was 25 ± 5 µm. Table 1 shows the average diameter of the zinc dust particles for the different series studied, the only difference between them was the grain size of zinc dust employed. Moreover, zinc particles in series C and D have a broader average diameter, which allows better packing in the polymer matrix.

Samples were immersed in 3% NaCl solution at the open circuit potential. Periodically, measurements of corrosion potential and EIS were performed at this potential. The typical three electrode arrangement was chosen for impedance measurements. The electrolyte was confined in a glass tube fixed on the painted surface by an O-ring; the surface area tested was 9 cm^2. A saturated calomel electrode (SCE) was used as reference electrode. The frequency domain scanned was 100 kHz to 1 mHz.

The impedance data were analysed using the simplex method described elsewhere [14–16]. The function to minimise, χ^2, is defined as:

$$\chi^2 = \sum_{i=1}^{N} \left[\left(\frac{a_{ei} - a_{ci}}{|Z_{ei}|} \right)^2 + \left(\frac{b_{ei} - b_{ci}}{|Z_{ei}|} \right)^2 \right] \tag{5}$$

where N is the total number of scanned frequencies, a_{ei} and b_{ei} are respectively the real and imaginary parts of the experimental impedance, Z_{ei}, at frequency i, and a_{ci} and b_{ci} the corresponding real and imaginary parts of the calculated impedance at frequency i.

3. Results and Discussion

3.1. Corrosion Potential–Time Behaviour

The potential–time behaviour of films is presented in Fig. 2 (p.128). The results obtained are similar to those in earlier work with high thickness paint [14]. Independently of the grain size of zinc particles, four main periods of time can be distinguished.

3.1.1. Period 0.

The corrosion potential shifts anodically. This displacement is a function of the system under study. It can be interpreted in terms of two phenomena occurring in parallel:

Table 1. *Average diameter, ϕ, of the zinc dust particles used in the tested paints*

Series	A	B	C	D
ϕ (μm)	2.5–3.5	3.5–4.5	4.5–6.0	6.0–9.0

(1) The surface of the steel substrate begins to be wetted, and thus, the electrode potential shifts towards positive values, near those typical for steel.

(2) The increase of the Zn/Fe surface ratio as a consequence of the reaction between the solution and the natural ZnO that covers the zinc particles, activates the zinc which then starts to be galvanically coupled to the steel substrate. The result is displacement of potential to more negative values.

Consequently, the corrosion potential results from the combination of both kinetics: increase of the Zn/Fe area ratio due to zinc particle activation and decrease of this area ratio due to steel wetting. The electrochemical behaviour of each of the different series will depend on the prevailing process and on the packing of the film. When the grain size and the average diameter of zinc particles increase, the activation of zinc kinetics is faster than the steel wetting kinetics. Series D shows continuous displacement towards more negative potential values in the first stage of exposure. Only in series A can a clear shift to the anodic direction be observed, i.e. there is a delay in zinc activation with respect to the steel substrate wetting. The higher average diameter for zinc particles in series D increases the pore tortuosity and thus the length of the path for the electrolyte to reach the steel substrate. This allows zinc activation prior to steel wetting and thus the initial anodic shift as observed for series A.

3.1.2. Period 1

The corrosion potential shifts in the cathodic direction to reach values close to those corresponding to a zinc specimen in this medium. This shift can be explained again as a function of the Zn/Fe area ratio variation. After a few minutes of immersion, the active steel surface does not increase in area because it is completely wetted, while the zinc particles having a thicker natural oxide layer continue to activate owing to the reaction with the electrolyte. This reaction tends to increase the Zn/Fe area ratio and explains the electrode potential variation in this period.

3.1.3. Periods 2 and 3

The corrosion potential slowly rises with time although in period 3 there are potential fluctuations not present in period 2. These variations in the electrode potential correspond to the decreasing Zn/Fe area ratio due to zinc corrosion.

As can be seen in Fig. 3, the paint film thickness increases with immersion time as a result of the accumulation of zinc corrosion products. Therefore, it is reasonable to consider that zinc particles are not completely fixed in their own positions but can move, albeit over a limited distance, around their original positions. The driving

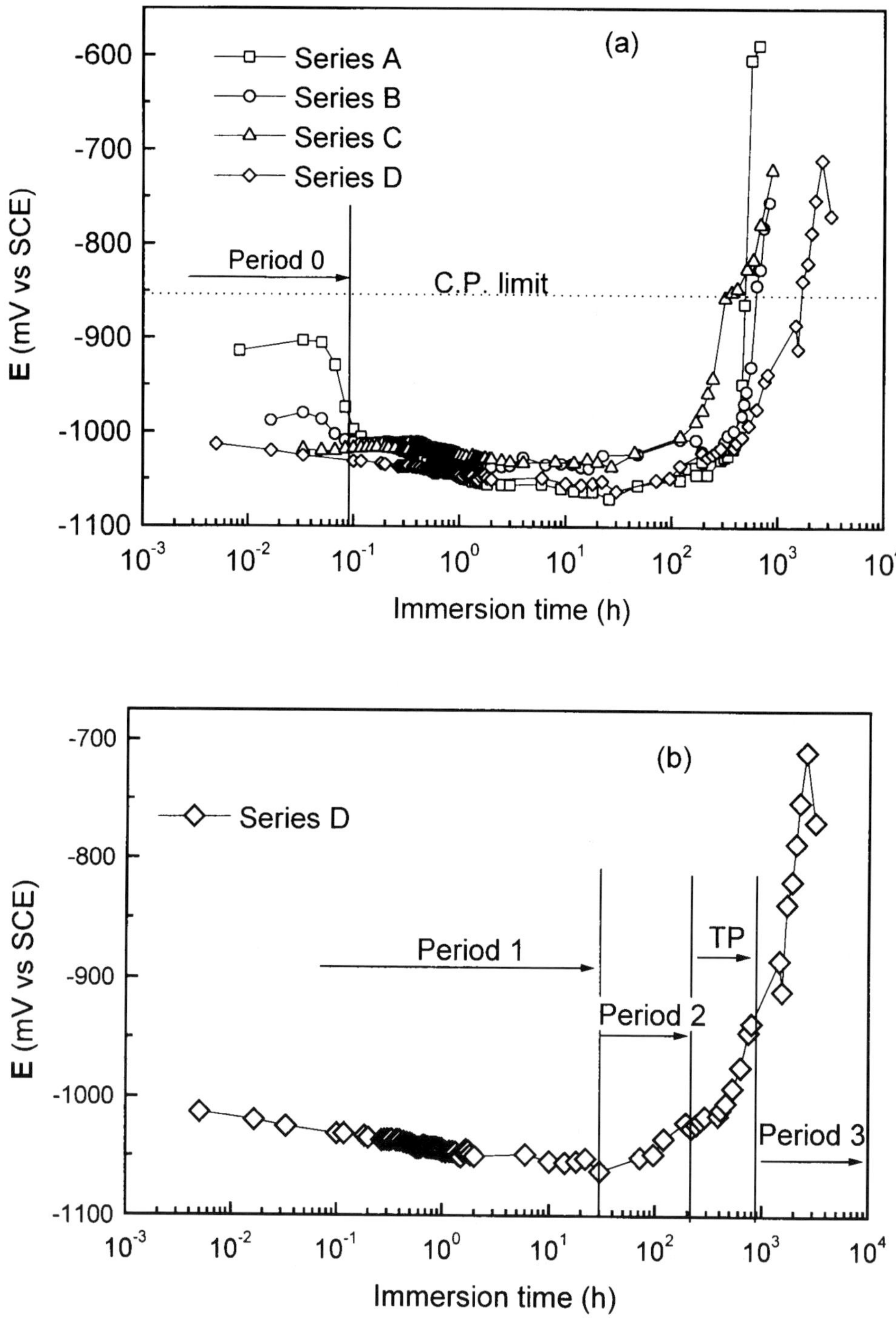

Fig. 2(a) Potential evolution of the thicker ZRP/steel systems with immersion time. The dotted line (–0.86 V) represents the thermodynamic Cathodic Protection (CP) limit. (b) Individual plot for series D. The different periods in the film life time are also depicted.

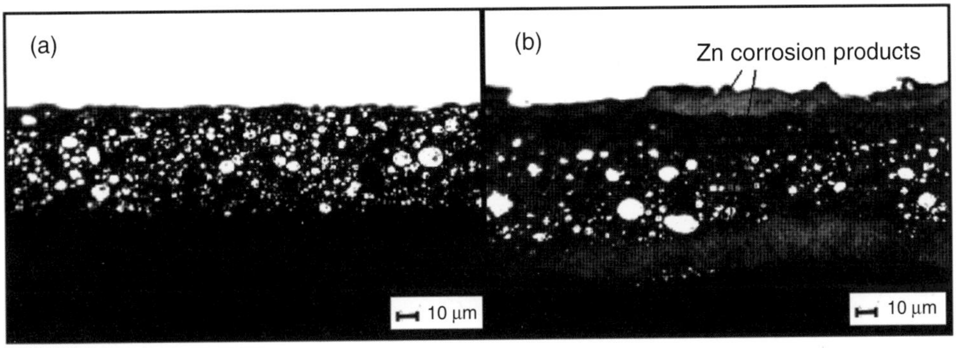

Fig. 3 *Cross-section of the ZRP. (a) Prior to exposure. The black vacancies are due to the zinc particles leaving the sample surface on polishing. (b) The same coating after 366 days of immersion in 3% NaCl. Substrate steel is seen as the white region at the top of micrographs.*

force for this dynamic effect is likely to arise from the more expanded volume of zinc corrosion products compared with that of zinc particles. Consequently, the number of zinc particles in contact with each other changes with time. The contact number and the active surface area of zinc decrease progressively with time due to the corrosion of the zinc particles. When this contact number is relatively high, the potential fluctuations are small since a large surface is involved in the system. Therefore, this phenomenon leads to a smooth change towards a more positive potential value (period 2). When the number of contacting particles decreases drastically, the change of this number generates potential fluctuations (the end of period 3) since the active surface of zinc particles is now small.

3.2. Impedance Spectra–Time Behaviour

Figure 4 corresponds to a typical impedance plot for periods 1 and 2: only two time constants are observed. However, three time constants are present at period 3, as depicted in Fig. 5. The third time constant, clearly differentiated in the medium frequency range at period 3, starts to develop at the transition period, TP, in Fig. 2.

Figures 4 and 5 show that a good agreement exists between experimental data and their fitting to the proposed equivalent circuit. Table 2 summarises the corresponding best fitting values. The development with time of the electric-equivalent parameters obtained for higher thickness films has been discussed in previous papers [15,16] in which the influence of grain size and zinc particle distribution in the behaviour of these systems was analysed. The present paper focuses on the results obtained for lower thickness films.

3.3. Systems Having Smaller Grain size (Series A and B)

For the lower thickness systems with small grain size (series A and B), as for the higher thickness A system, it is not possible to give a unique electrical model capable of assessing the paint behaviour over the immersion time. These systems behave randomly as flat or porous electrodes. Thus, it was not possible to obtain the electric-

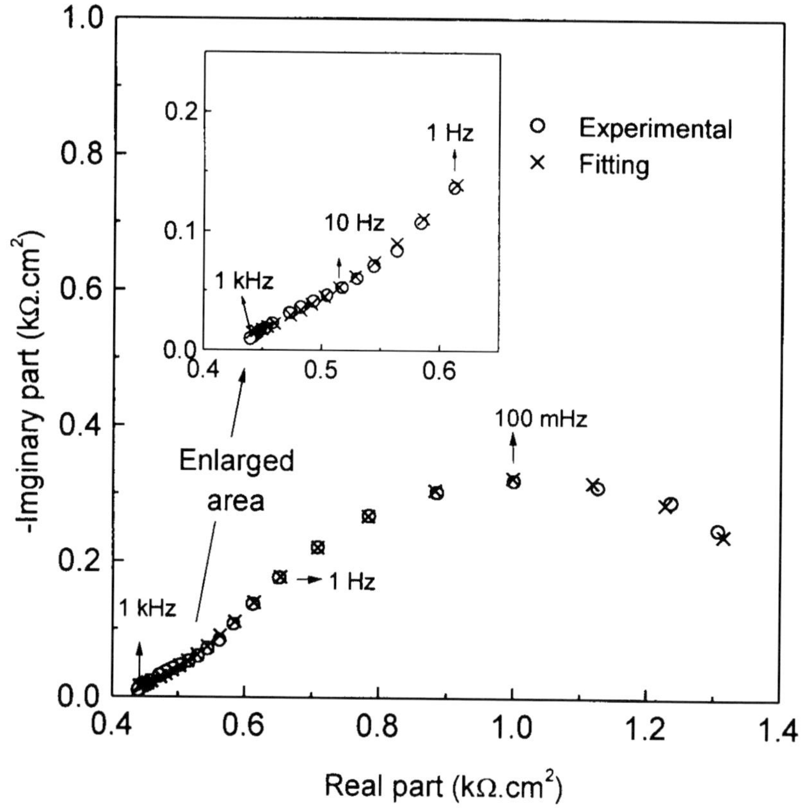

Fig. 4 Impedance spectrum for ZRP, series D, 23 µm, after 3 days immersion in the electrolyte. The corrosion potential was –1048 mV vs SCE. The values of the parameters for best fitting to eqn (1) are listed in Table 2.

equivalent parameters. In a flat electrode, the surface is open to the electric signal for all frequencies whereas in a porous electrode, the electric signal penetrates more into pores the lower the frequency.

The electric signal penetration depth λ, can be defined as $\lambda = \gamma^{-1/2}$. The parameter γ was defined in eqn (1). Thus, λ depends on the various impedances along the transmission line. Any change in one of them or in the electrode thickness can lead to a flat ($\lambda > L$) or porous ($\lambda < L$) electrode. Obviously, when the impedance values are such that $\lambda \approx L$ the small random changes due to the own system dynamics will result in random porous/flat electrode behaviour.

The flat electrode behaviour obtained for thinner series A and B means that the penetration depth of the electric signal in the film pores is higher than the pore length. Thus, all zinc particles participate in the galvanic protection of the steel substrate. However, for the systems showing porous electrode behaviour (series C and D) a part of the zinc in the film does not contribute to the galvanic action and is consumed in a self-corrosion process.

The observed flat/porous electrode behaviour can be explained in terms of the pore tortuosity, which increases with the zinc particle grain size and diameter

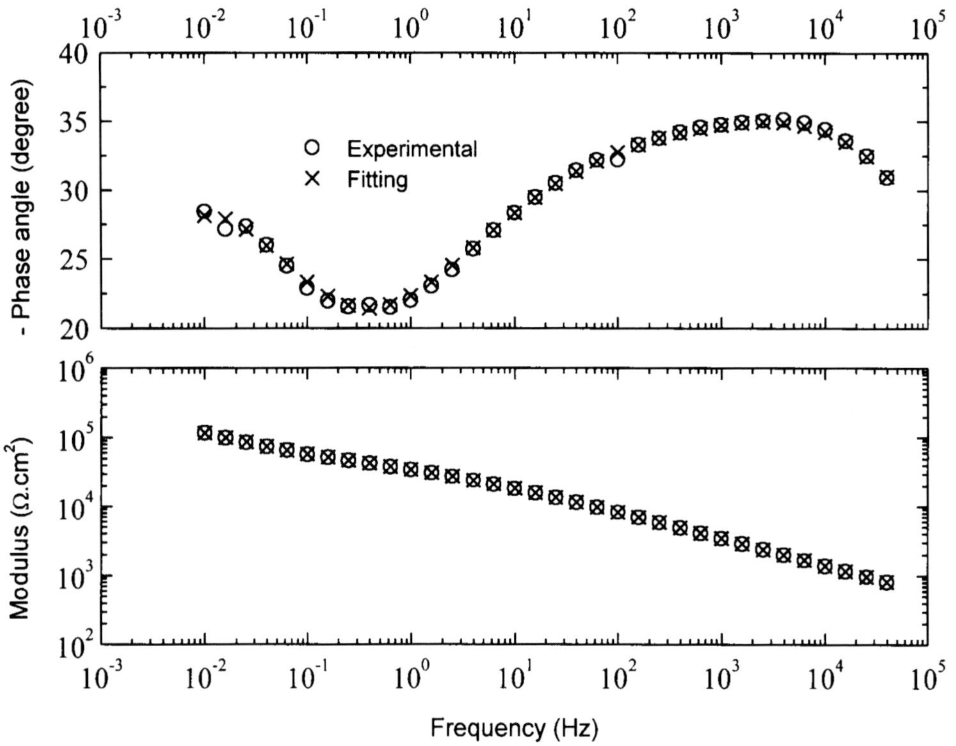

Fig. 5 Impedance spectrum for ZRP, series D 23 μm, after 60 days of immersion in the electrolyte. The corrosion potential was –885 mV vs SCE. The values of the parameters for best fitting to eqn (4) are listed in Table 2.

dispersion. The tortuosity, defined as the ratio, ρ, between the actual pore length, L_0, and the thickness of the film, L so that $\rho = L/L_0$. Thus only ideal pores normal to the steel surface will have $\rho = 1$.

Table 2. *Best fitting parameters to the data in Figs 4 and 5. The experimental data have been fitted to eqns (1) and (4) respectively*

	R_m(kΩ.cm)	C_m(nF.cm^{-1})	α_m	R_1(Ω.cm^{-3})	C_1(mF.cm^{-1})	α_1
Fig. 4	76	116	0.90	2.97	599	0.74
Fig. 5	31.1 10^3	0.25	0.71	1830	132	0.47
	R_e(kΩ.cm)	R_2(Ω.cm^2)	C_2(nF.cm^{-2})	α_2	Reo(Ω.cm^2)	L(μm)
Fig. 4	24.6	—	—	—	433	23
Fig. 5	2350	1749	10.6	0.60	293	23

3.4. Systems Having Standard Grain Size (Series C and D)

The evolution of the fitting parameters on immersion time (Figs 6–8) shows the same trend as described for the higher thickness systems [15,16], that is to say:

- The electrolyte resistance throughout the paint, R_e, remains practically constant for periods 1 and 2 (see Fig. 6), and rises sharply after the transition period.
- After period 2 the contact resistance, R_m, increases gradually with immersion time (see Fig. 7) due to the zinc dissolution process. The zinc corrosion products make zinc particles shift from their initial positions (see Fig. 3). This displacement leads to the observed increase in R_m.
- During period 2, important changes occur in the electric parameters that are related to the zinc dissolution process. The resistance associated with the charge transfer process for the zinc dissolution, R_1 rises and the associated double layer capacitance, C_1, decreases (Fig. 8). This evolution is consistent with a diminution of the zinc active area.
- In the transition period, important changes are observed for all parameters, especially those associated with ohmic resistance, R_e, and contact resistance, R_m (Figs 6 and 7). Moreover, a third time constant, not previously present, appears in the impedance spectra. This has been associated [14,15] with the

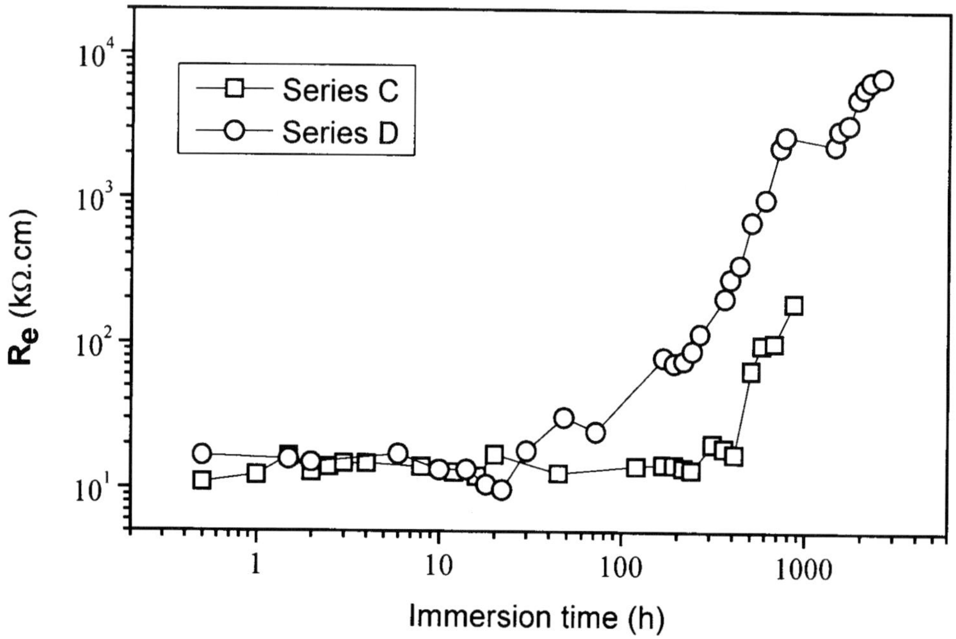

***Fig. 6** Evolution of the electrolyte resistance throughout the binder* R_e, *on immersion time for the lower thickness ZRP/steel systems.*

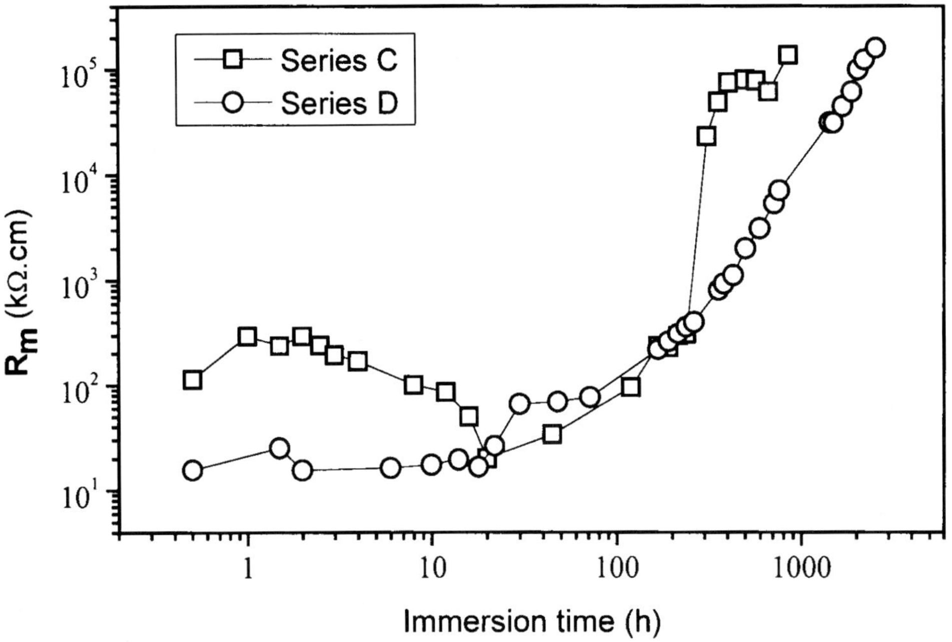

Fig. 7 Evolution of the resistance between zinc particles R_m*, on immersion time for the lower thickness ZRP/steel systems.*

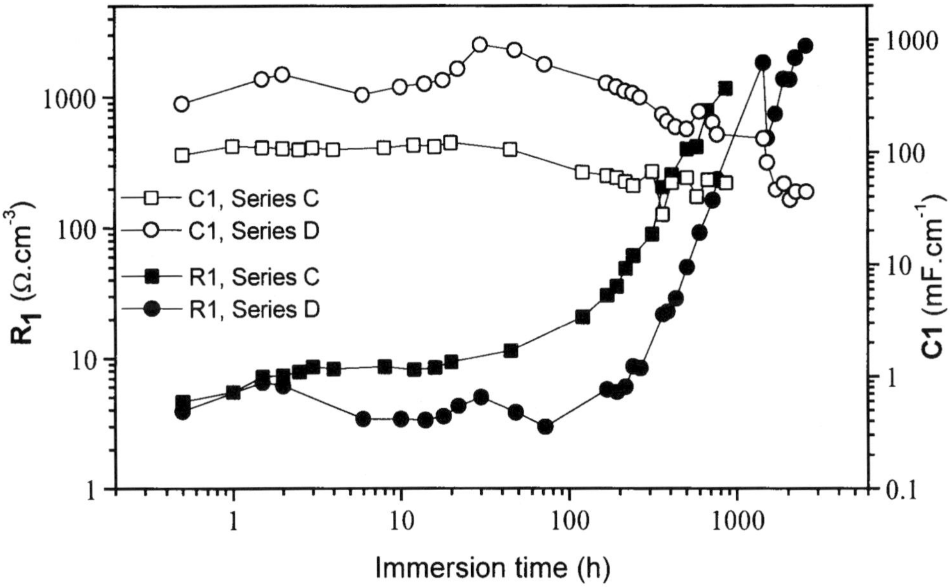

Fig. 8 *Evolution of the resistance* R_1 *and capacitance* C_1*, on immersion time for the lower thickness ZRP/steel systems.*

initiation of oxygen reduction at the outermost part of the film. This process makes the neighbouring solution more alkaline and zinc passivation starts to occur thereby increasing the contact impedance; moreover, ions in solution precipitate so making the solution throughout the film more resistive [15].

The third time constant that starts to appear in the impedance spectra at the transition period is completely defined at period 3. This change in the impedance spectra develops gradually with time. That is to say, the factors determining the cathodic protection decay (loss of electric contact between zinc particles and between them and the steel substrate, and the increase of the electrolyte resistance into the film) change gradually. Moreover, the barrier protection effect begins to operate as a result of the formation of insoluble corrosion zinc products (mainly oxides and carbonates).

3.5. Total Film Resistance

The analyses of the development with time of the various parameters relevant to paint behaviour (E_{corr}, R_m, R_1, C_1) are not able to differentiate clearly between the cathodic protection and barrier periods. It is evident that during periods 1 and 2 the galvanic action dominates and by the end of period 3 it disappears completely. However, this information has very little practical value because it cannot help in paint formulation.

Taking into account that the electric contact between zinc particles is necessary to maintain cathodic protection, and that the electrolyte resistance is the main parameter related to the barrier effect, a new parameter can be defined to evaluate the behaviour of the paint. This parameter assesses the Total Film Resistance TFR, and is defined as the parallel combination of the two resistances involved (eqn 6): the contact resistance between zinc particles R_m (galvanic parameter), and electrolyte resistance inside the film, R_e (barrier parameter):

$$TFR = \frac{R_m R_e}{R_m + R_e} \tag{6}$$

Figures 9 and 10 show the variation of TFR with testing time for the studied systems. Because of the reasons discussed above, only the systems showing porous electrode behaviour are considered.

In all cases the TFR parameter increases with immersion time. However, this evolution shows two different stages. Once the activation period (period 1) is overcome (approximately after one hour of immersion), R_e and R_m have low values, and thus the total resistance inside the film is small. For a certain immersion time the value of TFR is practically constant and the values approach an almost zero slope straight line. At a certain immersion time, an important increase in the straight-line slope can be observed (see Figs 9 and 10).

The variation of TFR permits a *break-point-time*, t_b to be defined that limits the end of the cathodic protection period and the beginning of the barrier type protection. In

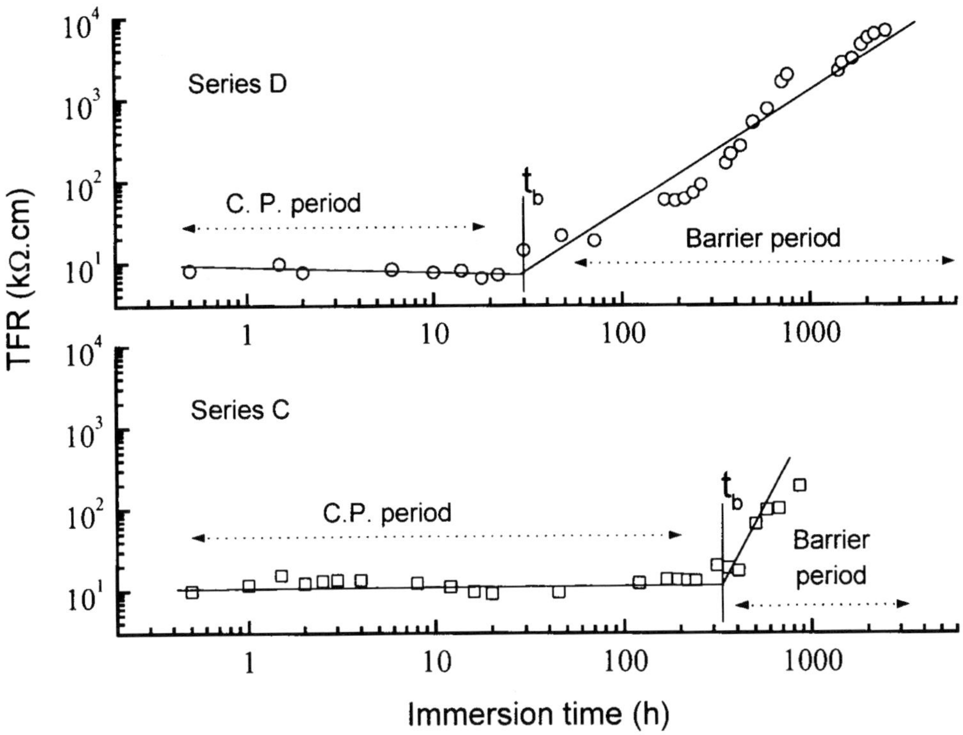

Fig. 9 Evolution of the Total Film Resistance, TFR, on immersion time for the lower thickness ZRP/steel systems. The break-point time, t_b, *is also shown for each system.*

an initial stage (low R_m and R_e values) the cathodic protection involves a low value for the total resistance throughout the film. After that, the displacement of zinc particles, and/or the corrosion products precipitation in the film, generates an important increase in R_m and R_e values. As usually $R_m > R_e$, TFR approaches R_e in the barrier period.

Table 3 summarises the slope values for the different straight lines in Figs 9 and 10. The cathodic protection period is characterised by a slope lower than 0.3 in the log–log plot, while in the barrier period this slope changes to values close to or higher than 1.

Furthermore, from Tables 1 and 3 it can be concluded that for the two thicknesses studied, the cathodic protection period shortens as the zinc particles grain size and their average diameter variation increase. The increases of grain size and diameter distribution increase the pore tortuosity (thus, the pore effective length) which increases the diffusion path for oxygen and ionic species. Once the galvanic couple starts to develop, the OH^- generated at the steel/coating interface and the Zn^{2+} produced in the paint must diffuse to balance the total charge throughout the film. The time constant for this diffusion process increases with the length of the diffusion path. The zinc corrosion products in long length diffusion path systems (series D in Figs 9 and 10) precipitate inside the film well before reaching the outer electrolyte, while for shorter diffusion paths (series B in Fig. 10) most of them diffuse into the

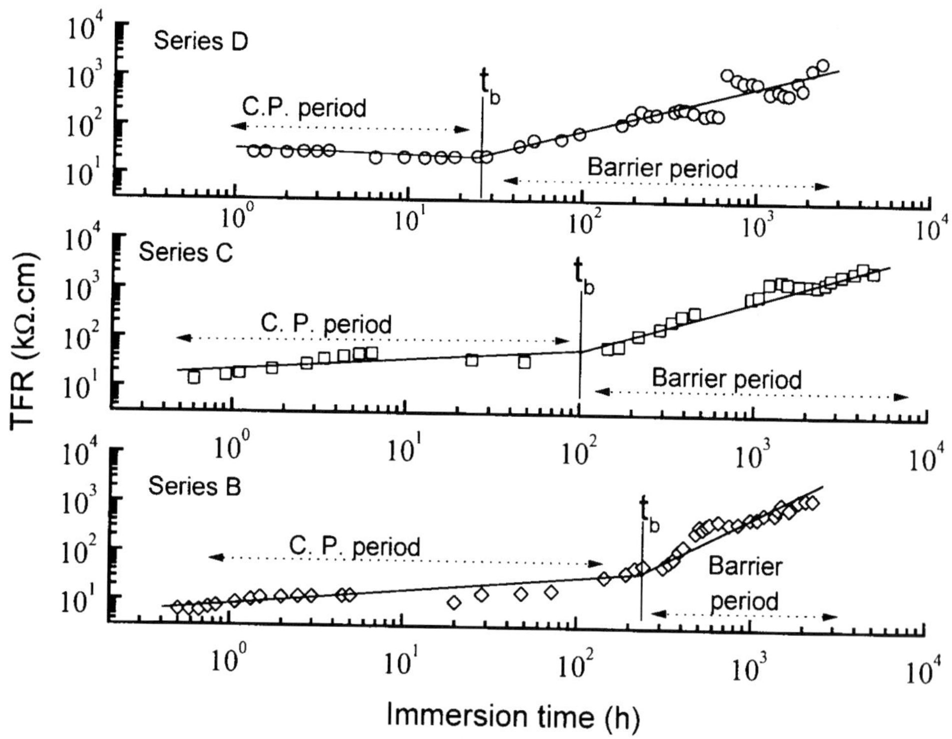

Fig. 10 Evolution of the resistance TFR with immersion time for the higher thickness ZRP/steel systems. The break-point time, t_b, *is also shown for each system.*

electrolyte. As a result, the pores of series D start to block before those of the B series giving the latter the longer cathodic protection period.

Considering now the barrier period, the data in Table 3 show that the slope of the log–log straight line increases from series D to B, i.e. as the diffusion path length decreases. Series D takes longer to achieve a certain barrier threshold than series B. This result is as expected if, again, the diffusion path length is considered. Once the oxygen in the fresh solution newly arriving to the steel/coating interface is consumed, the galvanic effect is supported/limited by the oxygen diffusing from the external solution. Systems with long diffusion paths (series D) will have low cathodic currents, i.e. low galvanic current and, thus, a low zinc corrosion rate. The film pores will be filled more slowly in those cases than for systems with shorter diffusion paths (series B for example).

Taking these results into account, zinc dust with a regular grain size that allows easy flow of species in and out of the film must be recommended both to increase the cathodic protection period and also rapidly produce a compact barrier. When the galvanic protection is exhausted, the barrier protection will be established more quickly the lower the grain size.

***Table 3.** Slope values (log–log plot) obtained by linear fitting in the resistance–immersion time curve for the studied systems (Figs 9 and 10). The break point time* t_b, *is also given*

		Systems with higher thickness (55 μm)
Series D	t_b = 25 h	$t < t_b$ Slope = –0.07 ⇒ Cathodic period $t > t_b$ Slope = 0.88 ⇒ Barrier period
Series C	t_b = 100 h	$t < t_b$ Slope = 0.25 ⇒ Cathodic period $t > t_b$ Slope = 1.07 ⇒ Barrier period
Series B	t_b = 236 h	$t < t_b$ Slope = 0.29 ⇒ Cathodic period $t > t_b$ Slope = 1.40 ⇒ Barrier period
		Systems with lower thickness (25 μm)
Series D	t_b = 30 h	$t < t_b$ Slope = –0.05 ⇒ Cathodic period $t > t_b$ Slope = 1.59 ⇒ Barrier period
Series C	t_b = 330 h	$t < t_b$ Slope = –0.03 ⇒ Cathodic period $t > t_b$ Slope = 2.55 ⇒ Barrier period

4. Conclusions

- The protection mechanism of zinc rich paints can be established on the basis of a Total Film Resistance, TFR. This resistance results from the parallel combination of the contact resistance between zinc particles and the resistance of the electrolyte throughout the film.
- A log–log plot of TFR vs immersion time allows a break-point-time, t_b, to be defined which identifies the transition between the galvanic and barrier periods.
- Zinc dust with a regular grain size must be recommended both to enlarge the cathodic protection period and to produce a compact barrier more rapidly.

References

1. R. Barajas, M. Morcillo, J. M. Bastidas and S. Feliú, *Pinturas y Acabados*, 1991, **33**, (188), 55–62.
2. S. Feliú, R. Barajas, J. M. Bastidas and M. Morcillo, *J. Coat. Technol.*, 1989, **61**, (775), 63–69.
3. S. Feliú, R. Barajas, J. M. Bastidas and M. Morcillo, *J. Coat. Technol.*, 1989, **61**, (775), 71–76.

4. S. G. Real, A. C. Eleías, J. R. Vilche, C. A. Gervasi and Di Sarli, *Electrochim. Acta*, 1993, **38**, (14), 2029–2035.

5. S. Feliú Jr., R. Barajas, J. M. Bastidas, M. Morcillo and S. Feliú, Study of Protection Mechanisms of Zinc–Rich Paints by Electrochemical Impedance Spectroscopy, in *Electrochemical Impedance: Analysis and Interpretation*, ASTM STP 1188, (J. R. Scully, P. C. Silverman and M. W. Kendig, eds.). A.S.T.M., Philadelphia, 1993, pp.438–449.

6. D. Pereira, J. D. Scantlebury, M. G. S. Ferreira and E. Almeida, *Corros. Sci.*, 1990, **30**, (11), 1135–1147.

7. E. Almeida, D. Pereira, A. M. Cabral and M. Morcillo, *J. Oil Colour Chem. Assocn*, 1991, **1**, 21–28.

8. S. E. Faidi, J. D. Scantlebury, P. Bullivant, N. T. Whittle and R. Savin, *Corros. Sci.*, 1993, **35**, (5–8), 1319–1328.

9. P. J. Frydrych, G. C. Farrington and H. E. Townsend, *Proc. Electrochem. Soc.*, 1987, 240–252.

10. X. R. Nóvoa, M. Izquierdo, P. Merino and L. Espada, *Mater. Sci. Forum*, 1989, **44–45**, 223–234.

11. M. Izquierdo, "Aplicación de la Espectroscopía de Impedancia Electroquímica al estudio de recubrimientos protectores". PhD Thesis Universidad de Santiago (http:\\www.usc.es). 1991.

12. C. A. Gervasi *et al.*, *Corros. Sci.*, 1994, **36**, (12), 1963–1972.

13. M. Izquierdo, X. R. Nóvoa, G. Pena and L. Espada, *Mater. Sci. Forum*, 1992, **111–112**, 257–268.

14. C. M. Abreu, "Pinturas ricas en zinc: Estudio de los factores que definen su acción protectora". PhD Thesis. University of Vigo (http:\\www.uvigo.es). 1996.

15. C. M. Abreu, M. Izquierdo, M. Keddam, X. R. Nóvoa and H. Takenouti, *Electrochim. Acta*, 1996, **41**, (15), 2405–2415.

16. C. M. Abreu, L. Espada, M. Izquierdo, P. Merino and X. R. Nóvoa, Factors affecting the electrochemical behaviour of zinc-rich paints, in *Organic and Inorganic Coatings for Corrosion Prevention. Research and Experiences.* Papers from *EUROCORR'96*, L. Fedrizzi and P. L. Bonora, Eds. Publication No. 20 in European Federation of Corrosion Series. The Institute of Materials, London, 1997, 23–32.

17. C. Gabrielli, F. Huet, A. Sahar and G. Valentin, *J. Appl. Electrochem.*, 1992, **22**, 801–808.

18. C. Gabrielli, F. Huet, A. Sahar and G. Valentin, *J. Appl. Electrochem.*, 1994, **24**, 481–489.

19. M. Pourbaix, "Atlas d'équilibres électrochimiques", ch. 4. Gautier-Villars Ed., Paris, France, 1963.

10
Coating Characterisation Using Pulsed and Gated Electrogenerated Chemiluminescence Imaging

F. E. BUTTREY, M. MALLACE* and H. N. McMURRAY†
The Engineering Doctorate Centre, Wales, University of Wales Swansea, Singleton Park, Swansea, SA2 8PP, UK
*British Steel Tinplate, Welsh Technology Centre, Port Talbot, SA13 2NG, UK
†Department of Materials Engineering, University of Wales Swansea, Singleton Park, Swansea, SA2 8PP, UK

ABSTRACT

A photoelectrochemical technique has been developed which permits a rapid, spatially resolved, evaluation of coating integrity for electrically non-conductive barrier coatings applied to metallic substrates. The technique is based on the anodically electrogenerated chemiluminescence (ECL) of luminol and relies on the generation of light at the site of penetrative coating defects. The coated metal is immersed in an alkaline aqueous electrolyte containing luminol and hydrogen peroxide (H_2O_2) and subjected to a series of anodic square wave voltage pulses. The voltage pulses produce a corresponding pulsatile light emission at the metal–solution interface. Complexing agents such as EDTA are shown to suppress background light emission due to homogeneous luminol–H_2O_2 chemiluminescence whilst minimally affecting ECL intensity. The maximum intensity and half-width of the ECL light pulse is systematically studied as a function of potential, electrolyte pH, H_2O_2 concentration and temperature. The relationships obtained are rationalised in terms of a proposed mechanism for the ECL process. They are also used to optimise the intensity, duration and contrast of light pulses for pulsed-and-gated ECL image acquisition using a low light digital (CCD) camera. The optimised system is used to study coating integrity in a steel based packaging material. Rapidly acquired (sub-second), high contrast, ECL images of defect populations are suitable for automatic image analysis providing information on the frequency, size distribution and location of penetrative defects.

1. Introduction

It is well known that the electrochemical oxidation of luminol (3-aminophthalhydrazide) in aqueous alkaline solution yields visible light, through a process known as electrogenerated chemiluminescence (ECL) [1–3]. Mechanistic studies have demonstrated that the intensity of ECL light emission [4–8], and luminol chemiluminescence generally [9,10], is dependent on solution pH and the concentrations of both hydrogen peroxide and luminol. Furthermore, a pulsed potential, or potential step has been found to increase the instantaneous light intensity generated at the electrode surface [8]. Steady state light intensities tend to be lower in poorly stirred solutions due to luminol depletion at the electrode surface [4]. Luminol ECL has been used to investigate the surface heterogeneity of electrode materials such as glassy carbon [7,11], platinum [12] and poly (phenylene oxide) coated platinum [13].

A "weak background" chemiluminescence is often observed in solution during luminol ECL experiments [14,15] due to homogeneous luminol chemiluminescence catalysed by trace quantities of transition metal cations [15] present as contaminants in reactants and or on equipment surfaces. The catalytic oxidation of luminol by transition metal cations in the presence of oxygen and/or hydrogen peroxide, and subsequent chemiluminescence is well known [16,17]. Furthermore, complexing agents are known to reduce this background emission by chelating the transition metal ions present in solution [15].

The aim of the work presented here was to optimise luminol ECL reaction kinetics for the detection of penetrative defects in anti-corrosion barrier coatings applied to steel substrates. Optimisation is achieved through (i) the use of complexants to reduce background chemiluminescence; (ii) by adjusting solution pH, reactant concentrations and temperature; and (iii) by control of the applied square wave pulsed potential. Images showing the distribution of coating defects are captured using a low-light video camera gated to match the potential pulse waveform. Results are presented demonstrating the suitability of this technique for detecting spontaneously occurring defect populations in experimental steel container systems after shelf-life testing.

2. Experimental Procedure

2.1. Reagents and Solutions

Luminol (3-aminophthalhydrazide, 97%) and hydrogen peroxide (unstabilised, aqueous 27.5% ww.) were obtained from the Aldrich Chemical Co. Ltd. All other chemicals were obtained from Aldrich at A.C.S. grade purity. All reagents were used without further purification. Doubly distilled water was used throughout for the preparation of solutions. The working electrode for ECL optimisation studies was composed of iron foil (purity 99.9+%), also obtained from Aldrich. The iron electrode was cleaned with 5 mm polishing alumina prior to use. Experimental coated steel food cans (dia. 73 mm, height 110 mm) were commercially produced for British Steel Plc. The internal coating consisted of a proprietary organic corrosion barrier layer, which was optically transparent. The coating thickness was 15 mm ± 2 mm. Cans were filled with green beans in 2.5% aqueous sodium chloride — an aggressive electrolyte allowing the simulation of commercial conditions. Cans were stored at 37°C for 2 years prior to opening, a period exceeding that of a typical shelf-life storage test.

All luminol solutions for ECL experiments contained 10^{-1}M aqueous Na_2HPO_4 as a pH buffer and supporting electrolyte. Solution pH values were adjusted by admixing equimolar aqueous solutions of NaOH and H_3PO_4. Concentrations of ethylenediaminetetracetic acid (EDTA) in luminol solutions were established using volumetric additions of 10^{-1}–10^{-3}M aqueous tetrasodium EDTA. Concentrations of H_2O_2 in luminol solutions for ECL were established using volumetric additions of 0.1–10×10^{-1}M aqueous H_2O_2 prepared by volumetric dilution of the 30% stock. All solutions containing H_2O_2 were used within 12 h of preparation. All experiments were conducted at room temperature, 22–23°C, unless otherwise stated.

2.2. Apparatus

2.2.1. Light Intensity Measurements

The measurement of ECL light intensity was performed using the apparatus shown in Fig. 1(a). The electrochemical cell comprised an iron working electrode, a counter electrode composed of platinum gauze contained in a light proof casing, and a saturated calomel reference electrode. The cell was connected to a Bank Wenking potentiostat LB75M controlled by a Topward 8201 pulse generator. An Electron Tubes 9125A photomultiplier was focused on the working electrode using a 5 cm dia. compound glass lens of 10 cm focal length. An Electron Tubes A1 transimpedance amplifier was used to convert the photomultiplier output current to a voltage signal, which was recorded using a Tektronix TDS 340A storage oscilloscope. The apparatus was contained inside a light tight housing. A Grant model L14 water bath allowed water to be circulated through a 650 mL water jacketed glass beaker, enabling the electrolyte temperature to be varied between 10 and 60°C.

2.2.2. Imaging Experiments

ECL imaging of can coating defect populations was performed using the apparatus shown in Fig. 1(b). The electrochemical cell comprised the can body acting as the working electrode, a counter electrode composed of platinum wire (dia. 0.25 mm), and a silver/silver chloride wire reference electrode (dia. 0.25 mm). Images were obtained using a Hamamatsu C5985–10 monochrome chilled CCD camera (Kern-Paillard f10 mm lens) and C5985 camera control unit. A Hamamatsu C2400–80H image intensifier and M4314-01 HV control unit preceded the camera. Images were digitised as 8 bit resolution grey scale images of 320 by 240 pixels using a video capture card and software supplied by Motion Picture Ltd. Image analysis was carried out using Graf Tek Optilab software.

(a)

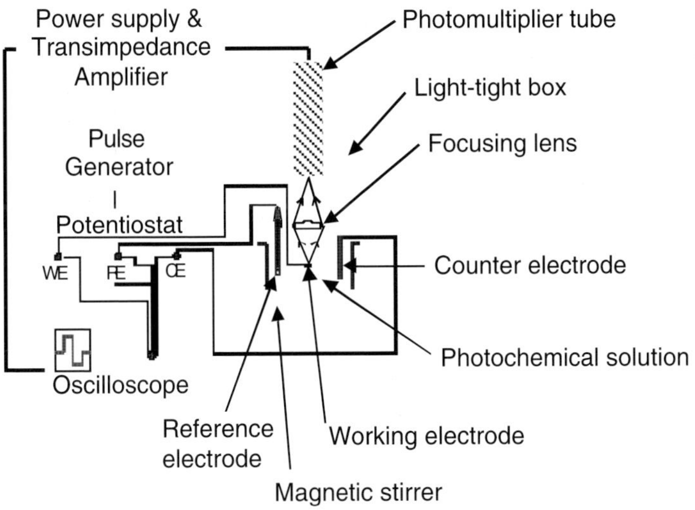

Fig. 1(a) *Experimental apparatus — Luminol ECL intensity measurements.*

(b)

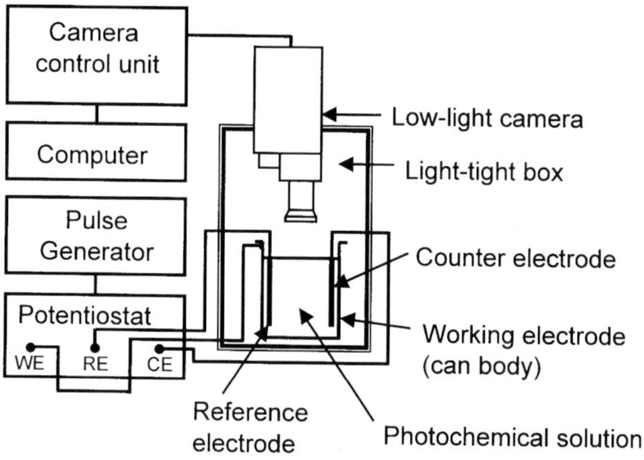

Fig. 1(b) Experimental apparatus — Luminol ECL imaging of penetrative coating defects.

2.3 Results

2.3.1. Effect of EDTA on ECL and background chemiluminescence

In these experiments, the iron working electrode was subject to a square wave anodic potential pulse in which the starting and final potential was 0 V (vs SCE). The pulse potential amplitude was 1.4 V and the pulse duration 670 ms. The application of the potential pulse resulted in a corresponding ECL emission in which ECL intensity rapidly rose to a peak value (I_p), and then decayed more slowly to zero. The relationship between the applied potential pulse and the resulting ECL waveform is shown schematically in Fig. 2.

The effect of EDTA concentration ([EDTA]) on I_p is shown in Fig. 3 for an electrolyte containing 3.5×10^{-2}M H_2O_2 and 1.125×10^{-3}M luminol at pH 11.70. Curve (a) indicates that I_p is reduced by around 20% as [EDTA] is increased from 0 to 10^{-4}M. A measure was also made of the effect of EDTA concentration on the background chemiluminescent light intensity of the electrolyte in the absence of the working

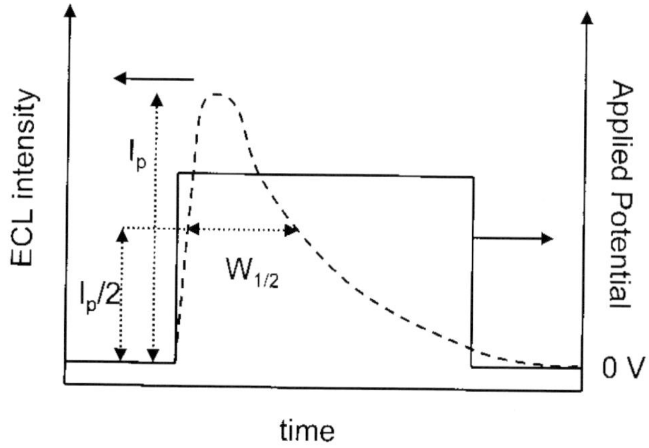

Fig. 2 Schematic plot of ECL light intensity against time: —— *applied potential pulse (vs SCE);* - - - - *ECL light waveform.* I_p *= peak ECL intensity,* $W_{1/2}$ *= half-width (width at half-height).*

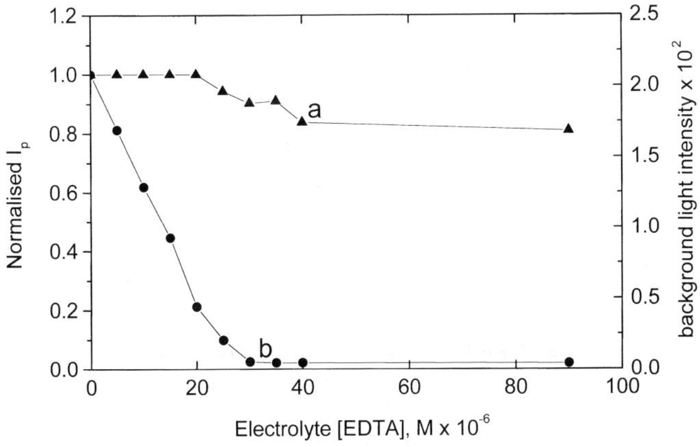

***Fig. 3** Effect of EDTA concentration on (a) normalised ECL I_p values at an Fe electrode, (b) background chemiluminescent light intensity scaled to maximum value of ECL I_p. Electrolyte composition: 2.25×10^{-3}M luminol, 0.1M sodium phosphate buffer pH 11.70. Applied square wave potential pulse amplitude 1.4 V vs SCE, pulse duration 670 ms. Rest potential = 0 V.*

electrode. Curve (b) in Fig. 3 shows how the background chemiluminescent light intensity was reduced by 98% when [EDTA] exceeded 3×10^{-5}M.

In order to confirm the hypothesis that background chemiluminescence arises from the transition metal catalysed oxidation of luminol, experiments were performed in which the experimental electrolyte was spiked with both uncomplexed transition metal ions, and transition metal ions complexed with an excess of EDTA. Curve (a) in Fig. 4 shows the characteristic change in emitted light intensity observed on addition of $CuSO_4$(aq) to give a concentration of 10^{-4}M Cu^{2+}. Curve (b) in Fig. 4 shows the effectively invarient emitted light intensity observed on addition of $CuSO_4$(aq)

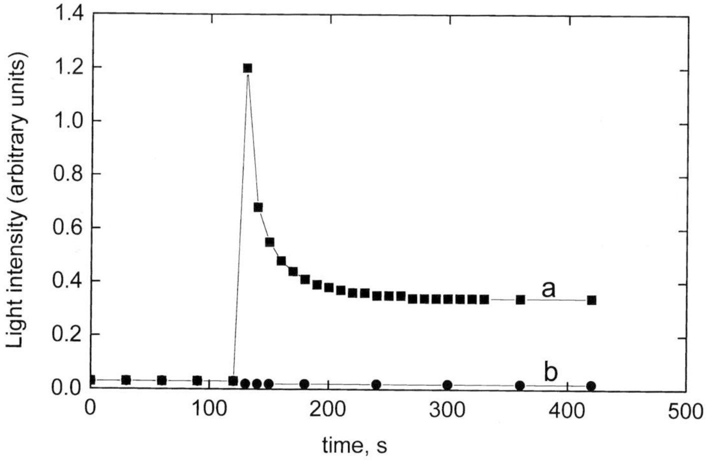

***Fig. 4** Background chemiluminescent light intensity as a function of time when (a) $CuSO_4$ added at 120 s, to give a concentration of 1×10^{-4}M Cu^{2+} (b) $CuSO_4$ + EDTA added at 120 s, to give a concentration of 1×10^{-4}M Cu^{2+} and 2×10^{-4}M EDTA. Electrolyte composition: 1.125×10^{-3}M luminol 3.5×10^{-2}M H_2O_2, 0.1M sodium phosphate buffer, pH 11.70.*

and EDTA to give a concentration of 1×10^{-4}M Cu^{2+} and 2×10^{-4}M EDTA to the same solution. Entirely similar results were obtained using additions of similar concentrations of Ni^{2+} ($NiSO_4$ (aq)).

2.3.2. Effect of pH and hydrogen peroxide concentration

Experiments to determine the effect of H_2O_2 concentration ($[H_2O_2]$) on the intensity of ECL were conducted over a pH range of 7–13. The experimental electrolyte contained 2.25×10^{-3}M luminol and 5×10^{-5}M EDTA. Concentrations of H_2O_2 were varied between 0 and 2.6×10^{-2}M. A 1.4V (vs SCE) square wave potential pulse of 670 ms was applied to the working electrode as described in section 2.3.1. and the resultant ECL waveform was recorded. Potential pulses were applied at 40 s intervals until reproducible ECL waveforms were attained (typically < 5 repetitions). Figure 5 shows I_p as a function of $[H_2O_2]$ and pH. It is clear that at any given H_2O_2 concentration, I_p values increase with electrolyte pH to a maximum at pH 11–12. However, at pH 13 I_p values are considerably reduced. It may also be seen from Fig. 5 that at pH < 9 I_p is relatively unaffected by $[H_2O_2]$, whereas the I_p vs $[H_2O_2]$ curves at pH10–13 show maxima at $[H_2O_2]$ between 2 × 10–3 and 5×10^{-3}M.

It was also possible to measure the half-width ($W_{1/2}$) of the ECL waveforms, as indicated in Fig. 2. Figure 6 shows $W_{1/2}$ values corresponding to the I_p data illustrated in Fig. 5. At $[H_2O_2] < 2 \times 10^{-3}$M, the relationship between $W_{1/2}$ and $[H_2O_2]$ appears complex and pH dependent. However at pH <12, $W_{1/2}$ is relatively independent of $[H_2O_2]$ for $[H_2O_2]$ between 4×10^{-3} and 3×10^{-2}M.

2.3.3. Effect of applied potential and temperature

Experiments to determine the effect of applied potential on I_p were conducted using the electrolyte described in section (ii), at pH 11.70, and $[H_2O_2] = 8 \times 10^{-3}$M. Square wave potential pulses of varying amplitudes between 0.5 V and 2.5 V were applied to the working electrode as described previously.

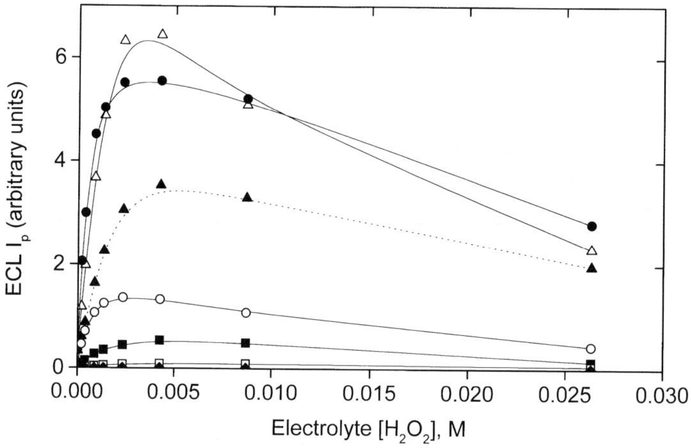

Fig. 5 *ECL I_p values at an Fe foil electrode as a function of electrolyte H_2O_2 concentration at ♦ pH 7, □ pH 8, ■ pH9, ○ pH10, ● pH11, Δ pH12, ▲ pH 13 (dotted line). Electrolyte composition: 2.25×10^{-3}M luminol, 0.1M sodium phosphate buffer, 5×10^{-5}M EDTA. Applied square wave potential pulse amplitude 1.4 V vs SCE duration 670 ms, rest potential 0 V vs SCE.*

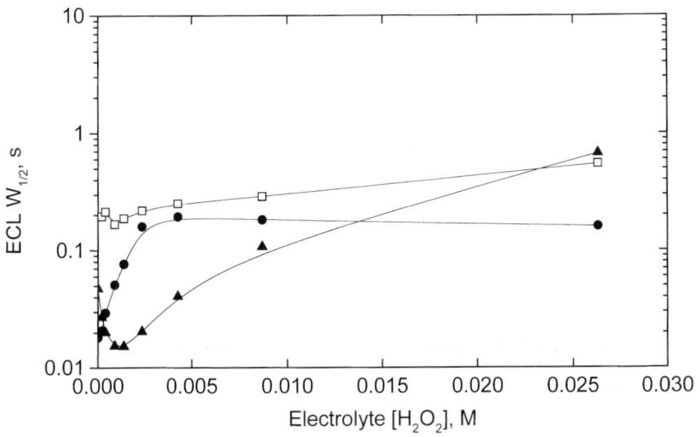

Fig. 6 *ECL emission peak half-width ($W_{1/2}$) values at an Fe foil electrode as a function of electrolyte H_2O_2 concentration at □ pH 8, ● pH11, ▲ pH 13. Electrolyte composition: 2.25 × 10^{-3}M luminol 0.1M sodium phosphate buffer, 5 × 10^{-5}M EDTA. Applied square wave potential pulse amplitude 1.4 V vs SCE duration 670 ms, rest potential 0 V vs SCE.*

The value of I_p was found to increase linearly with increasing potential pulse amplitude; a straight line plot of I_p vs potential pulse amplitude exhibited a linear correlation coefficient of 0.996. Variations in the electrolyte temperature had no measurable effect on the peak ECL output.

2.3.4. ECL imaging of penetrative defect populations in organic barrier coatings

Fourteen food cans previously filled with green beans in brine for 2 years were emptied, washed with distilled water and filled with approximately 330 ml electrolyte consisting of 1.4 × 10^{-2}M potassium chloride and 1.3 × 10^{-2}M hydrated potassium ferrocyanide. Cans were then anodically polarised to a potential of 6 V vs a nickel gauze counter electrode. The resultant current was recorded after a four-second stabilisation period. Typical currents observed were 10–50 mA per can, indicating a significant level of penetrative defects in the organic barrier coating i.e. areas in which electrolyte came into contact with the substrate metal. Subsequently, the anodic polarisation of the cans was resumed for approximately 40 s to permit the development of visible precipitates of Prussian Blue ($Fe_4[Fe(CN)_6]_3.xH_2O$) at defect sites. The interior of the cans were then examined using a hand-held magnifying lens, and populations (typically 2–20 per can) of small (5–50 μm dia.) penetrative defects were observed. (This method of defect detection is conventionally used by canmakers and fillers.)

In order to evaluate luminol ECL as a means of rapidly imaging defect populations, a second series of 14 cans was emptied, washed and filled with an electrolyte as described in section 2.3.3. A –1.3 V (vs Ag/AgCl) cathodic potential was applied to the can for 40 s to reduce or loosen any corrosion products present on exposed metal surfaces. This was followed by sonication in an ultrasound bath for 5 min. (Repeat application of the combined cathodic/ultrasonic cleaning procedure was not found to increase the number of detectable coating defects or the size of individual coating defects.) A 0.8 V (vs Ag/AgCl) square wave anodic potential pulse was applied to the can, and the resultant light output was recorded as a digital image using on-chip integration with an exposure time of 8 ms. The CCD camera shutter was electronically triggered such that the exposure period coincided with maximum ECL intensity. Figure 7 shows paired images of can interiors under normal illumination and during ECL emission. No variation in can images was observed after 5 successive cycles of cathodic polarisation and ultrasound. Computerised image analysis was applied to the ECL images derived from all 14 cans, in order to determine the spatial distribution of penetrative defects. Figure 8 is a histogram showing the distribution of penetrative coating defects as a function of distance from the centre of the ECL image.

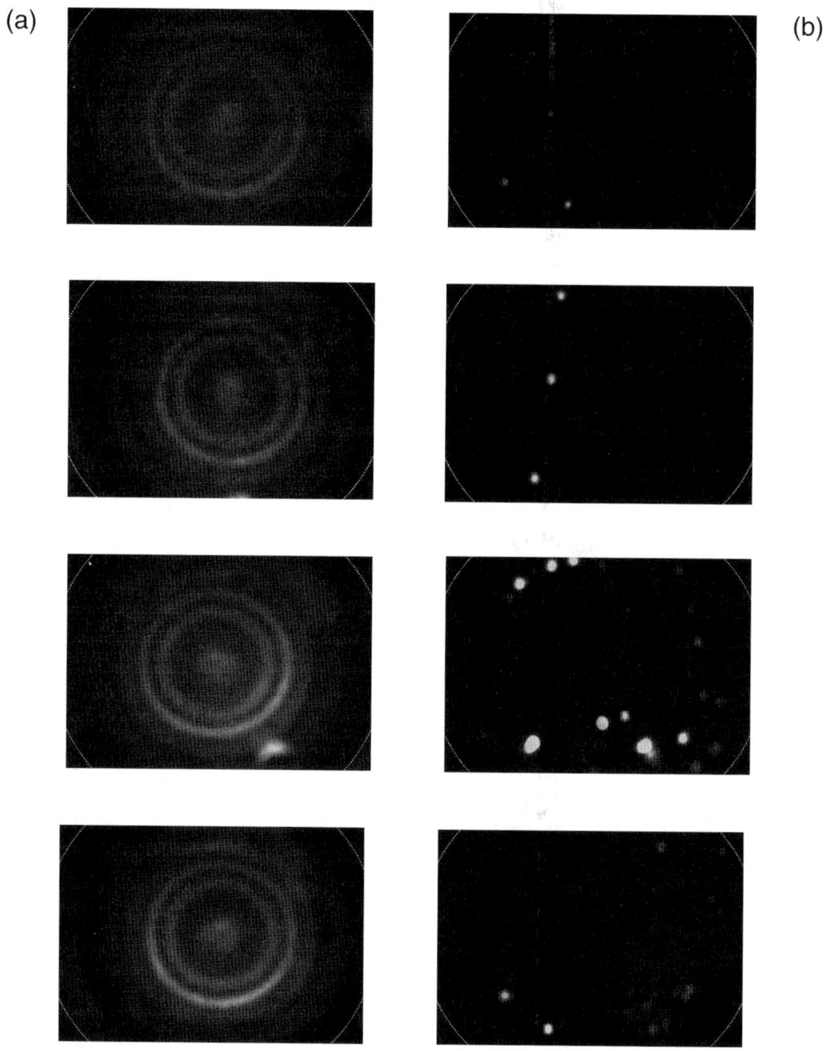

Fig. 7 Organic coated interior of steel food cans imaged using a co-axially mounted video camera. Dotted line indicates can rim. (a) normally illuminated image, (b) ECL image of penetrative defect population. Electrolyte composition : 2.25 × 10^{-3}M luminol, 0.1M sodium phosphate buffer, 8 × 10^{-3}M H_2O_2, 5 × 10^{-5}M EDTA, pH 11. Voltage pulse for ECL 670 ms, 0.8 V vs Ag/AgCl.

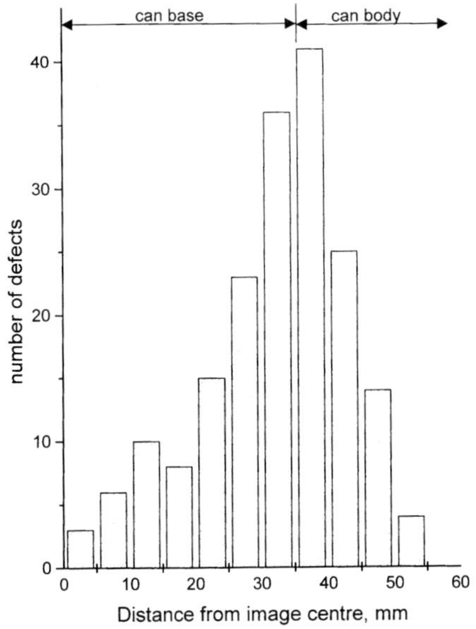

Fig. 8 *Image analysis data indicating the special distribution of penetrative defect sites on the organic coated interior of a population of 14 steel food cans.*

3. Discussion

Luminol is a dibasic acid (first and second pK_a 6.3 [18] and approximately 13 [7] respectively), and the process of chemiluminescence is initiated when the luminol monoanion is oxidised in the presence of O_2 or hydrogen peroxide H_2O_2 [3]. Light is emitted when the luminol radical anion so formed reacts with superoxide radical derived from O_2 or H_2O_2 to produce an hydroperoxide addition product. The hydroperoxide addition product is itself a weak acid (pK_a 10.4 [10]) and it has been shown [9] that it is only the monoanion form of this intermediate which decomposes to give an excited state aminophthalate molecule, subsequently relaxing with emission of photons [3] at 430 nm. The undissociated hydroperoxide regenerates the starting material (luminol) via a dark reaction [9].

It has further been shown [3] that, in the presence of hydrogen peroxide, electrogenerated luminol chemiluminescence (ECL) proceeds via the following mechanism:

Electrode reaction for luminol

$$LH^- + OH^- \xrightarrow{k_1} L^{\bullet -} + e^- + H_2O \tag{1}$$

$$L^{\bullet -} + 4OH^- \xrightarrow{k_2} AP^{2-} + N_2 + 3e^- + 2H_2O \tag{2}$$

where LH^- is the luminol monoanion, $L^{\bullet -}$ is the luminol radical anion, and AP^{2-} is the ground state aminophthalate dianion.

Electrode reaction for H_2O_2

$$HO_2^- + OH^- \xrightarrow{k_3} O_2^{\bullet -} + H_2O + e^- \quad (3)$$

$$O_2^{\bullet -} \xrightarrow{k_4} O_2 + e^- \quad (4)$$

where HO_2^- is the hydroxide anion (pK_a, H_2O_2 = 11.65) and $O_2^{\bullet -}$ is the superoxide radical.

Solution reaction

$$L^{\bullet -} + O_2^{\bullet -} \xrightarrow{k_5} \text{hydroperoxide} \xrightarrow{\text{FAST}} AP^{2-} + N_2 + h\upsilon_{430\text{ nm}} \quad (5)$$

It has been shown that under highly convective conditions (at a rotating disc electrode), the steady state intensity of ECL, is first order with respect to both luminol and $[H_2O_2]$. However, under the effectively quiescent conditions used here, time-dependent concentration polarisation of the working electrode is likely to be significant.

If it is assumed that the interfacial single electron transfer kinetics for reactions 1 and 3 are sufficiently facile that these processes are at all times mass transport controlled then the time dependent anodic currents for each process will be given by the Cottrell equation:

$$i = n F A D^{1/2} C_b \pi^{-1/2} t^{-1/2} \quad (6)$$

where n represents the number of electrons transferred, F is the Faraday constant, A is the electrode surface area, D is the diffusion coefficient for the species being oxidised, C_b is its bulk concentration, and t is time.

If it is further assumed that the multi-electron transfer kinetics for reactions 2 and 4 are sufficiently slow that both of these processes are entirely surface controlled then it may be shown that the time dependent surface ($[\,]_s$) concentrations of $L^{\bullet -}$ and $O_2^{\bullet -}$ are given by:

$$[L^{\bullet -}]_s = [LH^-]_b D_{LH^-}^{1/2} k_2^{-1/2} \bullet \exp(-k_2 t) \bullet \text{erf}(k_2 t)^{1/2} \quad (7)$$

and

$$[O_2^{\bullet -}]_s = [H_2O_2]_b D_{H_2O_2}^{1/2} k_4^{-1/2} \bullet \exp(-k_4 t) \bullet \text{erf}(k_4 t)^{1/2} \quad (8)$$

where k_2 and k_4 are electrochemical rate constants for reactions (2) and (4) respectively. The bimolecular kinetics of reaction 5 imply that the intensity of ECL emission (IECL) is given by

$$I_{ECL} = \text{const } [L^{\bullet -}]_s [O_2^{\bullet -}]_s \quad (9)$$

Combination of eqs (7), (8) and (9) produces a time-dependent function for ECL light intensity which is entirely consistent with the ECL light intensity peak shapes observed experimentally.

The notion that the efficiency of ECL light emission increases with increasing dissociation of the luminol hydroperoxide intermediate, and therefore solution pH, is consistent with the pH dependence of I_p data shown in Fig. 5, which indicate that maximum ECL I_p values are achieved at pH 11–12. The decrease in ECL I_p values seen at pH > 12 may be explained by the known decrease of aminophthalate luminescence quantum yields at high pH [9]. Haapakka and Kankare [6] have demonstrated similar findings for luminol ECL at a platinised carbon electrode.

The I_p data in Fig. 5 suggest that for electrolyte pH > 8 and $[H_2O_2] < 2 \times 10^{-3}$M ECL, emission intensity increases with $[H_2O_2]$ in an approximately linear fashion, which is consistent with eqs (7–9). However, Fig. 5 also shows that at higher $[H_2O_2]$ values I_p decreases with increasing $[H_2O_2]$. These observations suggest that some competitive process is occurring whereby higher $[H_2O_2]$ inhibits the production of $L^{\bullet -}$ via reaction 1, possibly due to increased surface coverage by chemisorbed oxygen produced via reactions 2 and 4. If reaction 1 is being inhibited this implies that reaction 1 (and possibly reaction 3) must exhibit some degree of surface control at higher values of electrolyte $[H_2O_2]$. This assumption is consistent with the observed increase in I_p values with increasing anodic voltage pulse amplitude at pH 11.7 and $[H_2O_2] = 8 \times 10^{-3}$M. However, we do not yet possess a solution to the luminol ECL rate equations when reactions 1 and 3 are under mixed surface and mass transport control. Consequently, we are unable to apply further quantitative interpretation to the I_p data in Fig. 5 or the $W_{1/2}$ data shown in Fig. 6. However, it should be noted that all the $W_{1/2}$ values shown in Fig. 6 are significantly less than the potential pulse width of 670 ms. This implies that, in all cases, the shape of the time dependent ECL intensity peak is determined by the electrochemical kinetics of the ECL process.

Let us now consider the influence of EDTA concentration on the intensity of background electrolyte chemiluminescence. The data shown in curve (a) of Fig. 4 indicate that the addition of uncomplexed divalent transition metal ions to the electrolyte homogeneously catalyses the oxidation of luminol producing an increase in background chemiluminescent light intensity. Divalent transition metal ions are well known to catalyse the chemiluminescence of luminol when in the presence of dissolved oxygen and/or hydrogen peroxide [17]. The mechanism for this catalysis is not entirely clear, but is thought to involve production of hydroxyl radicals ($OH^{\bullet}$) through a 'Fenton type' reaction, i.e.

$$M^{2+} + H_2O_2 \longrightarrow M^{3+}\ OH^- + OH^{\bullet} \qquad (10)$$

The observation that chelation of metal ions by complexants such as EDTA greatly reduces their catalytic activity for this process [15] is consistent with curve (b) in Fig. 4.

Figure 3 indicates that EDTA is extremely effective in minimising background chemiluminescent light emission, whilst minimally affecting the intensity of ECL generated at the working electrode. This finding tends to confirm that EDTA is not acting as a reductive quencher for radicals such as $O_2^{-\bullet}$ or $L^{-\bullet}$, and it is the complexant properties of EDTA which are important here.

Considering the ECL imaging of coating defect populations, it may be seen from Fig. 7(b) that the sample ECL images show areas of light corresponding to the location of penetrative defects in the organic barrier coating (approximately 2–20 per can). Furthermore, comparison with the normally illuminated can images in Fig. 7(a) indicated that the majority of these defects lie in the base of the can. These results are broadly consistent with the location of defects as determined by Prussian Blue precipitation. However, caution must be exercised in the interpretation of Fig. 7(b) due to the perspective distortion of the can image, specifically foreshortening of the can walls. Furthermore, the ECL image areas corresponding to individual defects appear significantly larger than the actual defect areas due to de-focusing and other artefacts introduced by the imaging technique. As such, the images in Fig. 7(b) provide information on the number and spatial distribution, but not the absolute size of coating defects.

Figure 8 shows that penetrative defects in the organic coating are located principally at the perimeter of the can base, and at the interface between the can base and the can wall. These findings are again consistent with observations of defect distribution made using the Prussian Blue precipitation technique. The organic barrier coating is applied to the steel substrate in sheet form, prior to can forming. The coated can base is therefore subjected to a higher degree of stretch during manufacture than other areas of the can body. Additional stresses are induced by the application of a rolling bead adjacent to the can base, which is intended to guide the can along conveyor systems during manufacture and filling processes. These factors combine to generate increased levels of porosity in the base of the can.

4. Conclusions

Application of 670 ms anodic potential pulses with amplitudes between 0.5 and 2.5V vs SCE (rest potential 0 V) to an iron electrode in an aqueous phosphate electrolyte containing luminol, EDTA and H_2O_2 produces a corresponding pulsatile ECL light output with a half width <<670 ms. Peak ECL emission (I_p) increases with pH up to pH11–12 but decreases at pH13. Values of I_p also increase with $[H_2O_2]$ up to $[H_2O_2]$ = approx. 5×10^{-3}M but decrease at higher $[H_2O_2]$ values. These findings are consistent with a mechanism in which light emission results from the decomposition of a weakly acidic hydroperoxide addition product resulting from a homogeneous reaction between the luminol radical anion and the superoxide radical anodically electrogenerated from the luminol monoanion and the hydrogen peroxide monoanion respectively. Background electrolyte chemiluminescence may be minimised by addition of EDTA, which inhibits the catalytic oxidation of luminol by trace transition metal ions without affecting ECL intensity.

The application of luminol ECL to the detection of defect populations in organically coated steel packaging materials allows the acquisition of high contrast images, which are suitable for automated image analysis. The ECL images provide information on the number and spatial distribution of penetrative coating defects. This technique shows promise as a tool for the evaluation of performance in coated steel products and exhibits considerable advantages over current techniques.

5. Acknowledgements

The authors would like to thank Dr B. J. Hewitt, Director – Technical, British Steel Strip products, Mr E. F. Walker, Manager, Technical Co-ordination, Welsh Technology Centre, British Steel Strip Products and Mr D. Jones, Manager, Research and Development, British Steel Tinplate for permission to publish this paper and gratefully acknowledge the support of the Engineering and Physical Sciences Research Council. In addition, the author would like to thank Hamamatsu Photonics UK Ltd. for their assistance in providing camera equipment.

References

1. N. Harvey, *J. Phys Chem.*, 1929, **33**, 1456–1459.
2. K. E. Haapakka and J. J. Kankare, *Anal. Chim. Acta*, 1982, **138**, 253–262.
3. J. E. Vitt, D. C. Johnson and R. C. Engstrom, *J. Electrochem. Soc.*, 1991, **138**, 6, 1637–1643.
4. B. Epstein and T. Kuwana, *Photochem. Photobiol.*, 1965, **4**, 1157–1173.
5. B. Epstein and T. Kuwana, *Photochem. Photobiol.*, 1967, **6**, 605–611.
6. K. E. Haapakka, and J. J. Kankare, *Anal. Chim. Acta*, 1982, **138**, 263–275.
7. S. Sakura, *Anal. Chim. Acta*, 1992, **262**, 49–57.
8. G. P. Jirka, A. F. Martin and T. A. Nieman, *Anal. Chim. Acta*, 1993, **284**, 345–349.
9.G. Merenyi and J. S. Lind, *J. Am. Chem. Soc.*, 1980, **102**, 5830–5835.
10. J. Lind, G. Merenyi and T. E. Erickson, *J. Am. Chem. Soc.*, 1983, **105**, 26, 7655–7661.
11. R. C. Engstrom, K. W. Johnson and S. DesJarlais, *Anal. Chem.*, 1987, **59**, 670–673.
12. R. C. Engstrom, C. M. Pharr and M. D. Koppang, *J. Electroanal. Chem.*, 1987, **221**, 251–255.
13. R. C. Engstrom, P. L. Nohr and J. E. Vitt, *Colloids and Surfaces A: Physiochemical and Engineering Aspects*, 1994, **93**, 221–227.
14. N. N. Rozhitskii, *J. Anal. Chem.*, 1992, **47**, 1288–1310.
15. L. J. Kricka *et al.*, *J. Immunoassay*, 1996, **17**, 1, 67–83.
16. O. M. Huta *et al.*, *Mater. Sci.*, 1995, 31, **3**, 307–315.
17. L. L. Klopf and T. A. Nieman, *Anal. Chem.*, 1983, **55**, 1080–1083.
18. L. Erdey, I Buzas and K. Vigh, *Talanta*, 1966, **13**, 463.

Part 3

Inorganic Coatings

11
Characterisation of Corrosion Behaviour of Tungsten-based Thin Film Hard Coatings on Steel Substrates

C. M. A. BRETT and A. CAVALEIRO*
Departamento de Química, Universidade de Coimbra, 3000 Coimbra, Portugal
*Departamento de Engenharia Mecânica

ABSTRACT
The corrosion of tungsten-based sputtered thin film hard coatings on high-speed steel substrates has been investigated in aqueous neutral chloride media. The coatings, of thickness between 3.5 and 8 μm, contain elements such as titanium, aluminium, nickel and nitrogen. The corrosion processes have been tested electrochemically and by surface analysis techniques. Despite differences in the macroscopic corrosion rate, in all cases it is found that immersion in chloride solution leads to points of localised corrosion, with corrosion pits of diameter typically up to 10 μm nearly always appearing within the first half hour after immersion. Penetration of the electrolyte through the coating leads to substrate corrosion and formation of a protuberance, which is often followed by its rupture with deposition of corrosion products in a ring around the pit; the mechanical and physical properties of the rest of the coating are unaltered. Salt spray tests have been carried out to study the corrosion process under more aggressive conditions and computerised image analysis of the pits has been performed in order to give a size distribution of the pits. The model for the mechanism of corrosion of the coatings is discussed in the light of all the experimental results.

1. Introduction

Industrial application of tools often leads to situations involving significant wear, which is often localised at specific areas, and possibly corrosion. The coating of components with highly wear-resistant films is one way to solve this problem without resorting to new bulk materials. Furthermore, by using sputtered thin-film coatings, it is possible to produce new metastable alloy compositions which are not stable in the bulk form. The composition of the sputtering target determines the ratio of the elements sputtered onto the substrate and the chemical composition of the film. Also, the introduction of gases into the sputtering chamber at low pressures, for example nitrogen, oxygen or methane, permits their incorporation into the growing coating and, under some conditions, the obtaining of a large range of compounds (in these cases, nitrides, oxides or carbides, respectively). The corrosion resistance of these films is clearly also of interest.

Investigations have been made into the wear and corrosion resistance of amorphous surface alloys [1] and of a number of nitride coatings prepared by physical vapour deposition, for example, TiN and CrN [2–6]. Aluminium–titanium alloys have been the object of research [7] as well as the interaction between Al films and W–Ti films [8].

We have recently conducted investigations on sputtered coatings of the W–Ni–N, W–Ti–N and W–Ti–Al–N types on M2* high speed steel (HSS) substrates with respect to their high temperature oxidation behaviour [9,10] and corrosion resistance in aqueous saline media [11–13]. This choice of substrate was made for two reasons. First, HSS is commonly used for tools and thus any results obtained have immediate practical consequences. Secondly, it undergoes corrosion fairly rapidly which has been attributed to the corrosion of martensite, leaving the carbide phases largely untouched [14]. Such a substrate can therefore represent an added benefit in studies of film integrity, since it is a good test of electrolyte penetration through the film.

In this article results obtained recently on the above films are analysed and compared. Observations concerning the corrosion process are assessed and new considerations concerning the corrosion mechanism proposed.

2. Experimental

Heat-treated discs of M2 high speed steel (W 6.4; Mo 5.0; Cr 4.2; V 1.9; C 0.86; Si < 0.4; Mn <0.4mass%), dia. 12 or 30 mm and thickness 3mm, were used as substrates. After polishing to a mirror finish on a polishing table with diamond paste they were sputter-coated in a Hartec d.c. diode sputtering apparatus on one face for electrochemical experiments (12 mm discs) and on both faces for salt spray tests (30 mm discs). After mounting in the sputtering chamber and achieving vacuum conditions, the substrate surface was cleaned by a heating and argon ion etching procedure, i.e. heating by electron bombardment up to temperatures close to 450°C followed by argon ion bombardment at increasing negative substrate bias from 0 to 120 V, with 15 s at each bias value in steps of 10 V. Deposition was then conducted from the appropriate target at a power density of 11 W cm^{-2} and a negative substrate bias of 70 V. Target compositions were: W+10mass%Ni, W+20mass%Ti and W+25mass%Ti+5mass%Al. The total gas pressure in the sputtering chamber during deposition was 0.3 Pa; the gas was a mixture of nitrogen and argon, the partial pressure of nitrogen being varied. Each sputtering deposition led to samples with uniform coating thickness of 3.5–4.0 μm, which was subsequently checked by electron microscopy observation of the cross–section morphology.

Some experiments were carried out with W–Ti–N coatings varying the substrate preparation procedure and sputtering conditions on discs of M2 steel, 12 mm and 30 mm dia. Five variants were studied as follows:

Type 1: As above, 30 min total deposition time, thickness 3.5 μm;

Type 2: As Type 1 but with 3 min cleaning at –120 V at the end of the substrate cleaning procedure (to increase substrate roughness), thickness 3.5 μm;

---

*See Experimental section for analysis.

Type 3: As Type 1 but with a layer of W–Ti for the first 10 min of the total 30 min deposition (no nitrogen in the sputtering chamber during the first 10 min), thickness 3.8 μm (thicker due to the higher sputtering rate of W–Ti);

Type 4: As Type 1 but with two layers, each corresponding to 15 min deposition of W–Ti–N. After depositing the first layer the process was stopped for a period of 15 min. The surface was then etched again, as for the bare substrate, and deposition carried out for a further 15 min, final thickness 3.2 μm (thinner as a result of the extra etching);

Type 5: As Type 1 but with 60 min deposition time, thickness 7.5 μm.

For transformation into electrodes, a copper wire was attached to the rear face of 12 mm dia. samples with silver epoxy and the whole assembly protected, except for the coated face, with epoxy resin and varnish. The electrochemical cell also contained a platinum foil counter electrode and a saturated calomel electrode (SCE) as reference. Corrosion potential measurements and voltammetric experiments were made with a PC-controlled EG&G PAR273A potentiostat using PAR Model 352 software. All electrochemical experiments were carried out at 25°C using a solution of 0.1M potassium chloride prepared from analytical grade reagent and ultrapure water (Millipore Milli-Q, resistivity >18 MΩ cm). Solutions were not de-aerated and all potentials are with respect to SCE.

Salt spray tests were carried out with a Uniclima CNS-100 chamber at 35°C for up to 5 days following standard procedures with a 5% (0.85M) sodium chloride solution injected at 1 bar pressure. The samples of 30 mm dia., coated on both faces, were suspended by a thin nylon thread within the chamber from a tiny hole near the edge. This area around the hole and the uncoated edges of the samples were protected by a thin coating of epoxy resin.

Scanning electron microscopy (SEM) was with a Jeol T330 Scanning Electron Microscope equipped with a Tracor Northern Microanalysis Accessory.

3. Results and Discussion

Electrochemical and morphological observations will be discussed taking into account the film structure and a model for pitting corrosion of such films [13] refined.

3.1. Corrosion Potential and Tafel Plots

Typical variations of corrosion potential and Tafel plots are shown in Figs 1 and 2, respectively, for different W-based coatings. The corrosion potential (Fig. 1), becomes more negative at a relatively high rate during the first ten to twenty minutes and then the variation with time becomes much less. There is not much difference in the values of the corrosion potential for different types of coating, as shown in Fig. 1. Eventually the potential of the M2 steel substrate in 0.1M KCl is reached, i.e. –0.60 V [14].

Figures 2(a) and 2(b) show the negative change in corrosion potential with time.

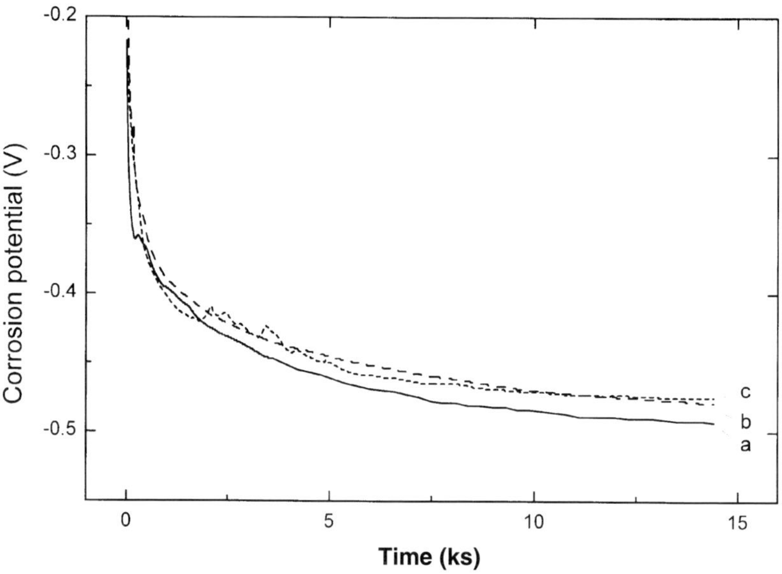

Fig. 1 *Variation of corrosion potential with immersion time in 0.1M KCl for coatings of (a) W–Ni, (b) W–Ti (c) W–Ti–Al.*

It can also be seen that the corrosion potential of W–Ti becomes close to that of the substrate after 4 h, whereas W–Ti–N shows a smaller change; this demonstrates the importance of incorporation of nitrogen, (see below). The large change between 10 min and 1 h immersion in Fig 2(a) can be ascribed to defects in the coating and appearance of corrosion sites. Corrosion currents, as deduced from Tafel analysis, are of the order of 10–20 μA cm^{-2}, with little influence from coating composition. Visual and microscopic examination shows that the corrosion is essentially point-like on all types of coating; with active, visible corrosion sites nearly all appearing within the first half-hour after immersion. Protecting these points of corrosion from contact with the solution by covering with epoxy resin essentially stops further corrosion, which lends support to the hypothesis that the corrosion occurs at microscopic or sub-microscopic flaws in the as-prepared films.

Scanning electron microscopy shows the formation of protuberances of ~10 μm dia. which are not present on the surface of the as-prepared films. There are also points of corrosion where there has been a sort of "explosion" and deposition of iron oxides (identified by X-ray diffraction and electron probe microanalysis [13]) in a ring at some distance from the pit, the open pit then undergoing corrosion more readily. Typically, pits are 5–10 μm dia. and the ring of corrosion products has a diameter of 150 μm.

Although the basic mechanism of corrosion appears to be the same for the different films, differences between the experimental results obtained can be interpreted in terms of variation in structure, which can vary from crystalline to amorphous, and cross-section morphology, which can vary from columnar to featureless, of the deposited coatings [15]. Structure and morphology depend on the deposition conditions, energy of the bombarding atoms, adatom mobility etc. In the case of the elements studied by us, compactness increases with increasing N content, and the

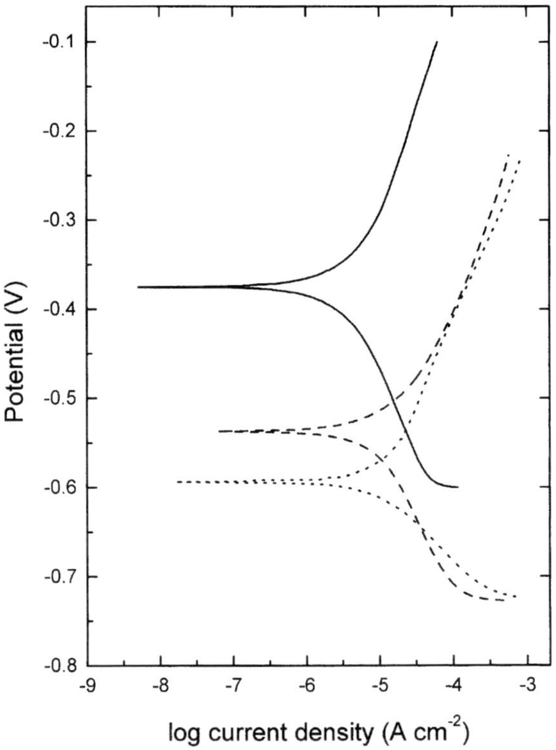

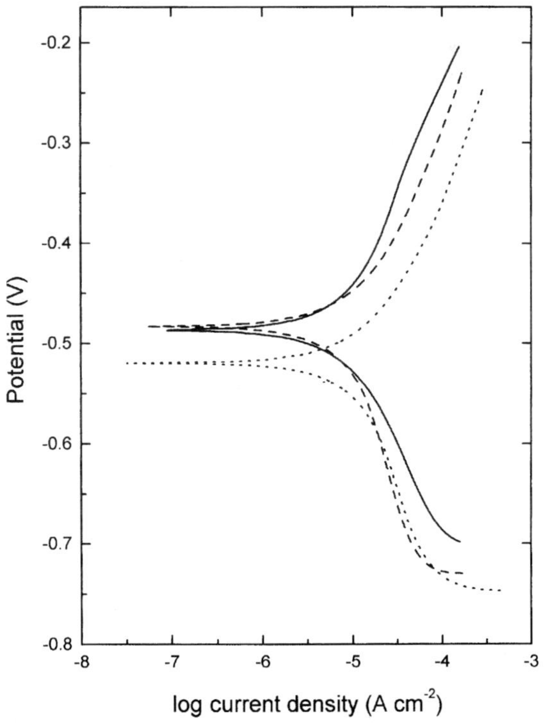

Fig. 2 *Tafel plots for samples coated with (a) W–Ti and (b) W–Ti–N, immersed for (——) 10 min (- - - -) 1 h (...........) 4 h. Scan rate 2 mV* s^{-1} *in positive direction; 0.1M KCl solution.*

resistivity of W–Ti–N coatings increases with increasing Ti content, both effects leading to higher corrosion resistance [13]. The addition of Al to W–Ti films leads to a very thin continuous passive layer several nanometres thick next to the substrate [13]. In the W–Ni–N system, W–Ni–N is amorphous whereas W–Ni is crystalline, the former having a higher corrosion resistance [12].

Under the experimental conditions employed, the above observations suggest that the most useful data for mechanistic interpretation are those obtained after short immersion times, for example 10 min, when only the most open paths to the substrate will have been penetrated by electrolyte, and after long immersion times, several hours, when all the possible corrosion sites have been accessed.

3.2. Influence of Sputtering Conditions

The role played by the structure of the film and its adhesion in determining the number of flaws has been investigated. Experiments were carried out with W–Ti–N films prepared under different sputtering process conditions in which film thickness, surface roughness and other sputtering parameters were varied. These different conditions are described under Experimental. Tafel plots from one of these conditions, Type 3, are shown in Fig.3. Results from samples prepared under the various sputtering conditions are in Table 1. Scanning electron microscope observations were also carried out.

A number of deductions can be made in relation to these results:

- corrosion is always localised and the formation of a protuberance and its rupture always occur in the same way.

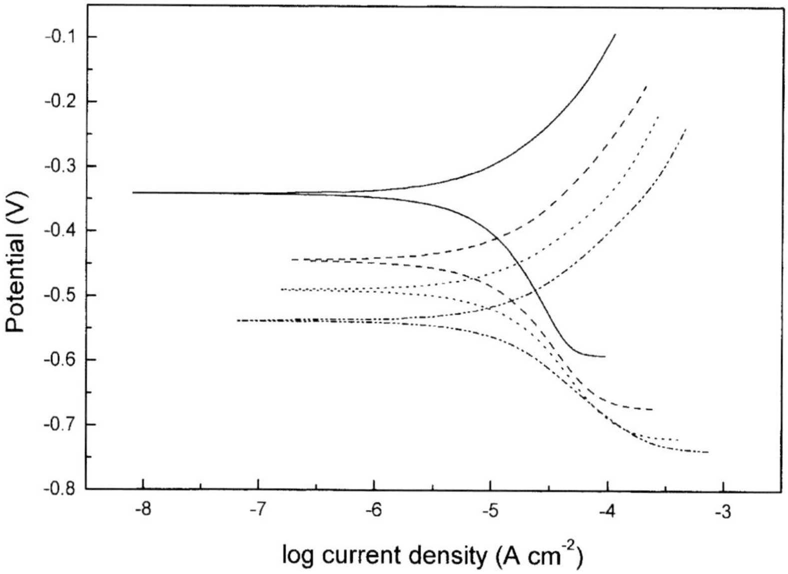

***Fig.* 3** *Tafel plots for Type 3 W–Ti–N coated samples immersed for (——) 10 min, (- - - -) 1 h, (..........) 4 h, (-..-..-..-..-) 24 h. Scan rate 2 mV s^{-1} in positive direction; 0.1M KCl solution.*

Table 1. *Tafel plot analysis for the various types of W–Ti–N coating (see Experimental section for details). All potentials vs SCE*

	Immersion time							
	10 min		60 min		240 min		1410 min	
Sample type	E_{corr} (V)	I_{corr} (μA cm^{-2})	E_{corr} (V)	I_{corr} (μA cm^{-2})	E_{corr} (V)	I_{corr} (μA cm^{-2})	E_{corr} (V)	I_{corr} (μA cm^{-2})
1	–0.461	15.8	–0.495	20.6	–0.511	24.5	–0.555	23.0
2	–0.370	13.7	–0.435	19.6	–0.474	21.4	–0.516	26.9
3	–0.341	17.6	–0.444	19.8	–0.490	20.3	–0.538	22.6
4	–0.384	11.2	–0.442	20.4	–0.480	37.5	–0.534	21.8
5	–0.392	9.5	–0.469	21.8	–0.494	23.1	–0.546	19.8

- corrosion currents are of the order of 10–20 μA cm^{-2}. After the first hour the change is small.

- values of E_{corr}, whilst following the trends in I_{corr} in general terms, do not differ significantly and, thus, it is not easy to use them for differentiating between the types of film.

- at long immersion times (1410 min) there seems to be a slightly lower corrosion rate from films of Type 5 (increased thickness) and a higher corrosion rate from the high surface roughness sample. Increased surface roughness and duplex films appear not to be advantageous.

Further information is forthcoming from salt spray tests, where samples were coated on both faces and suspended within the chamber. In these tests, the conditions are significantly more aggressive than in the electrochemical tests (see Experimental). In particular, in salt spray tests the temperature is higher and the access of oxygen to the surface is greater, and there is convection through movement of fluid vertically down the sample surfaces. However, some aspects are useful for comparative purposes. In particular, such tests show similar features to the electrochemical tests except that, owing to the convection the large amount of corrosion products precludes any formation of a ring and the pits themselves are significantly larger. Examples are shown in Fig.4 for Type 4 films.

After removal of the corrosion products, image analysis was used to determine the distribution of pit sizes on salt spray test samples for the various sputtering conditions; details are given in Reference [17]. An example of such an analysis is shown in Fig. 5 for Type 1 and Type 2 samples. The radius and number distribution of the pits is different in the two cases and there is a much larger corroded area for

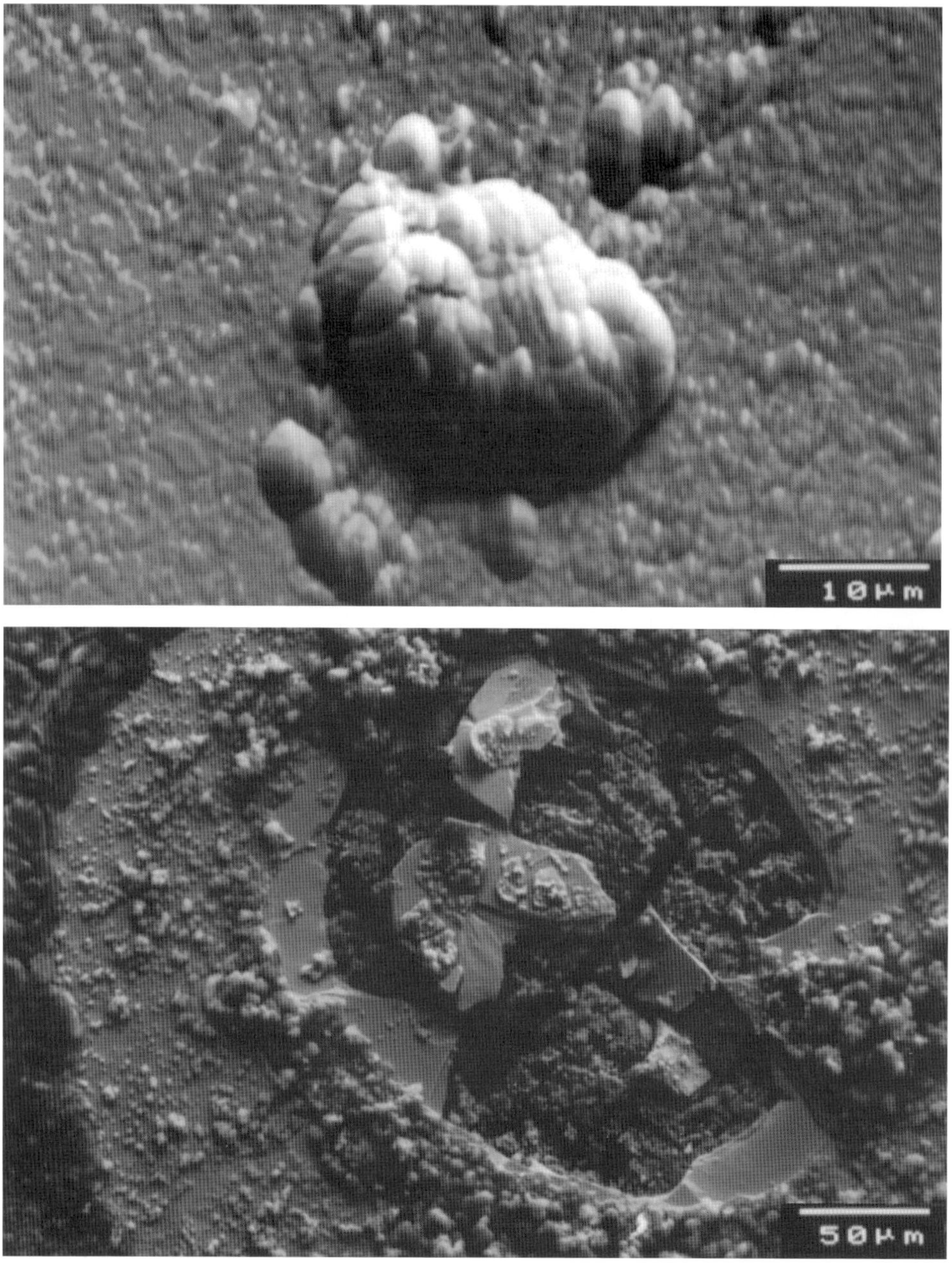

***Fig. 4** Scanning electron micrographs showing pitting corrosion of Type 4 W–Ti–N films after 5 days salt spray test: (a) bubble formation with formation of corrosion products between coating and substrate, (b) after rupture of coating showing the pit formed underneath.*

the Type 2 sample. This provides evidence for the negative influence of the increased surface roughness in Type 2 as compared to Type 1 samples, and parallels the electrochemical results. Results from the other sample types compared to the

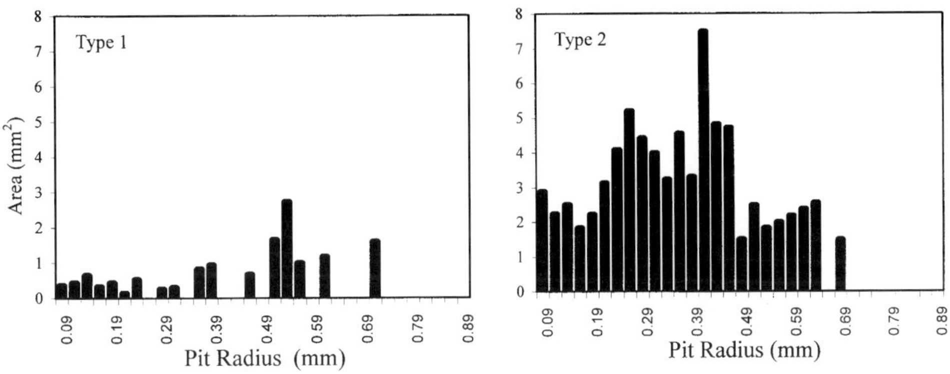

Fig. 5 Results from image analysis of W–Ti–N coatings prepared under different conditions showing pit area expressed as a function of pit radius for samples of (a) Type 1 and (b) Type 2.

reference, Type 1, can be summarised as follows. The two-layer structure (Type 3 and Type 4) allows easier penetration of electrolyte and more pitting, particularly for Type 3 where the first layer is W–Ti. Type 5 shows less pitting and the pits are of smaller area, reflecting the influence of greater film thickness.

3.3. Mechanism of Corrosion

The model for corrosion through these thin-film hard coatings stresses the similarities between the different types of film. After penetration of electrolyte solution through flaws to the substrate, the substrate undergoes corrosion, the initial rate depending on the size of the electrolyte penetration channel. Corrosion products occupy more volume than the substrate and this explains the lifting of the film in bubble-like formations observed by SEM. Eventually, the volume of products and the consequent pressure can lead to rupture of these protuberances, apparently with explosion and deposition of iron oxides in a ring around the pit, as described above, which then undergoes corrosion more rapidly. A scheme for this basic process at the point of film lifting is shown in Fig. 6. It has been demonstrated that the rest of the coating remains unaltered, including its mechanical properties [16], except for the occasional, rare indication of some uniform corrosion close to the pits.

Further refinements to the corrosion model can be introduced by taking into account the observations from the salt spray tests, since even under these aggressive conditions, some of the protuberances formed remained after 5 days' exposure in the chamber. It suggests that some coating compositions are able to absorb the mechanical stresses associated with protuberance formation more easily than others — and there will also be an influence on the stress from substrate adhesion. Thus, two situations can arise:

- corrosion products effectively internally seal the sub-microscopic channels to the external solution and further corrosion is blocked, equivalent to pit death. This is demonstrated in Fig. 6. It may be difficult to find direct evidence of such corrosion sites externally through posterior microscopic examination.

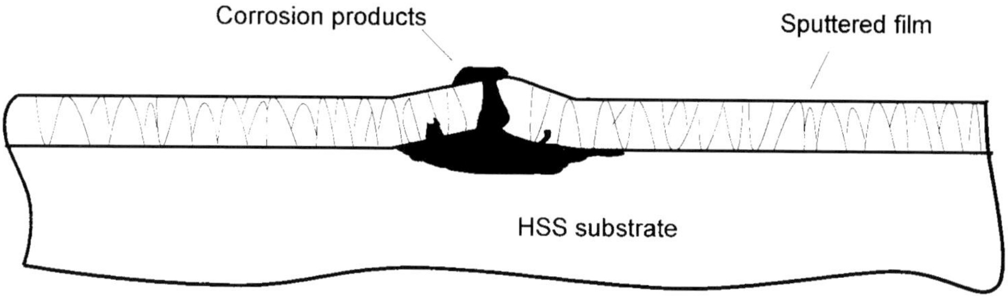

Fig. 6 Schematic showing the corrosion process when the corrosion products (mainly Fe_2O_3), begin lifting the film just before it explodes.

- where channels are larger and where the film cannot absorb the mechanical stresses, there will eventually be the rupture by explosion referred to above and the appearance of an open pit.

Further evidence for these mechanisms has to come from real-time microscopic observation of the corrosion processes. Such work is presently being undertaken. This should aid in enabling the necessary corrective action in the sputtering process to be undertaken to avoid the appearance of flaws during film preparation and the ensuing possibility of corrosion in aggressive media.

4. Conclusions

Detailed study of the corrosion resistance of W–M–N (M = Ni, Ti, Al) coatings shows that corrosion proceeds at specific flaws in the coating. The general mechanism is always the same with some influence from the film structure and morphology resulting from the identity of M and the presence or absence of N.

Electrochemical experiments and salt spray tests on W–Ti–N films prepared under different conditions demonstrate the importance of the sputtering process on film morphology. They also lead to refinements of the general model to explain these corrosion processes in which some corrosion sites are blocked by the corrosion products formed whereas others continue to be active, leading to localised film rupture and open pitting corrosion. The importance of film morphology is in turn influenced by film thickness and the ability of the film to absorb the mechanical stresses created during corrosion.

5. Acknowledgement

The authors thank Eng. F. Ramos, HUF Portuguesa, for carrying out the salt spray tests.

References

1. K. Hashimoto, *et al.*, *Corros. Sci.*, 1993, **35**, 363–370.
2. S. Berg, N. Eguchi, V. Grajewski, S. W. Kim and E. Fromm, *Surf. Coat. Technol.*, 1991, *49*, 336–341.
3. I. Montero, C. Jimènez and J. Perrière, *Surf. Sci.*, 1991, **251/252**, 1038–1043.
4. R. Brown, M. N. Alias and R. Fontana, *Surf. Coat. Technol.*, 1993, **62**, 467–473.
5. H. Barànková, L. Bárdos and S. Berg, *J. Electrochem. Soc.*, 1994, **141**, L8–L9.
6. S. D. Chyou, H. C. Shih and T. T. Chen, *Corros. Sci.*, 1993, **35**, 337–347.
7. H. Yoshioka, Q. Yan, K. Asami and K. Hashimoto, *Mater. Sci. Eng. A*, 1991, **A134**, 1054–1057.
8. H. J. Wondergem, A. Heger and J. J. van der Broek, *Thin Solid Films*, 1994, **249**, 6–10.
9. C. Louro and A. Cavaleiro, *Surf. Coat. Technol.*, 1995, **74/75**, 998–1004.
10. C. Louro and A. Cavaleiro, *J. Electrochem. Soc.*, 1997, **144**, 259–266.
11. C. M. A. Brett and A. Cavaleiro, *Mat. Sci. Forum*, 1995, **192–4**, 797–804.
12. C. M. A. Brett and C.-M. Nimigean, *Thin Solid Films*, 1997, **311**, 1–6.
13. C. M. A. Brett and A. Cavaleiro, *Thin Solid Films*, 1998, **322**, 263–273.
14. C. M. A. Brett and P. I. C. Melo, *J. Appl. Electrochem.*, 1997, **27**, 959–964.
15. A. Cavaleiro, M. T. Vieira and G. Lemperière, *Thin Solid Films*, 1992, **213**, 6–12.
16. A. Cavaleiro, C. Louro, J. V. Fernandes and C. M. A. Brett, *Vacuum*, 1999, **52**, 157–163.
17. J. C. Oliveira, A. Cavaleiro and C. M. A. Brett, *Corros. Sci.*, in press.

12
Stress Generation during the Growth of Anodic Oxide Films on Titanium–Aluminium Alloys

A. MARINOVIĆ, M. METIKOŠ-HUKOVIĆ and I. MILOŠEV*
Department of Electrochemistry, Faculty of Chemical Engineering and Technology, University of Zagreb, Savska 16, P.O. Box 177, 10000Zagreb, Croatia
*J. Stefan Institute, Jamova 39, 1001 Ljubljana, Slovenia

ABSTRACT
The formation of stable passive oxide films is critical to the design engineering of new biomaterials for orthopaedic implants. The growth of anodic oxide films on titanium, Ti6Al4V and Ti6Al6Nb alloys has been investigated in physiological solution at 37°C. During anodic film growth, the electric field strength within the film was of the order of 10^7 V cm^{-1}. This electric field exerts significant electrostriction on the film and yields high internal compressive stress of the order of a few GPa. The observed transformation from amorphous to crystalline structure at 2 V for all investigated materials and the onset of pitting corrosion and passivity breakdown for the Ti6Al4V alloy are accompanied by the appearance of compressive stress within the oxide film. The results are discussed with respect to the relevant reactions occurring at the metal/oxide and oxide/electrolyte interface during oxide growth.

1. Introduction

Titanium has become the metal of today. It is a highly corrosion-resistant metal used in a large number of applications. A good combination of mechanical strength, environmental properties and availability has made titanium useful for the construction of surgical implants. Extensive research has led to the production of titanium alloys that have even better properties than titanium itself. One of these studies resulted in a Ti6Al4V alloy which has recently become the most widely used material for orthopaedic implants [1,2]. Since the human body environment promotes corrosion in a similar way as warm aerated sea water, corrosion investigations of surgical implants are of great importance. The corrosion resistance of titanium and its alloys is attributed to a surface oxide film whose degree of protection depends on its integrity, thickness and the diffusion constant of metal ions in the oxide which are, in turn, influenced by the nature of the film formation [3]. One of the most important methods for oxide layer formation is anodisation. The thicker films are commercially important as well as scientifically interesting, but there is a limit to the thickness they can achieve. The ultimate thickness to which an oxide film may grow anodically before it cracks depends on its ability to withstand or relieve stresses that may arise during growth [4]. According to Sato [5], an electrostriction pressure of the order of 10^7 Pa due to the high electric field is a main cause of compressive stress. Stress in an anodic film can also arise for other reasons, for example, point defects at interface boundaries [6], interfacial tension of the film, internal stress caused by the film/metal volume ratio, internal stress due to partial hydration or dehydration of the film or local stress caused by impurities [7]. Sato showed that the electric field in passive oxide films was of the order 10^6–10^7 V cm^{-1}, large enough to produce considerable electrostriction pressure.

According to Archibald [8], the magnitude of the stress is dependent on the anodisation conditions such as surface preparation [9], electrolyte pH, and electrolyte composition. Other factors also partly, or sometimes wholly, determine the nature and the magnitude of the stress. These factors include the nature of the diffusing ionic species involved in the film formation, epitaxial surface roughness and residual stress in the metal. Leach and Pearson [10] found that internal compressive stress is conducive to the formation of crystals. This occurs only when conditions of growth are such that they induce compressive stress in the oxide. The sign and magnitude of oxide stress are related to the growth rate and the stress becomes more compressive as the growth rate falls.

In this work, we investigate the behaviour of the Ti, Ti6Al4V and a new Ti6Al6Nb alloy during anodic polarisation in Hank's solution. The surface stress generated during anodic oxide film growth on Ti and Ti-alloys was particularly analysed, since there is a strong dependence between stresses produced within the oxide film and the onset of passivity breakdown. The appearance of localised corrosion is closely related to the electric and dielectric properties of the oxide film as well as to point defect events at the metal/oxide film/electrolyte interfaces. The net stress in anodic oxide films produced by electrostriction is derived as a function of the electric field strength, oxide thickness, and dielectric properties of the oxide films.

2. Experimental

Rotating electrodes were used. Titanium (99.9% pure, Johnson-Matthey) Ti6Al4V and Ti6Al6Nb (Goodfellow) materials were machined into cylindrical form, pressed into a cylindrical Teflon holder and mounted on a "Tacussel" rotating disc electrode. The surface areas exposed to the electrolyte were 0.125 cm^2 for titanium and 0.196 cm^2 for Ti6Al4V and Ti6Al6Nb. Prior to each measurement, electrodes were abraded with silica papers 600 P, degreased in acetone and rinsed with redestilled water. The counter electrode consisted of two graphite rods positioned symmetrically to the working electrode. A silver chloride (Ag/AgCl/3M KCl) electrode served as the reference electrode (0.207 V vs SHE) to which all measured potentials were referred.

Measurements were performed in simulated physiological (Hank's) solution at 37°C consisting of: NaCl 8, KCl 0.4, $NaHCO_3$ 0.35, NaH_2PO_4 H_2O 0.25, Na_2HPO_4 $2H_2O$ 0.06, $CaCl_2$ 2 H_2O 0.19, $MgCl_2$ 0.19, $MgSO_4$ $7H_2O$ 0.06, glucose 1 g L^{-1}, at pH = 6.9.

Preliminary results showed that there is no dependence between the rotation speed and current density, therefore all the measurements were performed with a rotation rate of 2000 min^{-1}.

Potentiodynamic measurements were taken in a potential range from –1.4 V up to 10 V, with a sweep rate of 3–100 mV s^{-1} and a rotation rate of 2000 min^{-1}. During impedance measurements, the electrodes were rotated at the same rate in order to prevent attachment of oxygen bubbles since this disturbs impedance signal response. The constant perturbing a.c. signal amplitude was ±5 mV on the d.c. applied potential, over a frequency range of 10^5 to 2×10^{-3} Hz. Data analysis was made by a complex-non-linear-least-square (CNLS) method that is used to fit data to the equivalent electrical circuit (EEC) model using software developed by Boukamp [11,12]. An equivalent electrical circuit based on the information about the physical phenomena at the interface is selected.

The experimental set-up for impedance and polarisation consists of a Solartron SI 1287 Electrochemical Interface and a Solartron SI 1255 HF Frequency Response Analyser. An IBM compatible PC performed the system control and data acquisitions.

3. Results and Discussion

3.1. Polarisation measurements

Cyclic voltammograms obtained on a titanium rotating disc electrode for various scan rates, in Hank's solution in the potential range between –1.4 and 3 V vs Ag/AgCl are presented in Fig. 1. Before each cycle, the Ti-electrode was abraded using 600 grit emery paper, and polarised at –1.5 V to remove oxides formed at the surface as a result of contact with air, and during immersion in the Hank's solution. The potential sweep in the positive direction first shows a peak A1 corresponding to the formation of titanium sub-oxides TiOOH and Ti_2O_3 [13–15] by the following reactions:

$$2Ti + 3H_2O - 6e \rightarrow Ti_2O_3 + 6H^+ \quad (1)$$

$$Ti + 2H_2O - 3e \rightarrow TiOOH + 3H^+ \quad (2)$$

At about –0.25 V vs Ag/AgCl, the anodic film changes its composition from Ti(III) oxide to Ti(IV) oxide. An abrupt increase in the anodic current at about 1.5 V signifies the beginning of oxygen evolution and an electrochemical transformation, i.e. a structural change of the Ti(IV) amorphous oxy-hydroxide layer, formed in the potential range of the first current plateau. In the first and the second plateau regions the current is nearly constant indicating that a thickening of the anodic surface film takes place during linear polarisation under a constant electric field.

The electrochemical behaviour of Ti6Al4V and Ti6Al6Nb electrodes is qualitatively similar to that of the titanium electrode, as can be seen from the cyclic voltammograms shown in Fig. 2. Typical polarisation curves recorded in the wide polarisation range are presented in Fig. 3. The Ti6Al6Nb electrode exhibited steady current plateau up to 10 V, while the Ti6Al4V electrode underwent an anodic breakdown at the potential of 3.6 V and scan rate of 20 mV s^{-1}.

Numerous investigations have shown that the oxide film anodically formed on titanium has a composition ascribable to TiO_2. Surface spectroscopic characterisation of oxide films thermally or anodically formed on Ti, Ti6Al6Nb and Ti6Al4V alloys

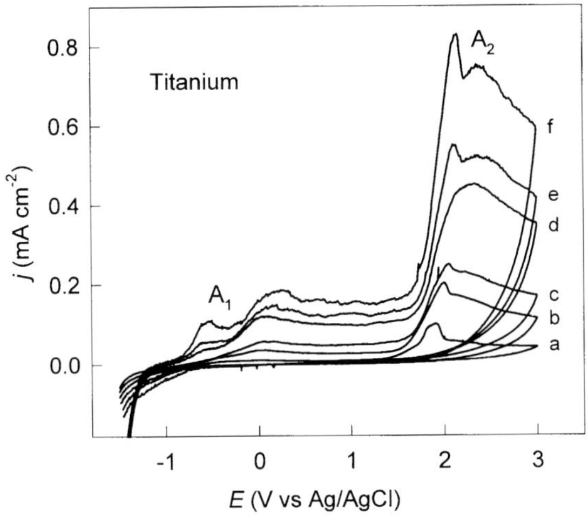

Fig. 1 *Cyclic voltammograms for a titanium rotating-disc electrode in Hank's solution (pH 6.9,* t *= 37°C) recorded at various scan rates: ν = 10 (a); 20 (b); 30 (c); 50 (d); 70 (e) and 100 (f) mV* s^{-1} *respectively and a rotation rate, ω = 2000* min^{-1}.

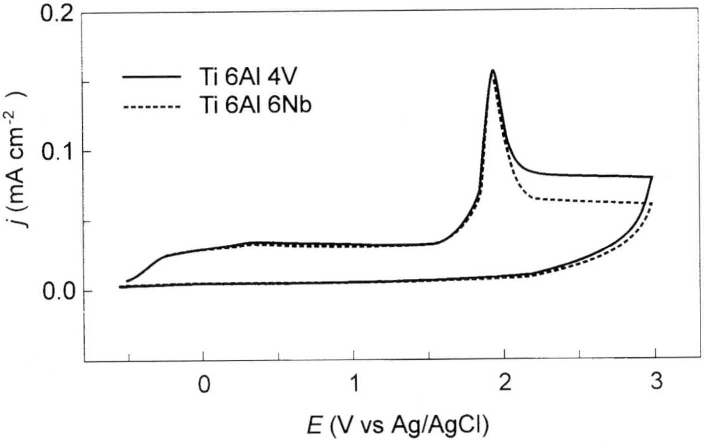

Fig. 2 *Cyclic voltammograms of Ti6Al4V and Ti6Al6Nb rotating-disc electrodes in Hank's solution (pH 6.9,* t *= 37°C), ν = 20 mV* s^{-1}, *ω = 2000* min^{-1}.

has shown that these films are predominately TiO_2. Enhanced contents of niobium and aluminium alloying elements with respect to the bulk stoichiometry of the alloy were found in the oxide, while the content of vanadium was reduced but still detectable (0.1%) [16–19].

The influence of rotation speed on the anodic profile of the cyclic voltammogram was not established. In contrast, the scan rate markedly influenced the shape of the cyclic voltammograms shown in Fig. 1. Since the current densities in the plateau region were independent of the rotation speed, the potentials of the anodic sweep limits equalled the voltage drop across the oxide film and a direct proportionality between the applied potential, *E* and oxide thickness, δ could be experimentally

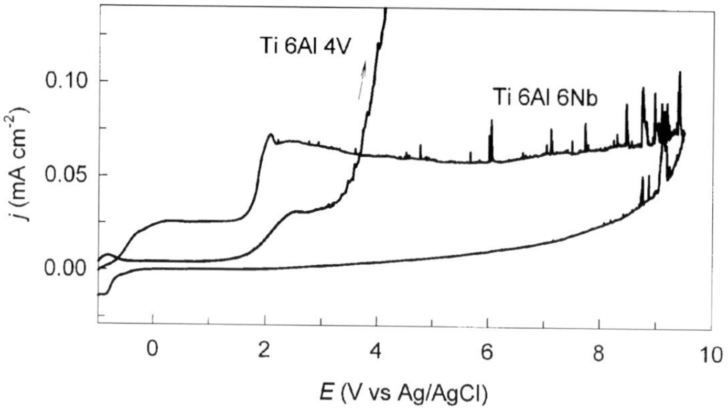

Fig. 3 *Cyclic voltammograms of Ti6Al4V and Ti6Al6Nb rotating-disc electrodes recorded in a wide potential range in Hank's solution (pH 6.9, t = 37°C), ν = 20 mV s^{-1}, ω = 2000 min^{-1}.*

verified. The effective field strength, H is defined as: $H = (dE/d\delta)$. In some respects titanium behaves as a typical valve metal (Ta, Al, Zr...) [4], and it is generally agreed that the growth of the anodic film involves a field-assisted migration of ions across the oxide film, according to the high-field law [20–23]:

$$j = A \exp B H \tag{3}$$

where j is the ionic current density, and A and B are the kinetic parameters of oxide growth. Thus, the effective field strength is expressed as:

$$H/\gamma = \left(\frac{dE}{dt} \times \frac{1}{j}\right)\left(\frac{1}{V}\right) \tag{4}$$

where V is the molar volume of the oxide formed per coulomb, γ is the roughness factor, (dE/dt) is the oxide growth rate, which corresponds to the sweep rate, ν under the potentiodynamic experimental conditions of oxide growth. In Fig. 4, a linear dependence between the plateau current density and the oxide growth rate is shown.

Using eqn (4) and taking the value of V to be 5.47×10^{-5} cm^3 C^{-1} [24], the electric field strength, H for all investigated materials was determined from the corresponding slope, as listed in Table 1.

The H values of the order of 10^7 V cm^{-1} suggest the validity of eqn (3), which can be described in the form:

$$R_j = \frac{1}{C} = \frac{1}{j}\left(\frac{dE}{dt}\right) = \frac{V}{\gamma B}\ln\left(\frac{j}{A}\right) \tag{5}$$

where R_j is the unitary formation rate or reciprocal capacity of the film. The kinetic parameters A and B were calculated and an estimate of the ionic conductivity of the oxide film, $\sigma = AB$ is presented in Table 1.

An electric field strength of the order of 10^7 V cm^{-1} (Table 1) yields considerable electrostriction pressure within the oxide film and is the most important source of

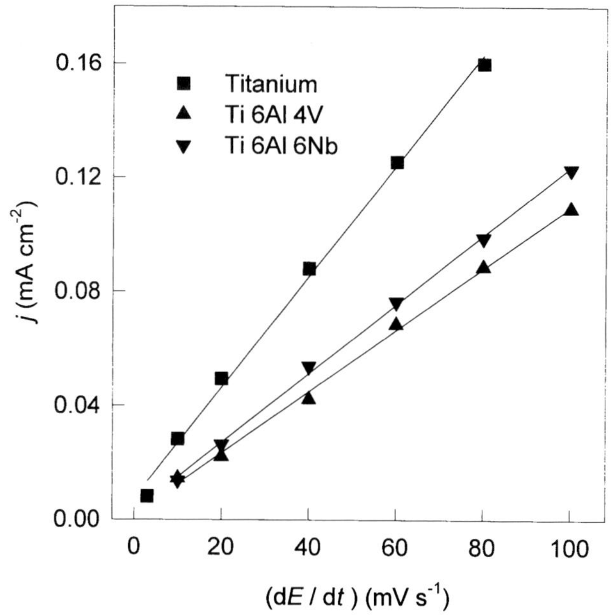

***Fig. 4** Dependence of the oxide growth rate (dE/dt) on the plateau current density for titanium, Ti6Al4V and Ti6Al6Nb electrodes.*

compressive stress in the film. The electrostriction pressure on a solid dielectric film is given by [5]:

$$p - p_o = \left(\frac{\varepsilon_o \varepsilon_r (\varepsilon_r - 1) H^2}{2} \right) - \left(\frac{\gamma}{\delta} \right) \tag{6}$$

where p is the electrostriction pressure, p_o is the atmospheric pressure, ε_r is the dielectric constant of the film, ε_o is the vacuum permitivity, γ is the interfacial tension of the film/electrolyte interface and δ is the film thickness. The stress pressures calculated by using $\gamma = 0.2$ kg s^{-2} [24] and $\varepsilon_r = 20$ [25] are listed in Table 1.

***Table 1.** Calculated values of oxide film conductivity, electric field and film pressure of the titanium, Ti6Al4V and Ti6Al6Nb rotating-disc electrode in Hank's solution during anodic polarisation from −1.4 V to 1.5 V vs Ag/AgCl, at ν = 20 mV s^{-1}, ω = 2000 min^{-1}*

	Conductivity, σ (10^{-12} S cm^{-1})	**Electric field, H (10^{6} V cm^{-1})**	**Film pressure, p (10^{9} Pa)**
Titanium	12.4	9.42	1.49
Ti6Al6Nb	9.52	15.38	3.98
Ti6Al4V	5.93	17.25	5.00

The experimental evidence from cyclic voltammetry measurements (Figs 2 and 3) shows that the oxide film appears to be crystalline under the above conditions. A sharp current peak at 2 V in Fig. 2 at the beginning of oxygen evolution [26] accompanies the structural changes within the oxide film, i.e. the transformation from an amorphous-glassy to a crystalline structure [27–30]. Crystallisation produces electron conductive paths through the anodic oxide film and oxygen can be sited in these crystalline regions during film growth so that oxygen evolution is attributed to an increased electronic conductivity of the oxide film [28,31].

According to Leach and Shibata [10,24], the crystallisation that occurs as a consequence of compressive stress is a way of releasing this stress. The question then arises as to why compressive stresses should favour the formation of a crystalline oxide instead of an amorphous, glassy structure. The amorphous oxides all occupy larger volumes than the same mass of crystalline material [10]. The growth of crystalline material will generally be favoured under those conditions when amorphous material is under compressive stress.

The results of polarisation measurements in Fig. 3 show that the oxide film formed on the Ti6Al6Nb electrode in Hank's solution, containing high concentration of Cl^- ions, is fairly homogeneous and that the electrode remains passive even under high compressive stresses up to 10 V. The remarkable corrosion resistance could be due to the formation of highly charged solutes of alloying elements in the barrier layers resulting in complexing of the immobile solutes with the mobile cation vacancies [32]. The other explanation is that of Nb(V) ion incorporation into the TiO_2 lattice because the Nb(V) ions possess very similar ionic radii to Ti(IV) ions. It has been suggested that Nb(V) ions enter the TiO_2 lattice substitutionally and replace the Ti(IV) ions [33].

Microstructural and morphological characterisations have demonstrated a large degree of heterogeneity of the anodic oxide film formed on Ti6Al4V alloy compared to that of anodic oxides on titanium [18]. Therefore, it is reasonable to assume that the compressive stress for the film growing on the Ti6Al4V electrode will be more likely to exceed the critical value at low anodic potentials and give rise to the film breakdown at 3.6 V (Fig. 3).

3.2. Impedance Spectroscopy

Wide frequency range impedance measurements (10^5 to 2×10^{-3} Hz) were applied to investigate the solid-state properties of the Ti, Ti-alloys/oxide film/electrolyte structure. Electric and dielectric behaviour of the growing oxide film in which electrostrictive stress takes place during film formation and thickening was investigated. For this study, the electrodes were potentiodynamically swept from –1.4 V to the desired potential where impedance data were collected. Thus, impedance spectra presented in Figs 5 and 6, as Nyquist and Bode plots, were taken at 0.5 and 2 V in the potential range of the first and the second current plateau, respectively (see Figs 1 and 2).

Because of the system complexity the impedance data modelling was undertaken by applying the electrical equivalent circuit (EEC) approach. The EEC used to fit the experimental data for Ti and Ti-alloys at 0.5 V is shown in Fig. 7.

Fit parameters are listed in Table 2. The shape of the impedance spectra at higher

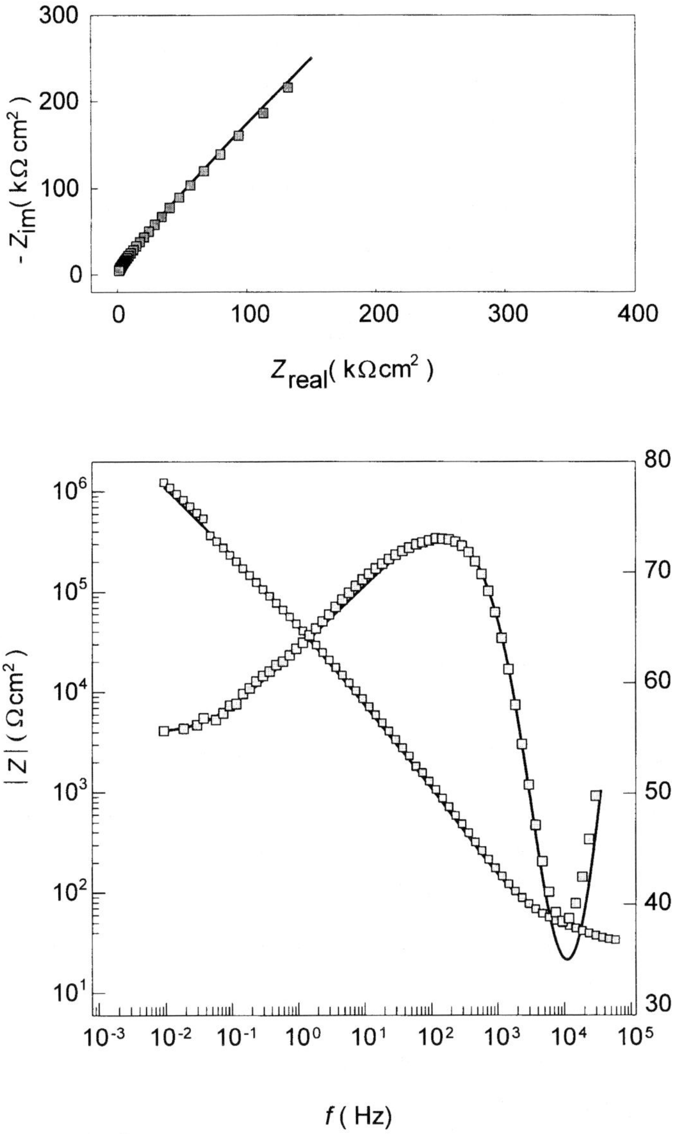

Fig. 5 *Measured (symbols) and simulated (full line) Nyquist and Bode spectra for a Ti6Al4V rotating-disc electrode in Hank's solution (pH 6.9* t = *37°C), obtained at* E = *0.5 V vs Ag/AgCl, and* $\omega = 2000$ *min*$^{-1}$.

frequencies is almost the same for all investigated materials but at low frequencies it varies for each investigated material. The high frequency part of the spectra is characterised by the oxide film capacitance in parallel with the oxide resistance that is associated with the ionic conduction through the amorphous barrier oxide film.

The low frequency part, Z_{low} of the impedance spectra is characterised by the inductance, L (for the titanium electrode), the Warburg diffusion element, W (for the Ti6Al4V electrode), thus indicating the different reaction mechanisms which take place at the metal/oxide interface. No inductive loop was obtained in the Nyquist plot of titanium and only a markedly depressed feature of the capacitance loop as frequency approaches zero was registered (shown as a linear dependence between Z_{real} and Z_{im}). The origin of inductive behaviour at low frequencies is not quite clear and may be interpreted as a rearrangement of the surface charge at the metal/oxide interface [34,35].

The changes at the low frequency response of the Warburg impedance for the Ti6Al4V electrode can be explained by ionic diffusion of vanadium ions through the barrier oxide film. Since the impedance spectra are obtained using the Ti6Al4V rotating disc electrode, the Warburg-like diffusion component must be related to the process under diffusion control in the solid phase [36]. Obviously, the existence of point defects (cation and/or anion vacancies) could markedly affect the molar volume of the oxide film and consequently the stress development during anodic oxidation of the Ti6Al4V electrode. According to Macdonald's point defect model (PDM) [32,37], the diffusion coefficient of vanadium ions is determined by:

$$D = 2\left(\frac{\sigma j_{ss}}{H}\right)^2 \tag{7}$$

where j_{ss} is the steady state current, and σ is the Warburg coefficient ($\sigma = dZ_{real}/d\omega^{1/2}$ = 181.345 kΩ cm^2 s$^{-1/2}$). The diffusion coefficient is calculated using the estimated values of j_{ss}, H and σ. A low value for the diffusion coefficient of $D_f = 5.25 \times 10^{-16}$ cm^{2}s^{-1} is obtained, through a finite region of approximately 1 nm in length, which is in good agreement with values of D_f usually reported for diffusion in solid films [38]. The extent of stress in the growing oxide film on the Ti6Al4V electrode depends not only on electrostrictive stress but also on the rate of competing annihilation or generation reactions of cation vacancies at the metal/oxide interface.

The diffusion of metal cations from the metal/film to the film/solution interface creates metal vacancies at the metal/film interface. These metal holes tend to submerge into the bulk of the metal and hence annihilate. However, if the cation diffusion rate inside the oxide film is greater than the metal hole submersion, the metal hole will pile up and form a void at the metal/film interface. In the region of the first current plateau, at the potential of 0.5 V, it is evident that the rate of annihilation is greater than the rate of cation diffusion. However, at potentials greater than 3 V, the generation of cations is faster than their annihilation, therefore, when the cation void grows to a certain critical size, the film suffers local collapse and breakdown of the film takes place. The presence of chloride ions makes this process faster. Chloride ions are the best-known agents of chemically induced breakdown and in Hank's solution they are present in a high concentration. In conclusion, the formation of cation vacancies at the alloy/oxide interface and their subsequent annihilation by an inward diffusion of vanadium ions are closely related to the stress developed during the anodic oxidation of the Ti6Al4V alloy.

In Fig. 6 typical Nyquist and Bode plots of impedance measurements with a Ti6Al4V electrode at 2 V are shown. Taking into account the complexity of the system under investigation, the impedance data were analysed in such a way to obtain the time constants for the whole set of data (from 10^5 to 2×10^{-3} Hz), using a minimum number of the equivalent circuit elements. The equivalent circuit presented in Fig. 8

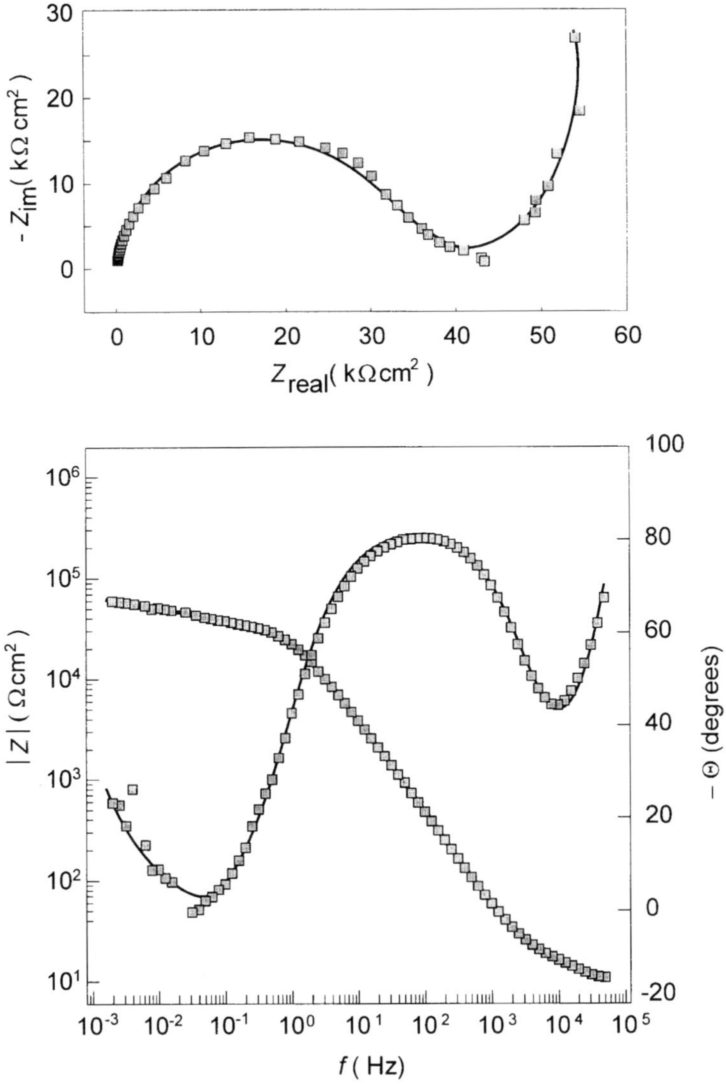

Fig. 6 *Measured (symbols) and simulated (full line) Nyquist and Bode spectra for a Ti6Al4V rotating-disc electrode in Hank's solution (pH 6.9,* t = *37°C), obtained at* E = *2 V vs Ag/AgCl, and* $\omega = 2000\ min^{-1}$.

predicts three time constants. Table 2 contains fit parameters. The quality of fitting to the equivalent circuit was judged firstly by the Chi-square value, and secondly by the relative error for all passive circuit elements comparing experimental and simulated data. A value of chi-square of 10^{-4} indicated a reasonably good fit, using a minimum number of passive elements with an error of 5%.

The fitting procedure showed that better agreement can be obtained between theoretical and experimental data if a frequency-dependent constant-phase element (CPE) is introduced instead of pure capacitance and Warburg impedance. The impedance of a constant-phase element is defined as:

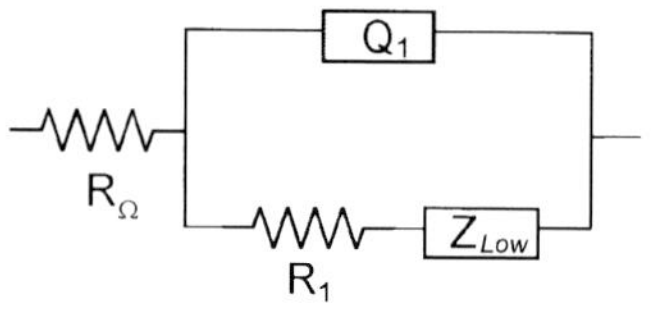

Fig. 7 Electrical equivalent circuit which describes experimental impedance data of a Ti6Al4V electrode at E = 0.5 V vs Ag/AgCl.

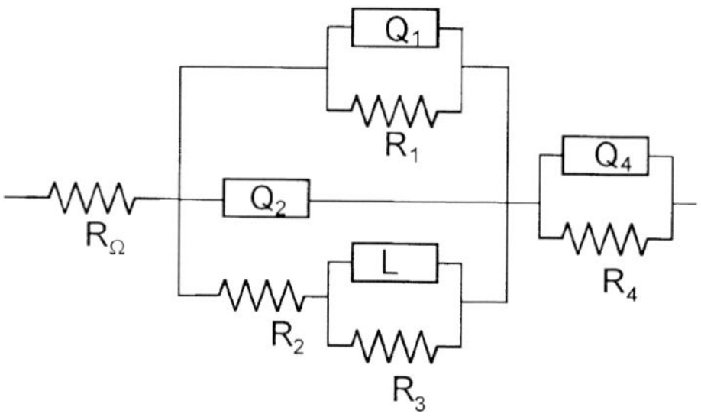

Fig. 8 Electrical equivalent circuit which describes experimental impedance data of a Ti6Al4V electrode at E = 2 V vs Ag/AgCl.

Table 2. *Values of the equivalent circuit elements (Figs 7,8) that fit the experimental impedance spectra of the Ti6Al4V electrode obtained at E = 0.5 V and E = 2 V vs Ag/AgCl*

Equivalent circuit elements	E = 2 V	E = 0.5 V
R_Ω (Ω cm^2)	8.87	40.00
Q_1 (Ω^{-1} s^n cm^{-2})	3.28×10^{-6}	3.28×10^{-6}
n_1	0.99	0.87
R_1 (Ω cm^2)	29	1.74×10^4
W (Ω^{-1} s$^{0.5}$ cm^{-2})	—	3.37×10^{-6}
Q_2 (Ω^{-1} s^n cm^{-2})	3.87×10^{-6}	—
n_2	0.69	—
R_2 (Ω cm^2)	4.49×10^4	—
R_3 (Ω cm^2)	-0.97×10^4	—
L (H cm^2)	2.90×10^4	—
R_4 (Ω cm^2)	-9.01×10^4	—
Q_4 (Ω^{-1} s^n cm^{-2})	5.70×10^{-4}	—
n_4	0.58	—

$$Z\,(\text{CPE}) = [Q\,(j\,\omega)^{\,n}]^{-1} \qquad (8)$$

where j is the imaginary number ($j^2 = -1$), Q is the constant, ω is the angular frequency, n is the exponent of the constant phase element. The exponent n has values between –1 and 1. A value of –1 is characteristic for inductance, a value of 1 corresponds to a capacitor, a value of 0 corresponds to a resistor and a value of 0.5 can be assigned to diffusion phenomena.The capacitive semicircle in the high and mid frequency ranges consists of two strongly overlapping time constants. The interpretation is based on a two-layer model of the oxide film, consisting of a barrier type inner layer at the metal/oxide interface and a porous outer layer at the oxide film/electrolyte interface. Although the anodic film which forms on titanium in aqueous solution is frequently described as a single TiO_2 layer, there is substantial evidence that this film exhibits a two layer structure under conditions such as exposure to phosphate-buffered saline solution, i.e. a dense inner barrier layer and porous outer layer [39,40]. The inductance at intermediate frequencies can be attributed to the adsorbed oxygen (chloride) ions at the oxide/solution interface. It can be proposed that adsorption is very strong since the inductive loop is very small in this case and will not be observed in the Nyquist plot.

The contribution of the third time constant is related to a slow process. The shape of the impedance spectra at very low frequencies tends to approach a negative value for the polarisation resistance $R_p = \lim_{\omega \to 0} [Z(j\omega)]$. For the lowest frequency << 0.002 Hz, the impedance turns to a negative value through a large semicircle around the origin of the Nyquist diagram. Extrapolation of $Z_{im} \to 0$ yields $R_4 = -9 \times 10^4\ \Omega$. The calculated value from the fitted data listed in the Table 1 is the same, i.e. $-9.01 \times 10^4\ \Omega$. It is worth emphasising that impedance spectra similar to those displayed in Fig. 6 were observed in the passivation phenomenon of iron [41–43] and during a stochastic process of $PbSO_4$ nucleation (crystallisation) and growth on lead in a sulfuric acid solution [44,45]. In both cases impedance spectra showed a capacitive arc, which changed the real component of impedance from a positive to a negative value in the Nyquist plot.

The appearance of negative resistances in the impedance spectrum on Fig. 6 is consistent with the negative slope of the current–potential characteristic in the potential range between 2 and 2.2 V, after the anodic peak in Fig. 2. The anodic peak around 2 V, as was discussed earlier, is tied to the stochastic process of crystallisation, in other words, to the evolution of a new phase in the matrix of the amorphous oxide and to an increase of electronic conductivity. This conductivity accompanies oxygen evolution sited at the crystalline regions within the oxide film. High internal compressive stresses within the oxide film of the order of a few GPa are inherent in the formation of crystals. Low frequencies reach deeper into the material so we can speculate that crystallisation, as well as stress, develops at the metal/oxide interface boundary which agrees with the conclusion of Pyun *et al.* [6] that stresses are limited to a narrow region of the aluminium/oxide film interface below 1 nm. Using reflection electron diffraction it was shown that the crystals appeared firstly at the metal–oxide interface and reached the oxide–liquid interface only after prolonged oxidation [10]. To obtain an accurate fit, an extra time constant (RQ-sub-circuit) was added in series with the EEC model in Figs 7 and 8, which can be observed from the phase angle of Bode plots at very high frequencies (f > 10 kHz) as a straight line after the main peak. This behaviour is similar to that found for porous electrodes [46–52] and accounts for the very rough surface of investigated electrodes which were abraded with silica paper 600 P.

4. Conclusions

- The combination of electrochemical methods with the structurally sensitive method of impedance spectroscopy has offered an insight into the structural, electric and dielectric properties of the oxide films anodically formed on titanium, Ti6Al4V and Ti6Al6Nb alloys in physiological Hank's solution at 37°C.

- The discussion of results confirms that the oxide layer formed anodically on Ti and Ti-alloys is predominately TiO_2, which contains sub-oxides closer to the metal/oxide interface.

- During anodic film growth, the electric field strength within the film was of the order of 10^7 V cm^{-1}. The validity of the high field law of film growth during linear polarisation of the electrode was established.

- Stress that amounted to 10^9 Pa was developed within the anodically formed oxide film of the Ti and Ti-alloys.

- The beginning of film crystallisation at 2 V for all investigated materials and the onset of passivity breakdown at 3.6 V for the Ti6Al4V alloy are attributed to high compressive stresses within the growing oxide film.

- Impedance results suggest that the formation of cation vacancies at the alloy/oxide interface and their subsequent annihilation by an inward diffusion of vanadium ions are closely related to the stress developed during the anodic oxidation of the Ti6Al4V alloy.

References

1. D. F. Williams, *Biocompatibility of Clinical Implant Materials*, Vol 1., Ch 2. CRC Press, Boca Raton, Florida, USA, 1981.
2. M. A. Khan, R. L. Williams and D. F. Williams, *Biomaterials*, 1996, **17**, 2117–2126.
3. D. H. Bradhurst and J. S. L. Leach, *J. Electrochem. Soc.*, 1966, **113**, (12), 1245–1249.
4. M. J. Dignam, *Oxides and Oxide Films*, Vol. 1. M. Dekker, New York, 1972.
5. N. Sato, *Electrochim. Acta*, 1971, **16**, 1683–1692.
6. S. Moon and S. Pyun, *Electrochim. Acta*, 1998, **43**, (21–22), 3117–3126.
7. J. C. Nelson and R. A. Oriani, *Corros. Sci.*, 1993, **34**, (2), 307–326.
8. L. C. Archibald, *Electrochim. Acta*, 1977, **22**, 657–659, and references therein.
9. D. A. Vermilyea, *J. Electrochem. Soc.*, 1963, **110**, 354.
10. J. S. L. Leach and B. R. Pearson, *Corros. Sci.*, 1988, **28**, (1), 43–56.
11. B. A. Boukamp, *Solid State Ionics*, 1986, **20**, 31–39.
12. B. A. Boukamp, *Equivalent Circuit*, International Report CT 89/214/128, University of Twente, 1989.
13. M. Pankuch, R. Bell and C. A. Meleders, *Electrochim. Acta*, 1993, **38**, (18), 2777–2779.
14. C. de Fonseca, S. Boudin and M. de Cunha Belo, *J. Electroanal. Chem.*, 1994, **379**, 173–180.
15. J. Pouilleau, D. Devilliers, F. Garrido, S. Durand–Vidal and E. Mahe, *Mater. Sci. Eng.*, 1997, **B47**, 235–243.

16. M. Wieland *et al.*, *Biomaterials, ECASIA '97*, 16–20 June 1997, (I. Olefjort, L. Nyborg and D. Briggs eds). John Wiley & Sons, Chichester, England, 1997.
17. M. Schmiedgen, B. Baretzky, A. Schminke and H. Schmidt, *Biomaterials, ECASIA '97*, 16–20 June 1997, (I. Olefjort, L. Nyborg and D. Briggs eds). John Wiley & Sons, Chichester, England, 1997.
18. M. Ask, J. Lausmaa and B. Kasemo, *Appl. Surf. Sci.*, 1988–89, **35**, 283–301.
19. D. G. Wiesler *et al.*, *Surf. Sci.*, 1994, **302**, 341–349.
20. A. Ammar, S. Darwish and M. W. Khalil, *Z. Werkstofftech.*, 1981, **12**, 174–183, 421–431.
21. A. T. Fromhold, *Passivity of Metals*, The Corrosion Monograph Series, The Electrochemical Society Inc., Princeton, NJ , USA, 1978.
22. S. Omanovic and M. Metikos-Hukovic, *Solid State Ionics*, 1995, **78**, 69–78.
23. M. M. Lohrengel, *Mater. Sci. Eng.*, 1993, (**R11**), 243–294.
24. T. Shibata and Y. C. Zhu, *Corros. Sci.*, 1995, **37**, (2), 253–270.
25. D. J. Blackwood and L. M. Peter, *Electrochim. Acta*, 1989, **34**, 1505–1511.
26. J. W. Schultze and S. Kudelka, *The Electrochem. Soc. Interface*, 1997, **6**, (2), 28–31.
27. C. K. Dyer and J. S. L. Leach, *J. Electrochem. Soc.*, 1978, **127**, (7), 1032–1038.
28. J. F. McAleer and L. M. Peter, *J. Electrochem. Soc.*, 1982, **129**, (6), 1252–1259.
29. G. Blondeau, M. Froehlicher, M. Froment and A. Hugot-le Goff, *Thin Solid Films*, 1976, **38**, 261–267.
30. Lj. Arsov, C. Kormann and W. Plieth, *J. Electrochem. Soc.*, 1991, **138**, 2964–2971.
31. J. Marsh and D. Gorse, *Electrochim. Acta*, 1998, **43**, (7), 659–670.
32. D. D. Macdonald, *J. Electrochem. Soc.*, 1992, **139**, (12), 3434–3449.
33. M. A. Malati and W. K. Wong, *Surf. Tech. Coat.*, 1984, **22**, 305–322.
34. J. H. W. Witt and H. J. W. Lenderik, *Electrochim. Acta*, 1996, **41**, (7/8), 1111–1119.
35. H. J. de Witt, C. Wijenberg and C. Crevecoeur, *J. Electrochem. Soc.*, 1979, **126**, (5), 779–785.
36. J. B. Bessone, D. R. Salinas, C. E. Mayer, M. Ebert and W. J. Lorenz, *Electrochim. Acta*, 1992, **37**, (12), 2283–2290.
37. C. Y. Chao, L. F. Lin and D. D. Macdonald, *J. Electrochem. Soc.*, 1982, **129**, (9), 1874–1878.
38. M. Metikos-Hukovic, R. Babic and A. Marinovic, *J. Electrochem. Soc.*, 1998, **145**, 4045-4051.
39. M. J. El Basiouny and A. A. Mazhar, *Corrosion*, 1982, **38**, (5), 237–242.
40. J. Pan, D. Thierry and C. Leygraf, *Electrochim. Acta*, 1996, **41**, (7/8), 1143–1153.
41. C. Gabrielli, Identification of Electrochemical Processes by Frequency Response Analysis, Monograph Reference 004/83, Solatron Instrumentation Group, Fransbourg, England, 1981.
42. M. Baddi, C. Gabrielli, M. Keddam and H. Takenouti, *Passivity of Metals*. The Electrochemical Soc. Inc., pp.625–645, 1978, and references therein.
43. M. Keddam, J.-F. Lizee, C. Pallota and H. J. Takenouti, *J. Electrochem. Soc.*, 1984, **131**, 2016–2024.
44. F. E. Varela, L. M. Gassa and J. R. Vilche, *J. Electroanal. Chem.*, 1993, **353**, 147–155.
45. S. Brinic, Metikos-Hukovic, R. Babic, *J. Power Sources*, 1995, **55**, 19–24.
46. M. Metikos-Hukovic, S. Omanovic, R. Babic and I. Milosev, *Ber. Bunsenges. Phys. Chem.*, 1994, **98**, 1243–1249.
47. L. Chen and A. Lasia, *J. Electrochem. Soc.*, 1992, **139**, 3214–3219.
48. N.A. Assuncao, M.J. de Giz, G. Tremiliosi-Filh and E.R.Gonzales, *J. Electrochem. Soc.*, 1997, **144**, 2794–2800.
49. R. de Levie, *Advances in Electrochemistry and Electrochemical Engineering*, Vol. 6., p.329. Interscience N.Y., 1967.
50. L. Chen and A. Lasia, *J. Electrochem. Soc.*, 1993, **140**, 2464–2472.
51. A. Lasia, *J. Electroanal. Chem.*, 1995, **397**, 27–33.
52. R. Karimi and A. Lasia, *J. Electrochem. Soc.*, 1997, **144**, 511–519.

13
Morphological Characterisation and Corrosion Resistance of Cr III- and Cr VI-Based Conversion Layers on Zinc Coatings

T. BELLEZZE, R. FRATESI and G. ROVENTI
Dipartimento di Scienze dei Materiali e della Terra, Università degli Studi di Ancona, via Brecce Bianche, 60131 Ancona, Italy

ABSTRACT

In the last few years, several conversion treatments have been developed by plating industries to improve the corrosion resistance of zinc and zinc alloy coatings. Conventional chromate treatments are based on baths containing Cr VI compounds. Recently, however, due to environmental concerns associated with the use of Cr VI, other baths have been proposed based on molybdates, permanganates, titanates, rare earth metals, Cr III compounds, etc.

In this work, zinc coatings obtained in an industrial plating plant from cyanide-free alkaline baths with a Cr III-based conversion treatment have been studied. Some samples were submitted to a sealing treatment, which forms a layer consisting primarily of an organic–inorganic matrix. For comparison, similar coatings with Cr VI-based conversion treatment were also examined.

The corrosion behaviour was studied by polarisation resistance (R_p) measurements and mainly by electrochemical impedance spectroscopy (EIS) during an exposure time of 27 days in a 5% NaCl solution (temp. 25°C). Additional investigations were performed by salt spray tests (ASTM B117). Finally, morphological investigations were carried out on the coating surface by means of scanning electron microscopy (SEM), before and after exposure to the corrosive environments.

In general, all the experimental results indicate a better performance of Cr VI-based coatings. However, it is noteworthy that the corrosion behaviour of the Cr III passivated coating with sealing treatment is comparable to that of the coatings submitted to Cr VI conversion treatment without sealant; therefore, the process based on Cr III salts can be used as a less polluting alternative to the traditional chromating treatment.

1. Introduction

In the last twenty years, zinc plating has developed remarkably because of the increasing demand from the automotive industry for coatings with better corrosion resistance [1]. Amongst the new processes proposed, one of the most important is that of zinc alloy (Zn–Fe, Zn–Co or Zn–Ni) deposition [2,3]; the corrosion resistance of alloys is better than that of pure zinc coatings because they are electrochemically more noble than pure zinc. Alloy coatings have also been developed because of the necessity to substitute cadmium, since this is considered to be a cancer-producing agent in various industrial applications [4].

Corrosion resistance of pure zinc coatings can be improved by means of suitable chemical conversion treatments, produced generally by immersion of the zinc coated articles in baths containing chromic acid or an acid solution of Cr VI salts [5–7]. After immersion, a chemical reaction on the coating surface leads to the formation of a conversion layer 10–1000 nm thick [5], containing zinc oxides and hydroxides, zinc chromate, mixed Cr III and Cr VI oxides and hydroxides [5,10]. In particular, $Cr(OH)_3$ is formed due to Cr VI reduction and to reduction of either water, hydronium ion, or dissolved oxygen to form hydroxyl ions at the metal surface [5–7]. A chromate conversion bath will also contain activators such as fluoride, sulfate, acetate, etc., which lead to thicker conversion layers [6,7]. These layers protect the zinc coating because they form a non-reactive barrier to humidity and air, but allow chromate ions to leach out and to act as corrosion inhibitors [8,9]. Furthermore, when scratched or mechanically damaged, enough water is absorbed by the layer to swell and mend the damaged areas (self-healing effect) [6–9].

In Europe, the use of Cr VI compounds, considered as cancer-producing agents [6,9], must be withdrawn within a short time, according to the directive 94/60/CEE 20–12–1994. Several alternative conversion treatments have been proposed by researchers [6,9–12]; these are processes based on molybdates, permanganates, titanates, rare earth metals and Cr III (considered to be non-carcinogenic) compounds.

In the last few years, zinc plating processes have also changed; thus the use of cyanide baths has strongly decreased, for example from 90% in 1970 to 20% in 1990. They have been replaced by chloride acid baths and alkaline cyanide-free baths [2,13].

In this work, zinc coatings obtained from NaOH based alkaline baths, with Cr III-based conversion treatments leading to Cr VI-free conversion layers, were studied. After passivation, some samples were submitted to sealing treatments, which form a layer of Si compounds in an organic–inorganic matrix; this treatment covers the porosity and the cracks in the conversion layer surface, with improvement in the corrosion behaviour even after heating [14].

The corrosion behaviour was studied by means of electrochemical impedance spectroscopy (EIS) and polarisation resistance measurements (R_p) during an exposure time of 27 days in 5% NaCl solution and also by salt spray tests (ASTM B117). A discussion of the analysis of EIS data was also carried out. Morphological investigations were performed on the coating surface by scanning electron microscopy (SEM) before and after exposure to the corrosive environments.

2. Experimental Details

Zinc coatings were obtained on mild steel sheets (total area 56.2 cm^2) in an industrial rack plating plant from a cyanide-free alkaline bath containing Zn^{2+} 11–19, NaOH 110–190 g L^{-1} and commercial addition agents (temp. 25°C; c.d. 1.5–3 A dm^{-2}). After zinc deposition, the samples were submitted to different passivation treatments, shown with respective identification symbols in Table 1. The experimental conditions of conversion and sealing treatments are reported in Table 2.

Table 1. *Treatments performed on zinc coatings and corresponding identification symbols*

Conversion treatment based on Cr III	C3
Conversion treatment based on Cr III + sealing treatment	C3S
Conversion treatment based on Cr VI	C6
Conversion treatment based on Cr VI + sealing treatment	C6S
Zinc coating without treatments	ZN

Table 2. *Experimental parameters of conversion and sealing baths*

Process type	**Concentration of the commercial product (mL L^{-1})**	**Bath temperature (°C)**	**Treatment immersion time (s)**	**pH**
Chemical conversion based on Cr III	25–35	18–25	20–40	1.9–2.1
Chemical conversion based on Cr VI	5–8	25	30–50	1.5–2.5
Organic–inorganic matrix sealing treatment	Ready to use	20–30	60	9–11

The compositions of the coatings were determined by means of inductively coupled plasma spectroscopy (ICP), after stripping the deposits in HCl 1:3 with hexamethylenetetramine to inhibit the dissolution of the iron substrate. The coating thickness was measured by the magnetic inductive method.

Before and after corrosion tests, morphological analyses of coatings were conducted by means of scanning electron microscopy (SEM) and energy dispersive X-ray spectroscopy (EDXS). To evaluate the conversion and sealing layer thickness, the samples were fractured in liquid N_2 and SEM analysis was conducted on the cross-sections.

The corrosion resistance was studied by means of electrochemical impedance spectroscopy (EIS) and polarisation resistance measurements (R_p), during an exposure time of 27 days. The electrochemical measurements were performed under potentiostatic control in quiescent aerated 5% NaCl solution at 25±1°C (pH 6.4). The cell was obtained by securing a plexiglass tube to the sample with epoxy resin (exposed area 13.7 cm^2). The reference electrode was saturated calomel (SCE) and the counter-electrode was a platinum sheet. The EIS measurements were performed using an EG&G P.A.R. 273 potentiostat, connected to a Solartron 1255 frequency response analyser. The frequency range used was varied depending on the coating studied, but was always kept between a maximum value of 50 or 60 kHz and a minimum value of 1.5 or 2 mHz (measurements/decade: 7 or 8, V_{ac} (rms), the sinusoidal signal amplitude: 5mV, V_{dc}: corrosion potential). The sample integrity was checked by measuring the corrosion potential after all the tests, to confirm that the change from the initial value was no more than ±5 mV. To correlate the admittance found at high frequencies with the conversion layer damage, some measurements were carried out within four hours from the start of immersion in the saline solution. In this case, the experimental conditions were as described above, but 5 measurements/decade in the frequency range 100 kHz–10 mHz were carried out. The polarisation resistance measurements were performed using an EG&G P.A.R. 273 potentiostat, with potentiodynamic scans from E_{corr}–10 mV to E_{corr}+10 mV. The scan rate was calculated according to the equation proposed by Mansfeld and Kendig [15], but with a different frequency choice; therefore before the R_p measurement, an EIS measurement was performed and from the Bode-|Z| diagram the f_{int} frequency, corresponding to the medium point of the horizontal part at low frequencies, was obtained. Then, the scan rate v_{scan} was calculated from the following equation:

$$\nu_{scan}(\mathrm{mV\ s^{-1}}) = 2\sqrt{2} \cdot V_{ac}\ (\mathrm{rms;mV}) \cdot \pi \cdot f_{int}\ (\mathrm{Hz})$$

If the horizontal part was not detected in the Bode diagram, because of the presence of diffusive processes or because the minimum values of the frequency range used were not low enough for the d.c.-limit, the scan rate used was 0.05 mV s^{-1}.

The corrosion resistance of the coatings was also studied by exposure to a salt spray chamber (ASTM B 117: NaCl 5%, temp. 38°C). The surface damage of the coatings was evaluated at various exposure times. Before exposure, some samples were submitted to heat treatment at 120 °C for 1 h.

The corrosion behaviour of the passivated materials depends significantly on the production procedure; therefore all the tests were repeated three times and the results obtained for the average behaviour of each series are reported in the Figures.

3. Results and Discussion

3.1. Chemical–Physical Properties

The chemical–physical characteristics of the studied coatings are reported in Table 3. The higher conversion layer thickness of samples C3S and C6S compared with that of C3 and C6 must be attributed to the presence of the sealant. In the coatings submitted to Cr III-based conversion treatment, the chromium content does not change when the sealant is present, while coatings submitted to Cr VI-based conversion treatment show a decrease in chromium content when the sealing treatment is performed. This fact indicates that during the sealing process some chromium is lost [14].

Table 3. Chemical–physical characteristics of the studied coatings

Coating	Coating thickness (μm)	Conversion layer thickness (μm)	Cr analysis (μg cm^{-2})	Si analysis (μg cm^{-2})
C3	14.7	0.07	4.9	—
C3S	12.1	1.20	4.6	327.3
C6	13.5	0.36	31.2	—
C6S	14.0	1.63	13.6	251.5
ZN	14.2	—	—	—

3.2. Morphology

The morphological analysis indicates that C3 samples have a rough surface (Fig. 1a) and that C3S samples show a network of micro-cracks on the surface (Fig. 1b). Figures 1(c) and 1(d) show the surface of the samples submitted to the Cr VI-based conversion treatment. The C6 surface has a network of very thin, parallel micro-cracks which are limited to the conversion layer. The C6S surface is almost similar to that of C3S, even if the conversion layer thickness is higher (Table 3); in general, a close cracks network corresponds to a higher thickness [13,16].

3.3. Corrosion Resistance

The corrosion behaviour of chromium-based conversion layers was mainly studied by EIS measurements; a comparison with R_p measurements was also made. Corrosion potential values for all coatings ranged between –1015 and –1057 mV SCE and remained almost stable during the overall exposure time.

In general, the major difficulty in EIS data analysis is to find an equivalent electrical circuit model and/or parameters that give, for the analysed systems, good corrosion behaviour indications. Some authors have reported circuit models which can be used to simulate EIS data from measurements on chromium- and molybdenum-based conversion layers [11,17–19]; other authors, rather than suggesting circuit models, report partial analysis based on Nyquist and Bode plots [20–23]. Normally, after EIS measurements, chromium-based conversion layers show, at least in the first stage of the electrolytic solution exposure, two time constants at different frequency ranges [17–20,23]; data analysis is particularly problematic in the high frequency region. Zeller and Savinell [20] limited their analysis to the distortions in this region of the spectrum. In a subsequent paper Pirnàt *et al.* [18] judged this interpretation incomplete and proposed an equivalent circuit with a high frequency time constant, attributed to the chromate layer and considered this layer as a capacitor with the plates formed by the metal substrate and electrolyte solution. Similar considerations are reported by other authors [11,17,19]. In other papers, the first loop at high frequency was not

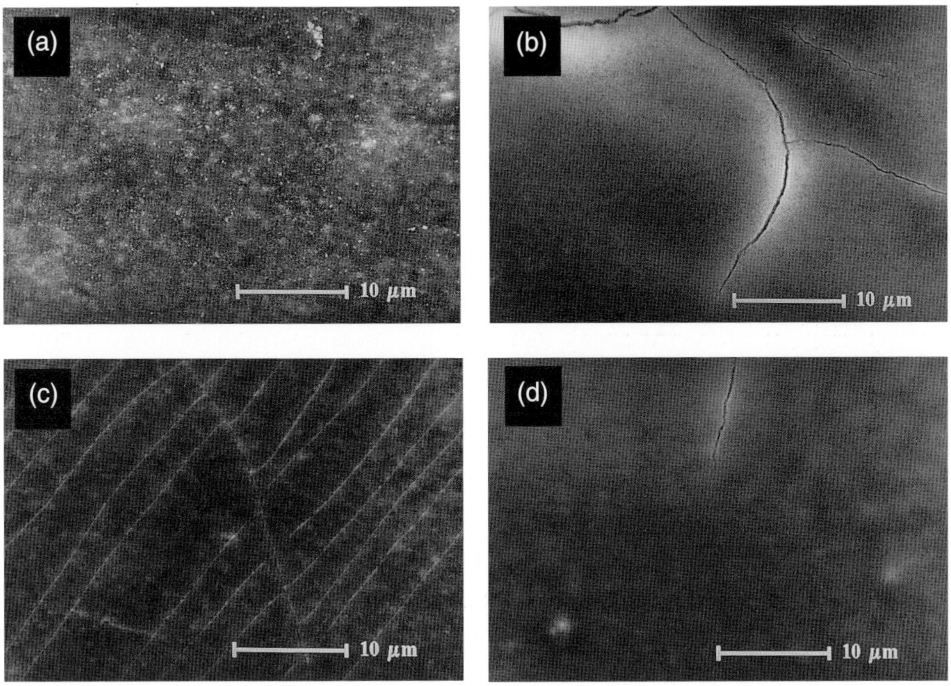

***Fig. 1** Microstructure of the studied coatings. (a) C3; (b) C3S; (c) C6; (d) C6S.*

attributed explicitly to the chromate layer, but the low value of 6 μF cm^{-2} calculated for the corresponding capacitance was assumed to be related to the presence of the chromated film [22,23]. Finally, Hazan [21] reported a loop at high frequency with a low capacitance (8.7 μF cm^{-2}) at the first stage of immersion, which was identified as "double layer capacitance" and related to a surface film.

In the present work, an equivalent circuit model that could describe completely the corrosion behaviour of the studied conversion layers during the overall observation period was not found and therefore a "corrosion parameter" R_{ac} was proposed. Barceló *et al.* [22] also used R_{ac} and measured the value corresponding to a chromated zinc coating, by reading the total resistance in the Nyquist plot. A parameter representing the corrosion trend was introduced, for example, by Kendig *et al.* [24] to study organic coatings and by Almeida *et al.* [11] to study conversion layers based on chromium and molybdenum on zinc; the former named this parameter R_{cor}, the latter R_{tot}. In both cases a circuit model and the corrosion parameter expressed by the sum of the series resistances, corresponding to the impedance at low frequency, was proposed. In the present work, R_{ac} was not measured using the Nyquist plot and it was impossible to calculate a sum of series resistances, because a circuit model was not available. In order to obtain an indication of impedance in the low frequency range for the conversion layer, R_{ac} can be evaluated from the Bode-$|Z|$ plot and taken as equal to the d.c.-limit, when present. When the d.c.-limit was not present (for example in those cases where a diffusion tail or a non-closed loop are present in the Nyquist plot), R_{ac} is taken as equal to the $|Z|$ value corresponding to the minimum frequency of the range employed in EIS measurements.

In Figs 2(a) and 2(b) the R_{ac}–time and R_p–time curves respectively are reported. In general they give similar information and in particular Cr VI-based coatings (C6 and C6S) have a better corrosion resistance than Cr III-based coatings (C3 and C3S). Coating C3S has good corrosion behaviour up to 15 days of salt solution exposure. For comparison, Fig. 2 also shows the non-passivated zinc corrosion behaviour. From one to five days of exposure, zinc shows extremely high resistance values (~40–80 kΩ cm^2) probably related to the presence of a layer of white corrosion products on its surface. This layer breaks off after seven days and the resistance then assumes lower values.

Subsequently, an EIS data analysis in the high frequency region of the spectrum was performed, in order to obtain a representative indication of the degree of coating

(a)

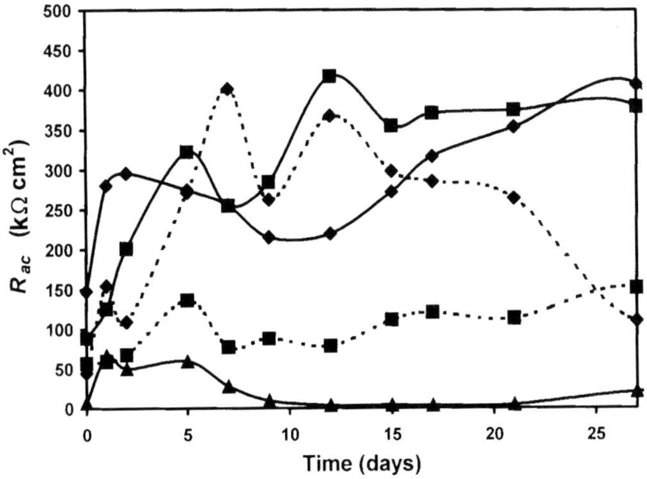

(b)

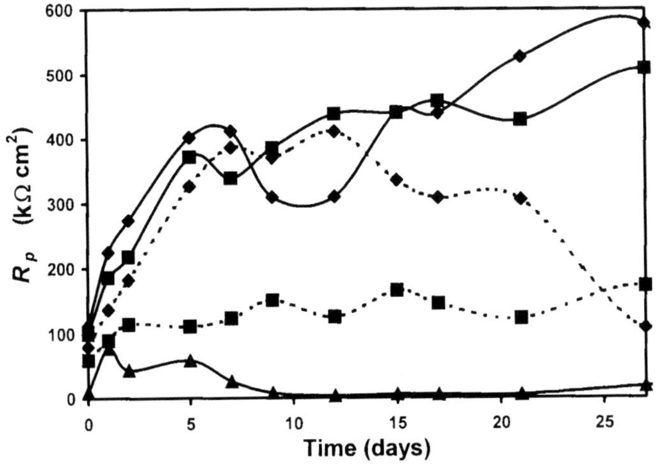

Fig. 2 *R_{ac}–time curve (a) and R_p–time curve (b) during exposure in quiescent aerated 5% NaCl solution. (—▲—) ZN; (···■···) C3; (···◆···) C3S; (—■—)C6; (—◆—)C6S.*

damage. Figure 3 shows the variation of the Bode-|Z| curves for the C3 coating during 4 h of exposure. In the high frequency region, it is clear that the oblique line shifts towards low frequencies. From the hypothesis that the Bode plot high frequency region is sensitive to the coating damage, a linear fitting procedure, which characterises quantitatively the line considered above, has been developed. At first, the linear trend at high frequency was assigned to a constant-phase element (CPE), whose admittance is $Y = Y_0(j\omega)^n$ [25]; from the slope and the intercept corresponding to the fitted line, it is possible to calculate n and Y_0, after correction for solution resistance.

This fitting procedure is similar to that applied by Shih and Mansfeld [26,27].

On plotting Y_0 and n constants as a function of time, it is possible to observe the increase in the first quantity and the decrease in the second quantity as indications of coating damage. This approach for coatings with passivation layers has been used by other authors [21,26,28], although they used the capacitance increase rather than the admittance increase.

The high frequency data analysis results, for an exposure time of 4 h in a NaCl 5% solution, are shown in Fig. 4(a). The admittance Y_0, for the C3 coating, increases sharply in comparison with the other coatings. This trend indicates a high degree of damage of the C3 coating. The n–time trends are not shown, however, they decreased throughout the period and assumed a value between 0.9 and 0.7.

At the early stage of exposure (5 min) n is close to 1 and it is possible to approximate Y_0 to a capacitance quantity C. At this stage C assumes increasing values equal to 0.21, 0.62, 0.91 and 1.93 μF cm^{-2} corresponding respectively to C6, C3, C6S and C3S.

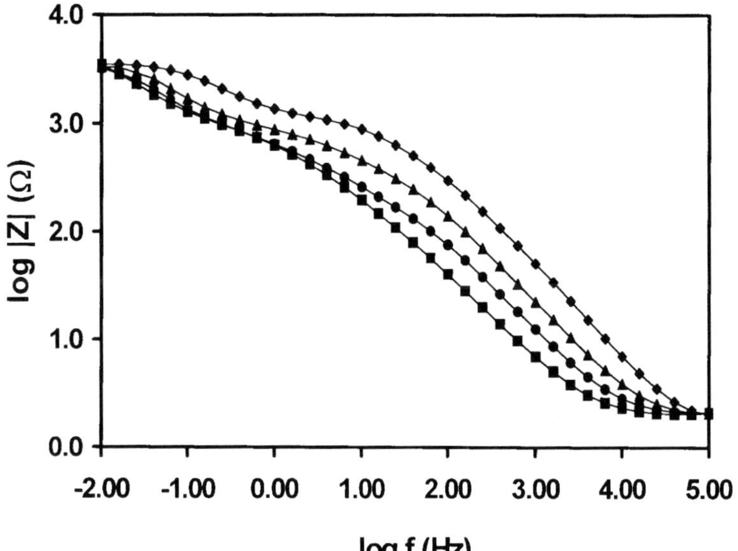

***Fig.** 3 Bode-|Z| plot for C3 coating at different exposure times in quiescent aerated 5% NaCl solution. (◆) 5 min; (▲) 40 min; (●) 100 min; (■) 240 min.*

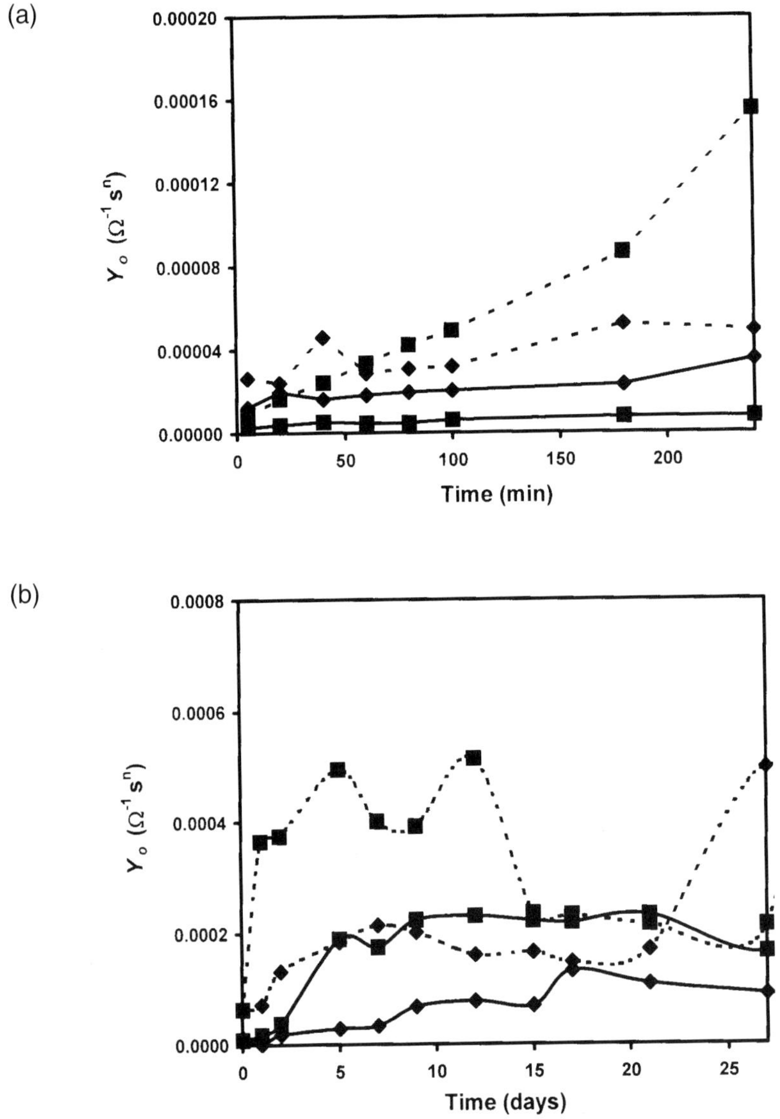

Fig. 4 *Y_o–time curves during short (a) and long (b) exposure in quiescent aerated 5% NaCl solution. (···■···) C3; (···◆···) C3S; (—■—) C6; (—◆—) C6S.*

These low values were reported in other papers [11,19,23]. If conversion layers without sealant (C3 and C6) have approximately the same dielectric constant and the conversion layers with sealant (C3S and C6S) have also approximately the same dielectric constant, but which is different from the C3 and C6 values, then, for the same surface area, a high conversion layer thickness (Table 3) corresponds to a low capacitance: $C_{C6} < C_{C3}$; $C_{C6S} < C_{C3S}$. Considering only the passivation film thickness, a decrease of capacitance could be observed passing from the conversion layer without sealant to the same conversion layer with sealant ($C_{C6S} < C_{C6}$ and $C_{C3S} < C_{C3}$); the exact inverse order was obtained, confirming the influence of the dielectric constant difference in the determination of the capacitance quantity.

In Fig. 4(b), the Y_0–time curves for long exposure times are reported. Again, the C3 coating has a high degree of damage; in comparison with C6, C3S has good corrosion behaviour, although this decreases towards the end of observation period. It is clear that C6S is the coating with the best resistance to corrosive agents. The value *n* (not shown) decreases with time and ranges between 1 and 0.6.

Shih and Mansfeld, for conversion layers obtained by $CeCl_3$ based treatments on aluminium, previously studied the capacitance increase with the pitting model [26], which used a pitting mechanism in the corrosion initiation. The corrosion of the coatings studied in the present work also initiates with a pitting mechanism and there is a high probability that this mechanism is revealed by the admittance increase, shown above, in the high frequency region of the spectrum. Visual observation of C3, after 2 days of exposure in NaCl 5%, shows small black spots on the surface which were not present in the other cases. When observed by SEM, these spots show a penetrating localised attack with corrosion products (Fig. 5a) in the central zone. The microanalysis highlights the presence of chlorides in the inner points of the pit and around it, indicating the diffusion of corrosion products below the conversion layer. In the case of C6 the Y_0–time curve is completely different from that of C3 (Fig. 4a) and correspondingly, from a morphological analysis, C6 has a surface where the conversion layer breaks off as shown in Fig. 5(c). Probably, the localised attack represents the sites of pitting corrosion initiation; no chloride was found and so no

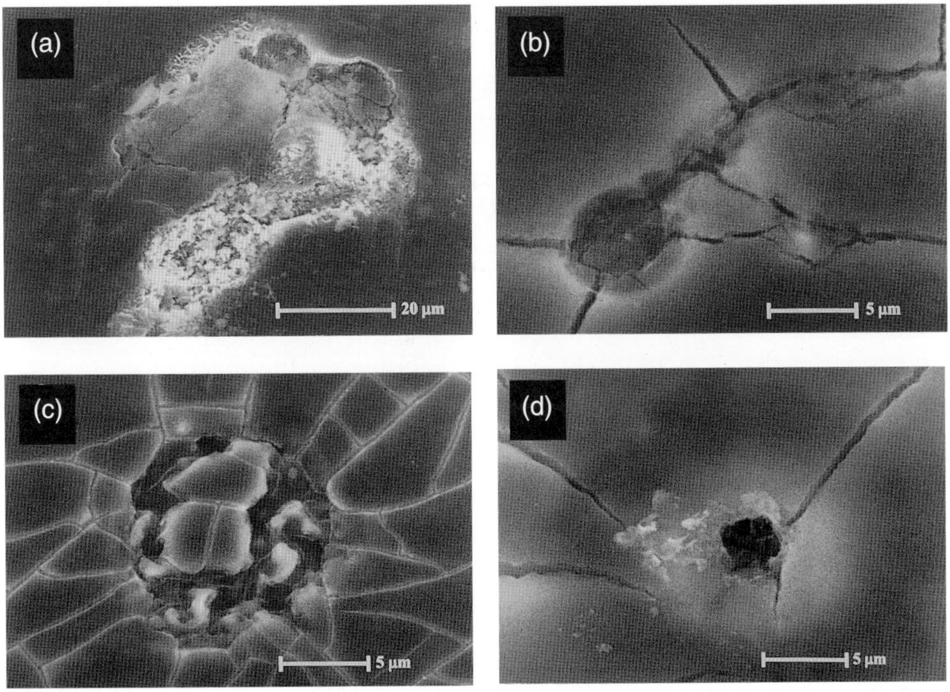

Fig. 5 *Microstructure of studied coatings after 2 days of exposure in quiescent aerated 5% NaCl solution. (a) C3; (b) C3S; (c) C6; (d) C6S.*

corrosion products are present. The C3S morphological aspect (Fig. 5b) is similar to that of C6 and in this case too no chlorides were found. This analogy is confirmed by similar R_{ac}–time, R_p–time and Y_o–time curves in the early stages of exposure (Figs 2 and 4). Coating C6S presents localised attack (Fig. 5d) similar to that observed on C3S (Fig. 5b) and C6 (Fig. 5c), but smaller. In conclusion, there is good correlation between admittance observations (Fig. 4a) and morphological observations (Fig. 5) after corrosion. Furthermore, pits or pores occluded by corrosion products can explain the initial R_{ac} increase (Fig. 2). For C6 and C6S coatings, the resistance increase can also be attributed to the effect of Cr VI inhibition [11].

Figure 6 shows the increase of the surface damage versus time during the salt spray exposure of the samples without (Fig. 6a) and with (Fig. 6b) heat treatment at 120 °C for 1 h. Coating C3S shows better corrosion resistance than C6, particularly when the coatings have been submitted to heat treatment (Fig. 6b). Coating C3S shows an improvement in corrosion behaviour after heat treatment and this, in agreement with other authors [14], an effect due to the sealant. Coating C3 shows very poor corrosion resistance, although a slight improvement was found after heat treatment. These trends agree quite well with the results obtained from electrochemical tests (Figs 2 and 4). In particular, it is noteworthy that the results of the salt spray exposure tests indicate that the increase in Y_o (Fig. 4) — related to the coating damage initiated by the pitting mechanism as considered above from a morphological point of view — is able to give rapid information on the corrosion behaviour of all the coatings. In fact, the rapid damage of C3 coating after salt fog exposure (Fig. 6) correlates well with the sharp increase in the corresponding Y_o–time curve during the early stage of immersion (Fig. 4a). In this period no damage on the C3S, C6 and C6S coatings was observed; they show the same trend, irrespective of their relative positions of the corresponding Y_o–time curves. To observe any difference between them, requires at least five days of exposure (Fig. 4b), when the sharp increase in the C6 Y_o–time curve becomes clear, i.e. in comparison to the curves of C3S and C6S. Again, this result correlates well with the salt fog exposure observations. On the other hand, R_{ac}, determined from EIS measurements (Fig. 2a), overestimates the performance of C6 coating, which shows the start of white corrosion after 240 h in the salt spray test, while the corrosion of C6S starts after 980 h. The trend of Y_o for long exposure times (Fig. 4b) shows that C6 coating begins to corrode before C6S and correlates well with the salt spray test results. This can be explained by considering that R_{ac} could include the contributions of the impedance due to the intact conversion layer, as well as to corrosion products, while Y_o is related only to the conversion layer. Values of R_{ac} and Y_o, even if they give similar information, are obtained from two different regions of the EIS spectrum: R_{ac} is determined from the low frequency range, whereas Y_o is calculated from the high frequency range.

4. Conclusions

Consideration of EIS data analysis has not been able to provide an equivalent circuit model capable of completely describing the corrosion behaviour of conversion layers on zinc coatings on steel as used in this work. Therefore, two quantities, R_{ac} and Y_o, were introduced; the first was obtained from the Bode-|Z| plot and the second from

(a)

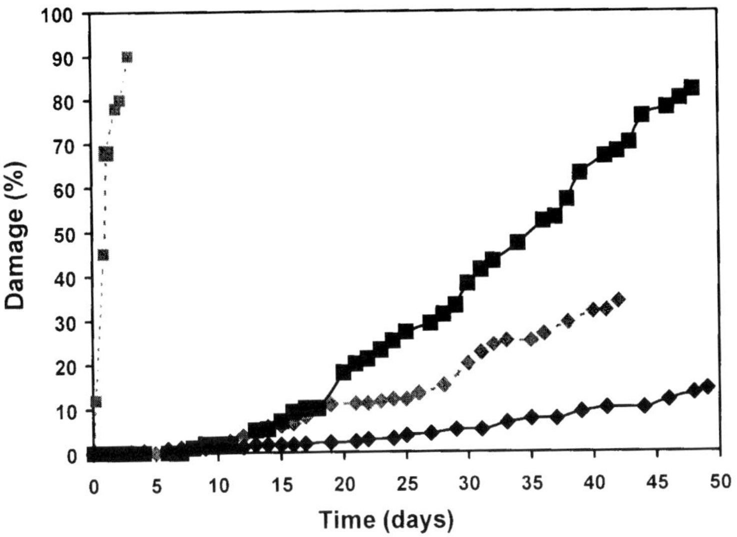

(b)

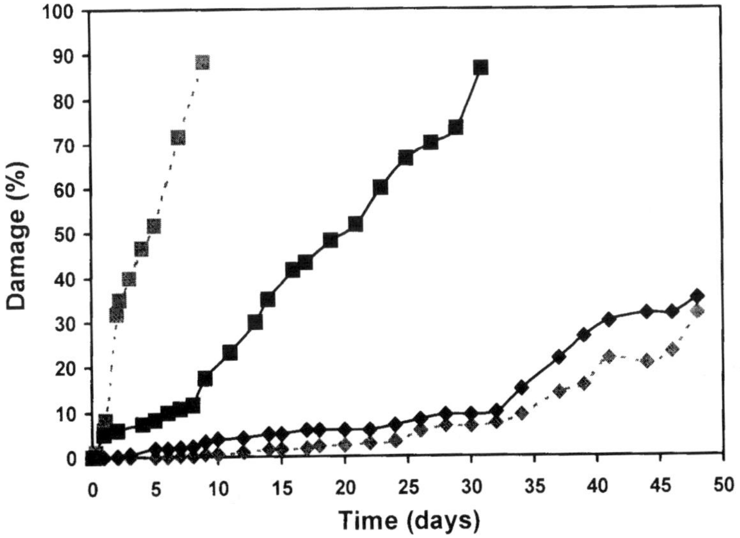

Fig. 6 Surface damage after salt fog exposure of samples without (a) and with (b) heat treatment at 120 °C for 1 h. (···■···) C3; (···◆···) C3S; (—■—) C6; (—◆—) C6S.

a linear fitting procedure that was developed in this work. These quantities R_{ac} and Y_0 give similar information, but R_{ac} overestimates the corrosion behaviour, for example, of the unsealed Cr VI coating (C6) in comparison with the corresponding sealed coating (C6S), whereas Y_0, for long exposure times, shows that C6 corrodes before C6S. The results obtained in this work show that Y_0 is a parameter able to give rapid indications of corrosion properties of the conversion layers and correlates well with salt spray test results. Probably, the Y_0 increase, as a function of exposure time, is related to the corrosion initiation occurring by a pitting mechanism. At least during the early stages of salt solution exposure, Y_0 can be attributed to the conversion layer and taken equal to the capacitance.

In general, the coatings based on Cr VI have a better corrosion behaviour compared with the coatings based on Cr III, but the sealed Cr III coating C3S shows a corrosion resistance which is comparable with that of the unsealed Cr VI coating C6. These results indicate that the presence of the sealant considerably improves the corrosion behaviour of the Cr III conversion layer. Therefore, Cr III passivation and subsequent sealing treatment can be used as alternative to the traditional chromating treatment.

References

1. H. E. Townsend, *Mater. Perform.*, 1991, **30**, 60.
2. T. J. Natorski, *Met. Finish.*, 1992, **90**, 15.
3. R. Fratesi, G. Lunazzi and G. Roventi, Organic and Inorganic Coatings for Corrosion Prevention- Research and Experiences. Papers from *Eurocorr '96* (L. Fedrizzi and P. L. Bonora, eds). Publication No. 20 in European Federation of Corrosion Series. Published by The Institute of Materials, London, 1997, pp.130–143.
4. G. F. Hsu, *Plat. Surf. Finish.*, 1984, **71**, 52.
5. K. A. Korinek, *ASM Metals Handbook Corrosion*, 1987, **13**, 389.
6. P. L. Hagans and C. M. Haas, *ASM Handbook Surface Engineering*, 1994, **5**, 405.
7. F. W. Eppensteiner and M. R. Jenkins, *Metal Finishing Guidebook and Directory Issue*, 1998, **96**(1), 489.
8. N. Zaki, *Met. Finish.*, 1988, **86**, (2), 75.
9. B. R. W. Hinton, *Met. Finish.*, 1991, **89**, (9) 55, (10) 15.
10. E. Almeida, T. C. Diamantino, M. O. Figueiredo and C. Sà, *Surf. Coat. Technol.*, 1998, **106**, 8.
11. E. Almeida, L. Fedrizzi and T. C. Diamantino, *Surf. Coat. Technol.*, 1998, **105**, 97.
12. C. B. Breslin and M. Geary, *Corrosion*, 1998, **54**, 964.
13. T. Bellezze, G. Roventi, R. Fratesi and G. Faranda, *AIFM GalvanoTecnica e Nuove Finiture*, 1999, **2**, 76.
14. Von G. Katen, *Galvanotechnik*, 1999, **90**, 650.
15. F. Mansfeld and M. Kendig, *Corrosion*, 1981, **37**, 545.
16. N. M. Martyak, J. E. McCaskie and R. J. Nichols, *Met. Finish.*, 1996, **94**, (1), 20.
17. B. Lengyel, Zs. Baróti-Lábár and R. Kahán, *Mater. Chem. Phys.*, 1982, **7**, 183.
18. A. Pirnát, L. Mészáros and B. Lengyel, *ACH- Models in Chemistry*, 1995, **132**, 647.
19. D. Ende, W. Kessler, D. Oelkrug and R. Fuchs, *Electrochim. Acta*, 1993, **38**, 2577.
20. R. L. Zeller III and R. F. Savinell, *Corros. Sci.*, 1986, **26**, 389.
21. J. Hazan and C. Coddet, *J. Appl. Electrochem.*, 1996, **26**, 203.
22. G. Barceló, M. Sarret, C. Müller and J. Pregonas, *Electrochim. Acta*, 1998, **43**, 13.
23. L. Fedrizzi, L. Ciaghi, P.L. Bonora, R. Fratesi and G. Roventi, *J. Appl. Electrochem.*, 1992, **22**, 247.
24. M. W. Kendig, H.-S. Ryang, T.- L. Liao and S. L. Jeanjaquet, *Corrosion*, 1999, **55**, 222.
25. B. A. Boukamp, Manual AC-imittance Data Analysis System Equivalent Circuit, University of Twente, Dept. Chemical Technology, Enschede, The Netherlands, 1989, p.12.
26. H. Shih and F. Mansfeld, *Corrosion*, 1989, **45**, 610.
27. H. Shih and F. Mansfeld, *Corros. Sci.*, 1989, **29**, 1235.
28. F. Mansfeld, *Electrochim. Acta*, 1990, **35**, 1533.

14
Structure, Corrosion Behaviour and Physical and Mechanical Properties of Composite Nickel Coatings with Sulfur Inclusions

V. MEDELIENĖ, K. LEINARTAS and E. MATULIONIS
Institute of chemistry, A.Gostauto 9, 2600-LT Vilnius, Lithuania

ABSTRACT

The structure, corrosion behaviour, physical and mechanical properties of Ni coatings deposited from an electrolyte with saccharin and microparticles of amorphous boron (B_{am}), B_4C and artificial diamond were studied. Inclusions of B_{am} and B_4C microparticles and sulfur-inclusions in the Ni matrix impeded free growth of crystals and led to the formation of an amorphous surface state. Non-metallic inclusions in electrode surfaces affected corrosion parameters as well as physical and mechanical properties of Ni. Correlation between morphology of coatings and corrosion properties was discussed.

1. Introduction

Interest in the production of composite materials as well as composite electrochemical coatings (CEC) has, up to now, been mainly concerned with their practical applications. Composite electrochemical coatings are heterogeneous systems consisting of a co-deposited metal and dispersed microparticles of various kinds, e.g. Al_2O_3, TiN, SiC, etc. The presence of a dispersed phase in the metal matrix improves the microhardness, high temperature strength, refractoriness, wear resistance etc. of the material. An important advantage of CEC lies in the possibility of obtaining protective layers which are ≥ 500 μm thick. Organic agents are included in the plating bath to maintain the elastic (in longitudinal elongation) properties of the Ni and Ni–CEC matrices [1,2].

The hardening effect is achieved by formation of a fine crystalline surface structure promoted by the presence of sulfur-containing compounds and by the incorporation of microparticles in the coatings. Non-metallic inclusions in the metal change its electron structure and break up the crystal lattice resulting in changes in the physical and mechanical properties of the electrodeposits. According to Polukarov [3] the degree of disruption of the crystal lattice of Ni increases with increase in the content of sulfur incorporated in the Ni deposits. The disruption of the lattice is associated with diminished sizes of block mosaic, increased concentration of packing defect deformations and twining types and increase in dislocations. When sulfur penetrates into the crystal lattice there is an inhibition of dislocation motion and recovery.

Inclusions of inert particles have an analogous effect on the fine structure parameters, viz. on the increase in density of dislocations and on their movement [4].

An undesirable effect is when the sulfur content is ≥0.05% since this results in a reduction in Ni corrosion resistance and overall strength at temperatures ≥250°C [1,5]. In contrast to the possible negative influence of sulfur, the incorporation of amorphous boron (B_{am}) in the Ni matrix improves the antifrictional properties of the coatings at high temperatures [6]. The introduction of B_{am} by the electrochemical method is an alternative to the usual process of metal saturation with boron which is used to improve the hardness, wear and corrosion resistance of metals and alloys. Co-deposition of Ni with diamond by plating from an electrolyte-suspension — as in the case of (Ni–B_{am}) — is also an alternative method for the production of diamond tools [7]. Microparticles of B_4C have abrasive properties and a high conductivity (2×10^7 S m^{-1}) [7] this being important in the production of modified composite coatings (Ni–B_4C).

Data on the influence of saccharin desulfuration product — sulfur — and inert microparticles of B_{am}, B_4C, artificial diamond on the structure, corrosion and physical, and mechanical properties of composite coatings are reported in this paper.

2. Experimental

2.1. Materials

Samples of electroplated 30 μm thick Ni coatings with microparticles of B_{am}, B_4C, artificial diamond and with sulfur were used. Nickel was electrodeposited on platinum, brass and copper substrates from an electrolyte containing nickel sulfate, 300; nickel chloride, 45; boric acid, 30g L^{-1}. Nickel coatings modified by microparticles were electroplated from electrolytes containing suspensions of microparticles intensively stirred by compressed air (400 L h^{-1}) at a cathodic current density i_c of 5 A dm^{-2} and at a temperature of 55°C. The concentration of microparticles was (in g L^{-1}): B_4C and artificial diamond — 40, amorphous boron (B_{am}) — 10. The amount of saccharin (o–sulfobenzoic imide) was 2 g L^{-1}. The microparticles of B_4C and of artificial diamond micropowders were dispersed by an industrial method down to 5 μm, i.e. the basic fraction M5, and an amorphous micropowder of boron was used. The pH of electrolytes was 4 and 3 in the case of electrolyte-suspension with micropowder of B_{am}.

The content of microparticles in the composite coatings was determined by the gravimetric method. The microparticles were weighed after dissolution of coatings in HCl (1:1), filtration and heating at 350°C. The coatings contained 4±0.5 without saccharin and 3±0.5 mass% of microparticles with saccharin.

The amount of sulfur in the Ni coatings was determined by an X-ray microanalyser of JXA-50A electron microscope (Japan). The source of the sulfur in the coatings is the saccharin but sulfur is also present in the bath composition in the form of sulfates (see Table 2, p.203). The sulfur as a product of saccharin decomposition at the cathode, i.e. as a desulfuration reaction product, is indicated with a sulfur symbol S.

2.2. Electron Microscopy Study of Structure of Ni and Ni–CEC

The morphology of coatings was studied and micrographed by a field-emission (SEI) type scaning electron microscope JXA-50A at an acceleration voltage of 25 kV and electron current = 5×10^{-12} A.

2.3. Corrosion Measurements

Corrosion of Ni and Ni–CEC was studied in naturally aerated 5% NaCl and 1.5% H_2SO_4 solutions. Solutions were prepared from 'analytical grade' salts and acid and triply distilled water. Temperature of solutions was 20 ± 1°C. The substrate under the nickel coatings in these corrosion studies was platinum.

Corrosion parameters of the coatings under study were measured by electrochemical methods. These methods are widely used for electrolytic corrosion measurements and provide information about not only corrosion rates, but also about peculiarities of the corrosion process.

Voltammetric measurements were taken by a PI–50–1.1 potentiostat. ΔE, log i dependencies were measured by the single potential step method. Steady-state values of current were used for the above dependencies (steady-state conditions were reached after 50–100 ms and measurements were taken after 5–10 s).

Corrosion current densities of the coatings studied were obtained by extrapolation of anode Tafel dependencies to the corrosion potential (E_{corr}).

Corrosion current densities were also determined according to a widely used formula [8,9]:

$$j_{corr} = b_a b_c / 2.3\,(b_a + b_c)\,R_p \quad (1)$$

where b_a, b_c are anodic and cathodic Tafel slopes, respectively, and R_p is a polarisation resistance. The R_p values were determined on the basis of potentiodynamic curves in the region near to the E_{corr} according to:

$$R_p = (\Delta E / \Delta I)_{\Delta E \to 0} \quad (2)$$

where ΔE is the electrode polarisation. The potential sweep rate was 0.1 mVs^{-1}.

Cathodic polarisation curves were recorded potentiodynamically at a sweep rate of 1 mVs^{-1}.

2.4. Determination of Physical and Mechanical Properties

Vickers microhardness (HV) was measured by an indenter PMT-3 on the mechanically polished 30 μm thick surface by continuous 50 g indentation load and was calculated by the formula:

$$HV_{50} = 1857.4\, P\, d^{-2}, \text{(Pa)} \quad (3)$$

where P = the load, g; d = the diagonal of diamond prism, mm.

An average value was calculated after ten measurements.

Internal stresses (σ) were measured by a conventional bent-cathode technique [10] and calculated by a formula:

$$\sigma = Ed_1^2 z/3d_2l^2 \text{ (Pa)} \tag{4}$$

where E = Young's modulus for the cathode substrate; d_1 = the thickness of the brass substrate, cm; d_2 = the coating thickness, cm; z = the deflection from initial position, cm; l^2 = area of cathode, cm^2.

Elasticity was evaluated by determining the relative extension of coatings (Δl) by the bending method using an elasticity scale NIILK OST 10086-39.M.I.22 [10]. In this coatings on brass substrates of various diameters (1–20 mm) rods are bent until the appearance of the first crack. A relative extension (Δl) was calculated by the formula:

$$\Delta l = d\,(2R + d)^{-1}\,100\%, \tag{5}$$

where d = the total thickness of coatings and substrate, mm; R = the radius of the smallest rod before the appearance of the first crack in the coating by a number divisible by 7.

3. Results

3.1. Structure of Surface

Investigation of the microstructure of the surfaces shows that the matrix of an electrolytic Ni surface consists of pyramidal crystals with pronounced crystallographic polyhedral forms and with equigranular texture (Fig. 1a–c). The surface of Ni with artificial diamond inclusions (Fig. 2a–c) has the same crystallographic forms as matrix crystals.

The surface of (Ni–B_4C) coatings consists of coalesced spherulites (Fig. 3a–c). Electroconductive B_4C microparticles changed the conditions of electrocrystallisation of Ni and changed the morphology of the matrix surface: deposition of Ni directly on the electroconductive microparticles of B_4C made a relief structure with accumulation of crystallites of Ni, from which spherulite with dendritic structure were formed (Fig. 3). X-ray analysis showed that the surface had an amorphous state.

It is known that saccharin actively inhibits the free growth of orientated crystals of Ni and leads to a reduction of their sizes [11,12]. The fine crystal structure was formed in this case (Fig.1d–f). Microparticles of an artificial diamond grow by compaction into the finely disperse sulfur-containing Ni matrix surface (Fig. 2d–f). The influence of saccharin on the structure of (Ni–B_4C) coatings is weaker in comparison with its influence on Ni and (Ni–diamond) coatings (Fig. 3d–f). It can be mentioned that the surface of (Ni–B_4C–S) has lost a dendritic structure and the crystallites have taken a smoother shape (compare Fig. 3a–c and d–f).

The surface of (Ni–B_{am}–S) coatings does not differ from (Ni–B_{am}) coating (Fig. 4a–d). In both cases, surfaces were formed by crystallites of Ni.

3.2. Corrosion of Ni–CEC

It is known that the surface of Ni is covered with a stable passive ~ 1 nm thick layer which provides a natural corrosion resistance of Ni electrodes. A study of Ni electrodes modified by microparticles of Al_2O_3, SiC and B_4C has shown that the corrosion

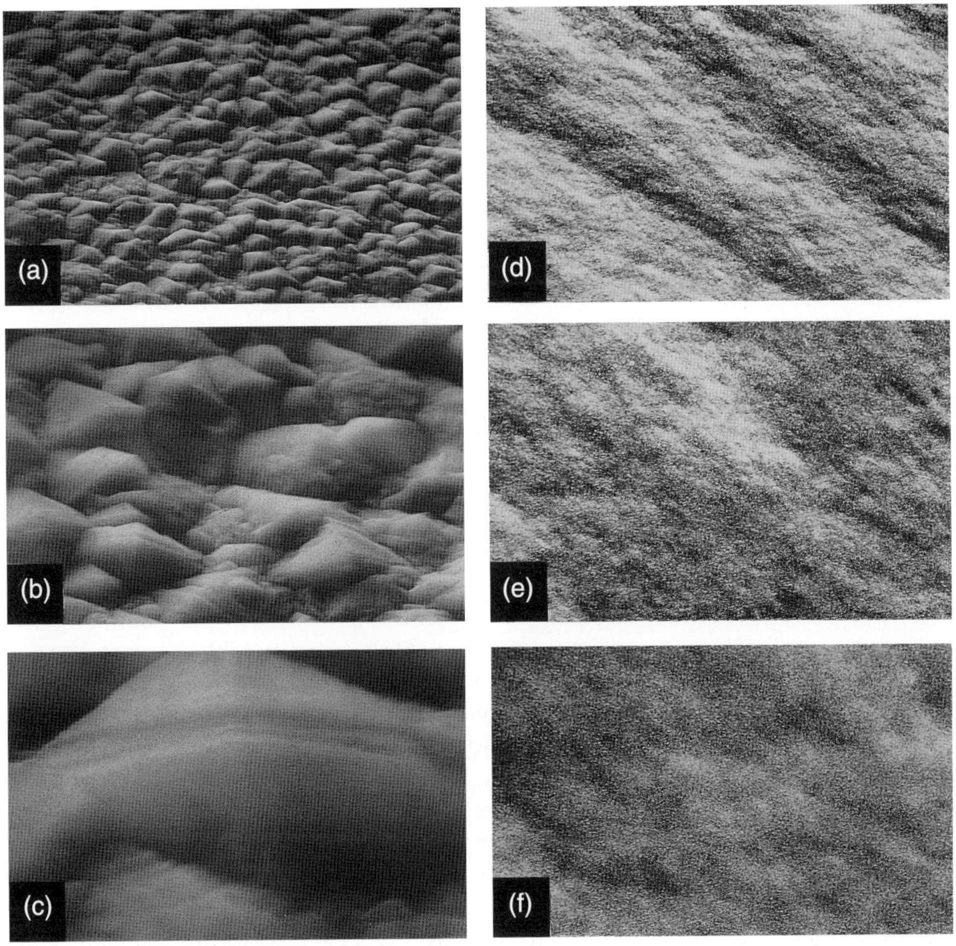

Fig. 1 SEM micrographs of the surface of electroplated Ni without saccharin (a–c) and with saccharin (d–f) (Magnifications: a,d — ×1560, b, e — ×4710, c — ×16 500, f — ×15 750.)

behaviour of the above electrodes depends on both the degree of blocking of the surface and the structure and physical properties of inclusions, e.g. their conductivity [13,14]. The correlation between b_a values and structural properties of coatings has been mentioned: lower values of b_a are more characteristic of crystalline Ni.

Experimental data of corrosion parameters of (Ni–B_{am}), (Ni–B_4C) and (Ni–diamond) and sulfur-containing CECs are listed in Table 1 (p.201) and shown in Figs 5–8.

According to the data listed in Table 1, the E_{corr} values of sulfur–containing Ni–CEC are more positive than those of Ni–CEC without sulfur in both acid and neutral media. The biggest shift of E_{corr} of about 0.2 V was measured for (Ni–B_4C) electrodes. A negative shift of E_{corr} in 1.5% H_2SO_4 solution was determined for (Ni–B_{am}) electrodes. The values of b_a for sulfur-containing (Ni–B_{am}) electrodes were close to those of (Ni–B_{am}) without sulfur both in acid medium and, especially, in 5% NaCl. According to structure measurements, the morphology of (Ni–B_{am}–S) surface did not differ from that of (Ni–B_{am}) (Fig. 4a–f).

Fig. 2 SEM micrographs of surface of (Ni–diamond) and (Ni–diamond) coatings with saccharin. (Magnifications: a, d — ×1600, b, e — ×4800, c, f — ×16 000.)

The strongest effect of sulfur on b_a values was found with (Ni–B_4C) coatings: b_a decreased from ~ 0.160 V to 0.090 V in 1.5% H_2SO_4 and from ~ 0.200 V to ~ 0.150 V in a neutral medium. The above effect may be caused by changes in the surface structure: as was confirmed by an X-ray analysis of the coatings under study, the microparticles promote formation of the amorphous state, and sulfur promotes a compact fine crystalline structure which is also an amorphous state. In the case of (Ni–B_4C–S) coatings, it was found that the dendritic structure was lost and crystallites of Ni had a smoother shape (Fig. 3d–f).

Contrary to (Ni–B_{am}) and (Ni–B_4C) coatings, higher values of b_a were measured for sulfur-containing (Ni-diamond) electrodes in both media. According to structural measurements, microparticles of an artificial diamond grew into a compact finely dispersed sulfur-containing Ni matrix (Fig. 2d–f).

The measurements of corrosion current densities of electrolytic Ni forming the basis for Ni–CEC matrix have shown that they were the same in naturally aerated

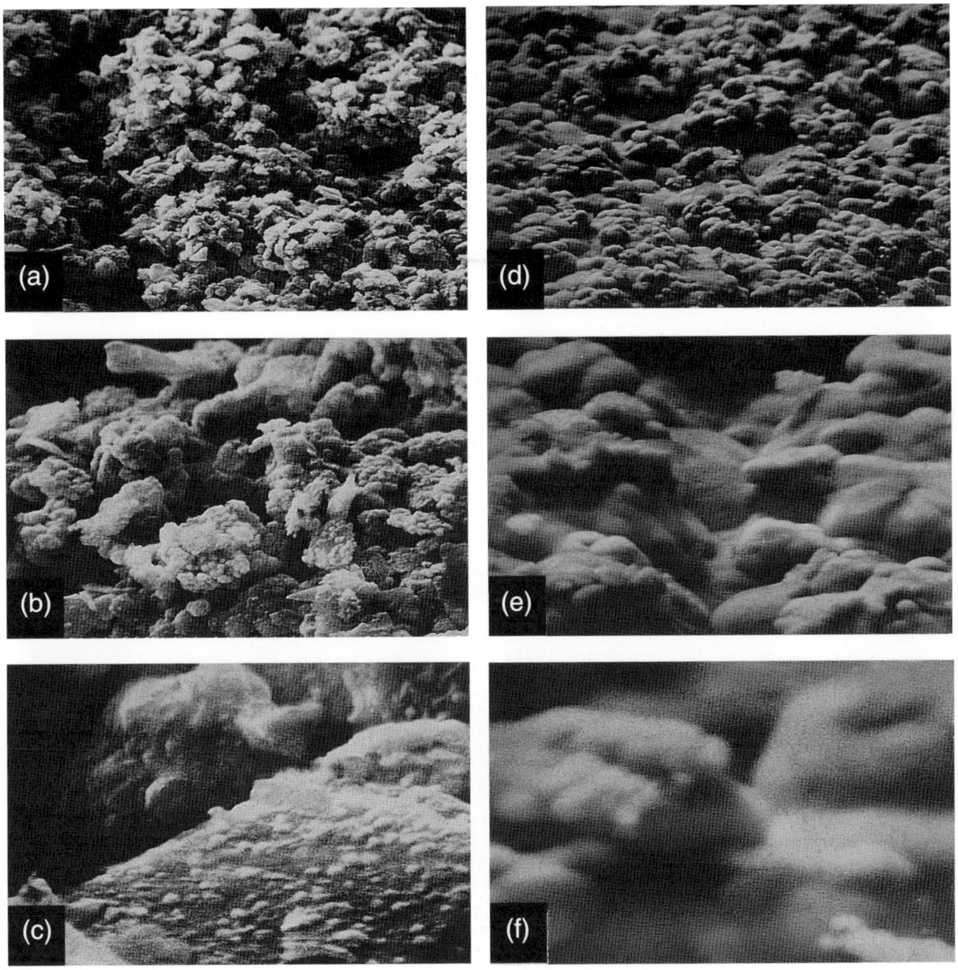

Fig. 3 SEM micrographs of the surface of (Ni–B_4C) and (Ni–B_4C) coatings with saccharin (Magnifications: a, d — ×795, b, c — ×2400, c, f — ×4800.)

1.5% H_2SO_4 and 5% NaCl solutions (Figs 5a,b). The value of j_{corr} was independent of exposure time in both solutions. It should be mentioned that the same values of j_{corr} were determined for the investigated Ni electrodes by extrapolation of Tafel dependencies method and by calculation according to formula (1) (Fig. 5(a), curve 2).

It was found that j_{corr} of (Ni–B_{am}) and (Ni–diamond) electrodes slightly increased in comparison with that of Ni (Figs 6,7). Especially high values of j_{corr} were determined for the (Ni–B_4C) electrode.

Voltammetric measurements of corrosion current densities in 5% NaCl solution showed that sulfur-containing Ni, (Ni–B_{am}) and (Ni–diamond) electrodes were slightly more resistant to corrosion than when sulfur-free (Fig. 6(a,b), curves 1,1'–3,3'). A significant decrease of j_{corr} was determined only for the Ni electrode with electroconductive B_4C microparticles (curves 4,4'). Corrosion currents increased for both (Ni–B_4C) and (Ni–B_4C–S) electrodes depending on the exposure time.

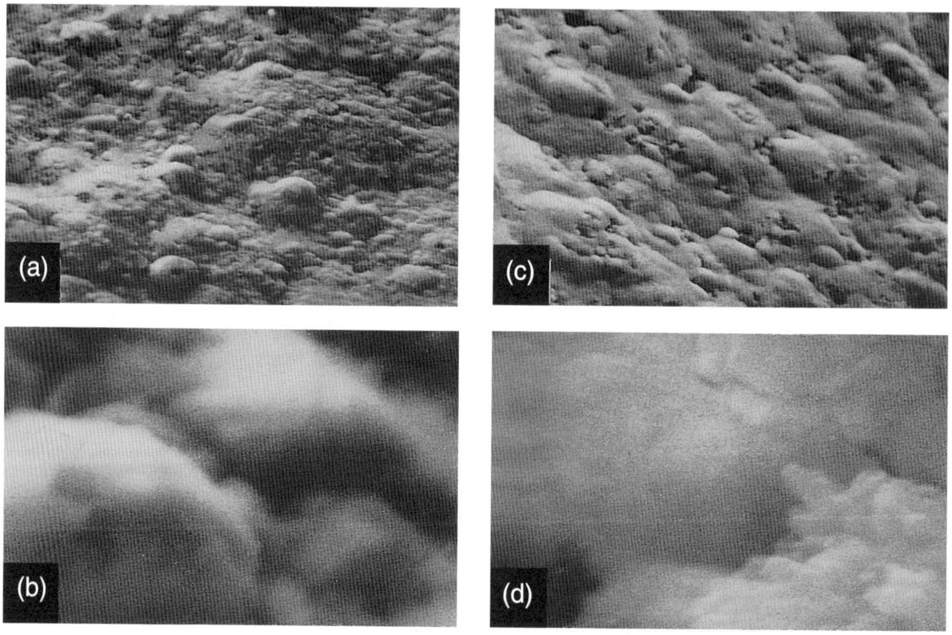

Fig. 4 SEM micrographs of the surface of (Ni–B_{am}) and (Ni–B_{am}) coatings with saccharin. (Magnifications: a, c — ×1550, b — ×16 000, d — ×15 500.)

Corrosion current densities of sulfur-containing Ni, (Ni–B_{am}) and (Ni–diamond) electrodes in naturally aerated 1.5% H_2SO_4 solution were also slightly decreased in comparison with those without sulfur and had a tendency to increase depending on the exposure time (Fig. 7(a,b), curves 1,1'–3,3'). As in the former case, a masked decrease in j_{corr} was also found for sulfur-containing (Ni–B_4C) electrode (curves 4,4'). A decrease of j_{corr} depending on the exposure time was observed for both electrodes.

It had been determined earlier [15] that the area of the Ni electrode surface that was covered by B_4C microparticles was ~ 22% and that of covered by B_{am} was ~ 10%. Sulfur-containing Ni–CEC electrodes had ~ 8% of surface occupied by microparticles. Microsmoothness of surface [16] due to sulfur influence was accompanied by a decrease in the limiting cathodic current (Fig. 12a–d). Decrease in oxygen reduction rate was especially great for the (Ni–B_4C–S) electrode (curves 3,3').

It is obvious that the above influence of saccharin decomposition product — sulfur — on corrosion parameters of Ni–CEC could not be caused only by the microsmoothness of surface and blocking of the electrode active area by sulfur. The change of the surface state of the electrodes and the physical properties of the microparticles also have an influence on corrosion behaviour.

3.3. Physical and Mechanical Properties

The inclusions of sulfur, amorphous boron and microparticles of B_4C in the Ni matrix formed a qualitatively new microheterogeneous Ni matrix. It obviously affected corrosion and the physical and mechanical properties of Ni and Ni–CEC (Table 2, p.203).

Table 1. *Values of* E_{corr} *and* b_a *for Ni and Ni–CEC in 1.5%* H_2SO_4 *and 5% NaCl solutions*

Types of coatings	t_{exp} **(min)**	**Solutions: 1.5% H_2SO_4** E_{corr} **(V)**	b_a **(V)**	**5% NaCl→** E_{corr} **(V)**	b_a **(V)**
Electroplated Ni	15	–0.050	0.100	–0.180	0.140
	30	–0.030	0.090	–0.160	0.130
	60	–0.010	0.090	–0.140	0.140
($Ni–B_{am}$)	15	–0.080	0.100	–0.170	0.160
	30	–0.070	0.110	–0.160	0.160
	60	–0.060	0.100	–0.140	0.160
($Ni–B_{am}–S$)	15	–0.070	0.080	–0.100	0.160
	30	–0.100	0.090	–0.090	0.170
	60	–0.100	0.090	–0.070	0.170
($Ni–B_4C$)	15	–0.160	0.160	–0.320	0.190
	30	–0.150	0.170	–0.320	0.200
	60	–0.150	0.170	–0.320	0.180
($Ni–B_4C–S$)	15	–0.060	0.090	–0.100	0.150
	30	–0.060	0.090	–0.100	0.150
	60	–0.080	0.090	–0.100	0.140
(Ni–diamond)	15	–0.170	0.140	–0.170	0.140
	30	–0.160	0.140	–0.150	0.140
	60	–0.130	0.130	–0.130	0.130
(Ni–diamond–S)	15	–0.100	0.160	–0.100	0.160
	30	–0.100	0.160	–0.100	0.160
	60	–0.100	0.150	–0.100	0.150

The data of Table 2 show a hardening effect on the Ni matrix brought about by microparticles as well as by sulfur-inclusions; HV increasing from 2450×10^6 to 4900×10^6 Pa.

Internal stresses (σ) of coatings become compressive, this being a characteristic of the plasticity of metallic deposits. Increase in plasticity of coatings is connected with

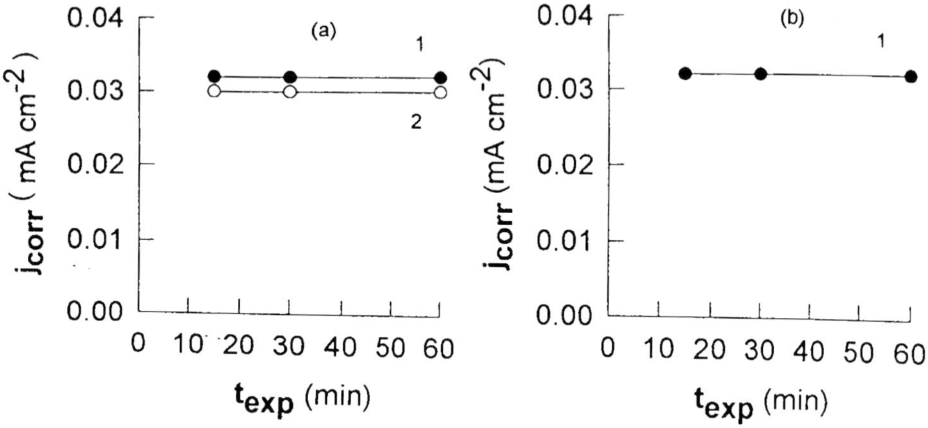

Fig. 5 *Dependencies of corrosion current densities of electrolytic Ni on exposure time. Curves: 1 — measured by extrapolation method (a, b); 2 — calculated according to formula (1) (a). Solutions: a — 1.5%* H_2SO_4*, b — 5% NaCl.*

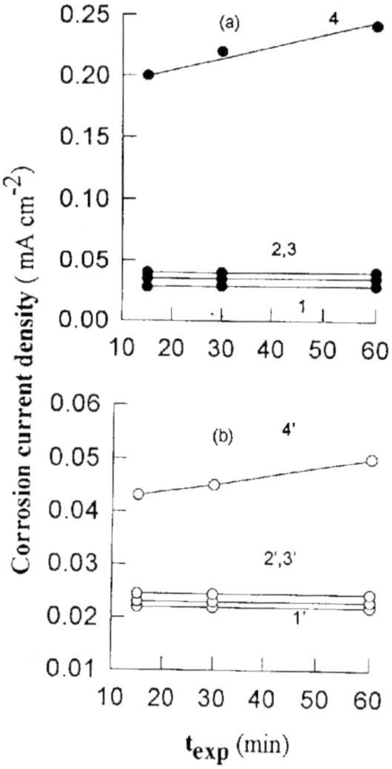

Fig. 6 *Dependencies of corrosion current density of Ni electrodes in 5% NaCl solution on exposure time. Electrodes: 1 — Electroplated Ni, 2 — Ni–diamond, 3 — Ni–*B_{am}*, 4 — Ni–*B_4C *(a). Electrodes with sulfur inclusions — 1′–4′ (b).*

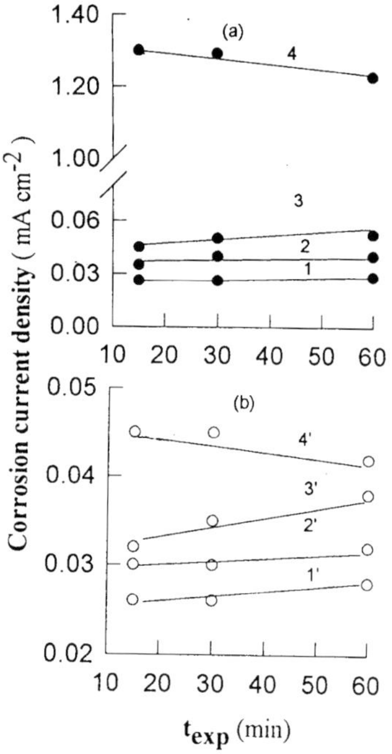

Fig. 7 *Dependencies of corrosion current density of Ni electrodes in 1.5%* H_2SO_4 *solution on exposure time. Electrodes: 1 — Electroplated Ni, 2 — Ni–diamond, 3 — Ni–*B_{am}*, 4 — Ni–*B_4C *(a). Electrodes with sulfur inclusions — 1′–4′ (b).*

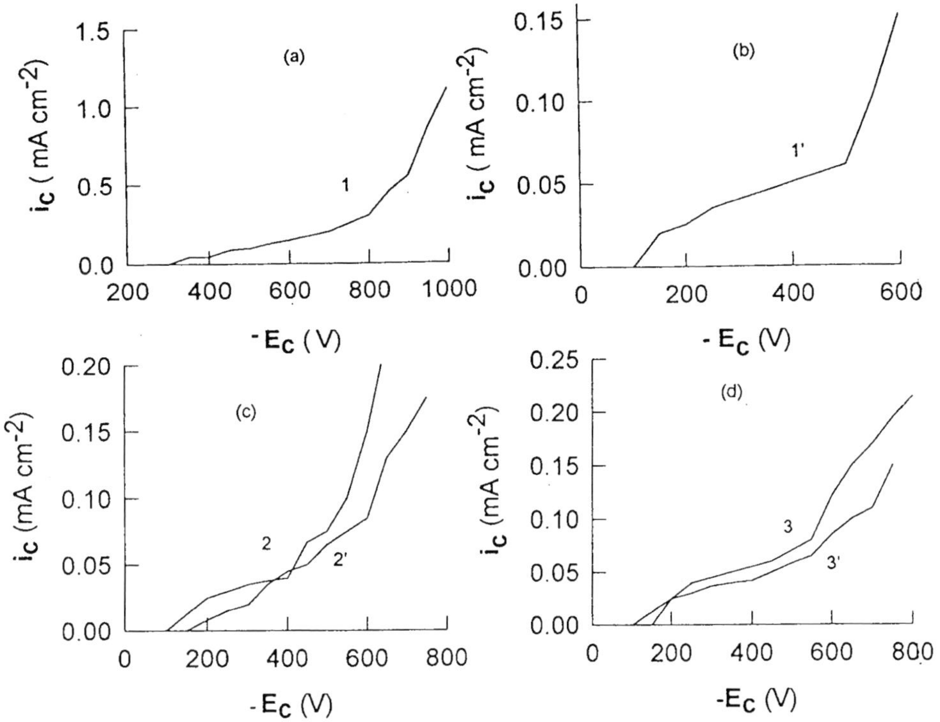

***Fig. 8** Cathodic potentiodynamic curves for Ni and Ni–CEC electrodes in 5% NaCl solution. Exposure time — 30 min. Curves: 1 — (Ni–B_4C) (a), 1′ — (Ni–B_4C–S) (b); 2 — (Ni–B_{am}), 2′ — (Ni–B_{am}–S) (c); 3 — (Ni–diamond), 3′ — (Ni–diamond–S) (d).*

***Table 2.** Influence of microparticles and sulfur on the microhardness (HV), internal stresses (σ) of Ni and Ni–CEC*

Types of coatings	**Content of sulfur (mass%)**	**HV_{50} (10^6 Pa)**	**σ (10^8 Pa) (– compressive stresses)**
Electroplated Ni	0.011*	2450	+117 600
(Ni–S)	0.039	3830	–53 900
(Ni–B_{am})	0.003*	2940	–39 200
(Ni–B_{am}–S)	0.035	3330	–37 240
(Ni–B_4C)	0.004*	2940	–39 240
(NiB_4C–S)	0.033	3430	–42 630
(Ni–diamond)	0.005*	4410	–42 140
(Ni–diamond–S)	0.037	4900	–48 020

the locking of crystal dislocations [17]; (Ni–B_{am}) coatings for example, are highly elastic under normal conditions. Internal stresses (σ) of deposits diminish three times in comparison with Ni when B_{am} is incorporated, and relative elongation was 34%. Therefore, both saccharin and B_{am} may be used to obtain elastic anticorrosive composite Ni coatings.

4. Conclusions

- The inclusions of B_{am}, B_4C and sulfur in the Ni matrix impeded free growth of texturing crystals and formed a qualitatively new microheterogenic Ni matrix with an amorphous surface state.
- The inclusion of sulfur in electrode surfaces affected corrosion parameters as well as the physical and mechanical properties of Ni and Ni–CEC.
- The decrease in corrosion rates of sulfur-containing Ni, (Ni–B_{am}) and (Ni–diamond) coatings was apparently due to a decrease in the active surface area. The corrosion behaviour of (Ni–B_4C) electrodes is also highly dependent on the electroconductivity of B_4C particles.

5. Acknowledgements

The authors thank Dr R. Juškėnas for conducting the X-ray structure analysis for identification of the amorphous state of surfaces.

References

1. W. H. Safranek, *The Properties of Electodeposited Metals and Alloys*. American Electroplaters and Surface Finishers Society, Orlando, Florida (1980).
2. V. Medelienė, Action of stimulators to codeposition of microparticles with Ni and some sulfur-containing compounds at electrodeposition of Ni composite coatings (in Russian), Doctorate thesis, Institute of Chemistry, Vilnius (1992).
3. J. M. Polukarov and Z. V. Semionova, *Elektrokhim.*, 1976, **12** (7), 1153 (in Russian).
4. I. N. Borodin, *Hardening of Components with Composite Coatings*. Publ. Mashinostroenie, Moscow, 1982 (in Russian).
5. K. E Volk, *Nickel und Nickellegierungen*. Publ. Springler-Verlag (1970) (in German).
6. I. M. Fedortchenko, J. A. Guslijenko and P. I. Epick, *Poroshk. Metall. (Powder Metallurgy)*, 1972, **8** (116), 31 (in Russian).
7. R. S. Saifullin, *Composite Coatings and Materials*. Publ. Khimiya, Moscow (1977) (in Russian).
8. M. Stern and A. L. Geary, *J. Electrochem. Soc.*, 1957, **104**, 56.
9. H. Kaesche, *Die Korrosion der Metalle*. Publ. Springer-Verlag, Berlin, 1979 (in German).
10. S. V. Jakubovitch, *Testing of Lacquer-Colour Materials and Coatings*. Publ. Goskhimizdat, Moscow, St. Petersburg (1952) (in Russian).
11. A. P. Dziūvė and J. J. Matulis, *Collected articles: The study of metals electrodeposition*. Publ. Inst. of Chemistry and Chemical Technology of Lithuanian Academy of Sciences, Vilnius, 1971 (in Russian).

12. D. Mockute, *Chemija* (in English), Vilnius, 1997, **4**, 14.
13. V. Medelienė, K. Leinartas, E. Juzeliūnas, *Zash. Met.*, 1995, **31** (1), 98 (in Russian).
14. V. Medelienė, K. Leinartas, E. Juzeliūnas, *Chemija*, Vilnius, 1996, **4**, 25 (in English).
15. V. Medelienė, M. Kurtinaitiene, K. Leinartas, *Cheminė Technologija*, Kaunas 1998, **3** (9), 14 (in Russian).
16. M. Trojon and L. Wang, *Appl. Surf. Sci.*, 1996 **103**, 517.
17. A. R. West, *Solid State Chemistry and its Applications*. University of Aberdeen (1984).

15
Ellipsometric Study of Tungsten Oxide (WO_3) Film Dissolution in Acid Electrolytes

M.A.PÉREZ and M. LÓPEZ TEIJELO
INFIQC, Depto. de Fisicoquímica, Facultad de Ciencias Químicas, Universidad Nacional de Córdoba, Pabellón Argentina, Ala 1, Ciudad Universitaria, 5000 Córdoba, Argentina

ABSTRACT
Ellipsometry is used as a monitoring technique in a study of the anodic growth and dissolution of WO_3 films in sulfuric acid solutions. The optical properties of the oxide are influenced by the presence or absence of the electric field across the film. The optical data allowed the conditions to be obtained in which either the chemical dissolution behaves as a single process (film thinning) or simultaneously with other processes such as surface roughening, generation of a hydrated layer, etc. For the first case, the evolution of thickness with time during the dissolution is linear in agreement with a zero-order rate law, i.e. the dissolution rate is constant. For the second case, the initial dissolution rates can be estimated.

1. Introduction

Anodically formed tungsten oxides have been found to be of interest for use in electric display devices but the main problem to overcome for commercial use is the short device life that results from WO_3 film dissolution and erosion during cycling [1]. The dissolution of oxides of valve metals like Al, Ti, Bi, Hf, Zr and W [2–9] has been studied in different media mainly by impedance measurements and both first-order kinetics for the dissolution rate law (i.e. dissolution rates proportional to the oxide film thickness) as well as zero-order kinetics have been reported.

The optical properties of anodic WO_3 films grown in both aqueous and acetic acid-based electrolytes as well as the electrochromic behaviour of the oxide studied by ellipsometry have been previously reported [10–13]. The field dependence of the dielectric constant during the anodic oxidation of the metal in a sulfuric acid electrolyte, has been analysed in terms of isotropic optical models while the effect of electrostriction in the non-aqueous electrolyte has been explained through optical anisotropy of the film.

The anodic WO_3 dissolution process has been studied in different acidic media. According to impedance measurements, oxide dissolution in sulfuric or acetic acid-based electrolytes follows a first-order law, while in phosphate-containing electrolytes, the rate is of zero-order [8,9].

The aim of this work is to characterise the WO_3 anodic growth and the oxide dissolution process in aqueous sulfuric acid solutions. Ellipsometry *in situ* is used as a monitoring technique during the galvanostatic growth of the anodic oxide films and in the dissolution at the open circuit potential in the formation electrolyte.

2. Experimental

The working electrode consisted of a polycrystalline tungsten rod (Koch-Light, 99.999% purity) of 4 mm dia. mounted in a Teflon holder to expose a circular area. The pre-treatment of this electrode was as follows: the surface was first abraded with emery paper and mechanically polished with 9 μm diamond paste. After the mechanical polishing, a WO_3 film was grown in a 0.005M H_2SO_4 + 0.995M Na_2SO_4 solution as indicated below and the variation of the ellipsometric angles Ψ and Δ (defined as the ratio of the electric field amplitudes and phase angles of the perpendicular and parallel components of the reflected polarised light), was followed during the growth. Then the electroformed film was dissolved chemically in a 0.1M NaOH solution and the procedure was repeated several times until a reproducible growth curve was obtained. A platinum sheet was used as counter electrode. All potentials were measured against a $Hg/Hg_2SO_4/Na_2SO_4$(1M) reference electrode.

Measurements were performed in *x*M H_2SO_4 + (1–*x*)M Na_2SO_4 ($0 \le x \le 1$) aqueous solutions of constant ionic strength, prepared from AR chemicals and purified water (Milli Ro- Milli Q system) under nitrogen gas saturation at 25°C.

The anodic growth of the WO_3 films was performed galvanostatically up to *ca.* 50 V by applying a constant current density of 2 mA cm^{-2}. In order to analyse the differences between the optical behaviour when the anodising field is removed, the circuit was opened at regular intervals during the growth of the oxide film formed at a constant current density. This procedure makes it possible to determine the values of the optical constants of the oxide at the value of the anodising field or at zero field and the change in the film thickness when the field is switched on or off. Chemical dissolution of the anodic WO_3 films was followed ellipsometrically at the open circuit potential in the formation electrolyte.

Ellipsometric data were obtained by using a three-compartment cell made of Pyrex glass. The working electrode was mounted in a Teflon holder and placed horizontally in the cell compartment provided with two plane glass windows adequate for optical measurements. The electrolysis cell was mounted in a Rudolph Research rotating-analyser automatic ellipsometer (vertical type, 2000 FT model) equipped with a 75 W tungsten lamp as light source and a filter (546.1 nm). The light reflected by the surface of the electrode passed through a rotating analyser to a photomultiplier tube (PMT) for detection. The sinusoidal output of the PMT was digitised by a HP 216 microcomputer and the values of Ψ and Δ were calculated by the computer from the three Fourier coefficients of the PMT signal. The PMT signal was averaged twice and data were collected in continuous mode. The time interval between measurements was *ca.*1 s. All measurements were made at an incidence angle of 70.00 degrees. The experimental data were fitted employing a model for a single anisotropic film, using the Simplex method [14] to minimise the function $G = \Sigma (\Delta^{exp} - \Delta^{th})^2 + (\Psi^{exp} - \Psi^{th})^2$.

3. Results and Discussion

Figure 1 shows the Ψ–Δ response obtained during the growth of an anodic WO_3 film at a constant current density up to a potential value of *ca.* 50 V where oxygen evolution starts, as well as the fitting curve for the growth of an anisotropic film on an isotropic substrate. The optical analysis was done, with no assumptions of the values of the optical constants, by varying six parameters in the Simplex algorithm: the refractive index (n_s) and extinction coefficient (k_s) for the substrate and the corresponding values in the transverse (n_x, k_x) and in the field direction (n_z, k_z) for the film. The results can be reasonably fitted with the anisotropic model provided that with increasing thickness the film is homogeneous and of constant refractive indices and extinction coefficients. Typical values obtained for the galvanostatic WO_3 growth at j = 2 mA cm^{-2} are n_x = 2.147; k_x = 0.0001; n_z = 2.083; k_z = 0.0016. The optical constants corresponding to the tungsten substrate are n = 4.45 and k = 2.78. Extinction coefficients are very small, indicating that the film is practically transparent.

As reported by Ord and De Smet [10] for anodic tungsten oxide grown in acetic acid-based solutions, the optical properties of WO_3 are influenced by the presence or absence of the electric field within the film. We have employed previously [15] a procedure of current interruption during the growth to obtain the optical properties of WO_3 anodic films in the absence of the electric field. Figure 2 shows the Ψ–Δ response obtained during the galvanostatic growth of an anodic oxide film and the

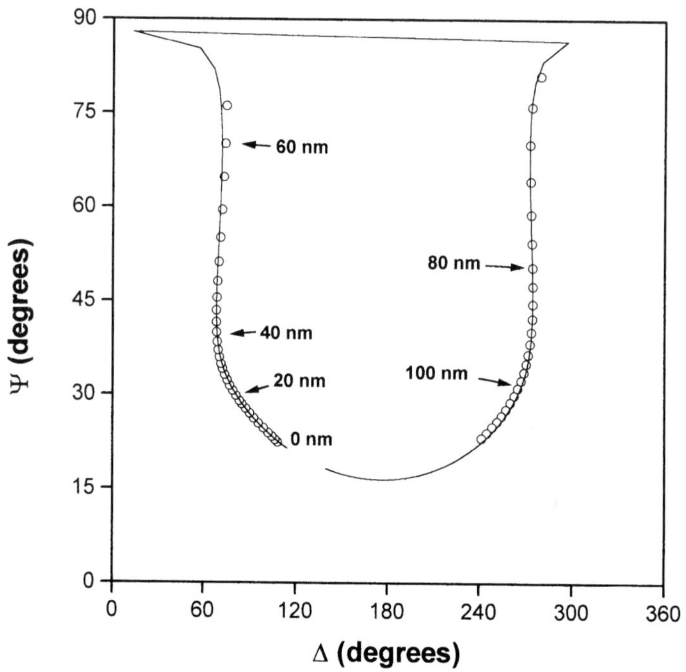

Fig. 1 *Ψ–Δ response for WO_3 anodic growth in a 0.001M H_2SO_4 + 0.999M Na_2SO_4 solution at j = 2 mA cm^{-2}. (o) anodic growth, (—) simulation with n_x = 2.147, k_x = 0.0001, n_z = 2.083, k_z = 0.0016.*

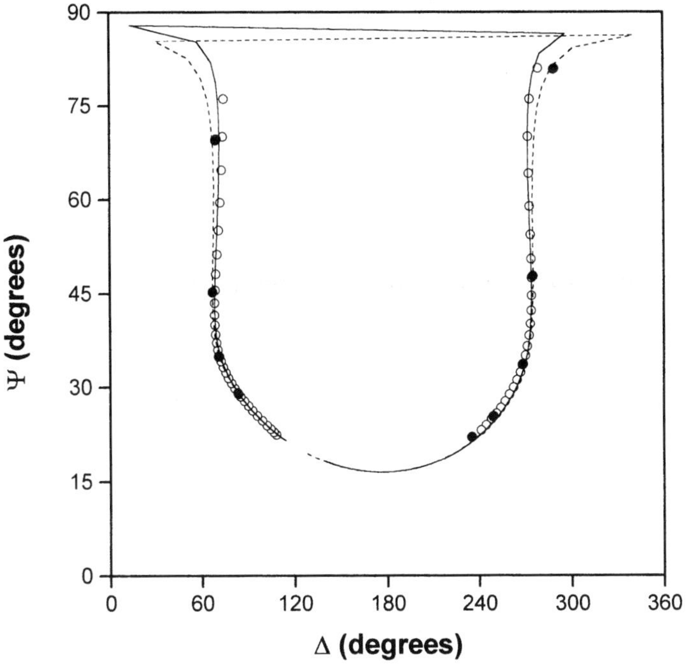

Fig. 2 *Ψ–Δ response for WO_3 anodic growth at j = 2 mA cm^{-2} (o) and after current interruption (•) in a 0.001M H_2SO_4 + 0.999M Na_2SO_4 solution; (—), (---) theoretical simulations.*

corresponding values when the circuit is open at regular intervals (different thickness) to analyse the differences of optical behaviour when the field is removed. The open points correspond to the WO_3 growth curve at j = 2 mA cm^{-2} and define the optical properties of a film of increasing thickness under these conditions. After opening the circuit, a rapid change (2–3 s) in the optical parameters Ψ and Δ is obtained (not shown in the figure) which corresponds to a rapid electrostrictive process governed by the rate of change of the electric field in the oxide film [12]. The Ψ–Δ values after *ca.* 5 s are practically constant and determine the optical behaviour of the film at zero field. When the anodic current is reconnected, the Ψ–Δ values return to the growth curve. At longer times, the change in the Ψ–Δ values corresponds to the chemical dissolution of the film.

The points obtained for different thicknesses after the current is switched off allow the optical properties of the oxide at zero field to be obtained. The fitting procedure was the same as described before, employing the optical constants for the tungsten substrate obtained from the fitting procedure for the growth of the anodic oxide. The values obtained for the oxide when the anodising field is switched off, are n_x = 2.173; k_x = 0.0001; n_z = 2.134; k_z = 0.0015. When the field is removed, the refractive index in the transverse direction increases by an average value of 0.025 while that in the direction of the field increases by 0.050, in agreement with results reported previously for WO_3 grown in acetic acid electrolytes [10]. The corresponding values of the extinction coefficients remain practically constant. These results show that the optical properties of WO_3 at open circuit are different from those in the presence of the anodising field, which means that if the dissolution process occurs only by thinning of the film (i.e. the reverse process of the growth), the evolution of the Ψ–Δ values with time must retrace the curve obtained for the oxide growth at zero field.

The Ψ–Δ response during chemical dissolution of WO_3 under open circuit conditions was followed in solutions of different composition as previously described. The results obtained can be classified in two groups. The first group shows a Ψ–Δ evolution during dissolution which retraces the points for the oxide growth curve at open circuit, while the second group presents changes in Ψ–Δ with hysteresis between the curves. Figure 3 shows the Ψ–Δ dependence for WO_3 galvanostatic growth and during current interruption as well as the Ψ–Δ evolution during chemical dissolution in a diluted acid solution (0.001M H_2SO_4 + 0.999M Na_2SO_4). It can be seen that the dissolution curve retraces practically the points of the curve obtained for the growth under open circuit conditions, indicating that no other process apart from film thinning takes place during the dissolution. For this case, the Ψ–Δ curve during the dissolution can be considered as a parametric function of the thickness value, from which the fitting procedure assuming a single anisotropic film thinning process allows the calculation of the thickness–time dependence. The optical properties of the substrate were obtained from the fitting procedure for the anodic oxide growth. Figure 4 shows the thickness–time relationship obtained in the 0.001M H_2SO_4 + 0.999M Na_2SO_4 solution. The dependence is linear indicating that the dissolution rate follows a zero-order law (i.e. the dissolution rate is independent of film thickness).

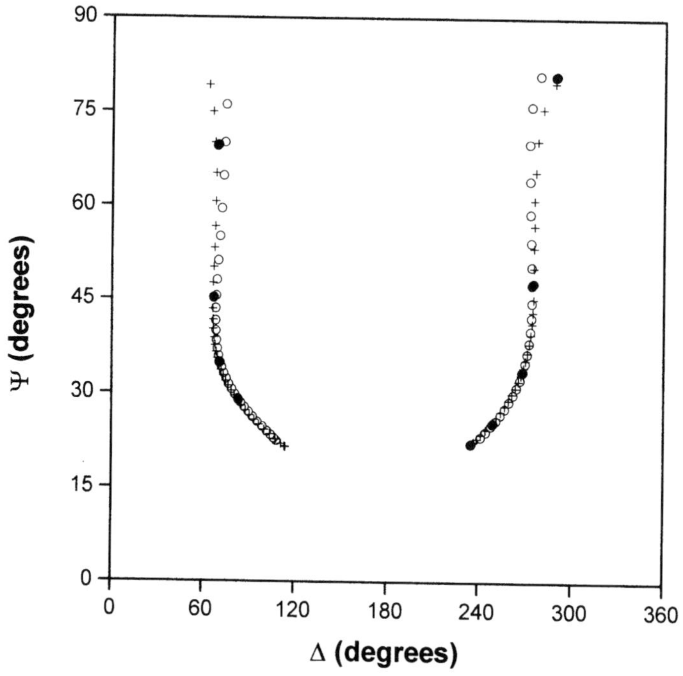

Fig. 3 *Ψ–Δ response during WO_3 galvanostatic growth (o), with current interruption (•) and chemical dissolution at open circuit (+). 0.001M H_2SO_4 + 0.999M Na_2SO_4, j = 2 mA cm^{-2}.*

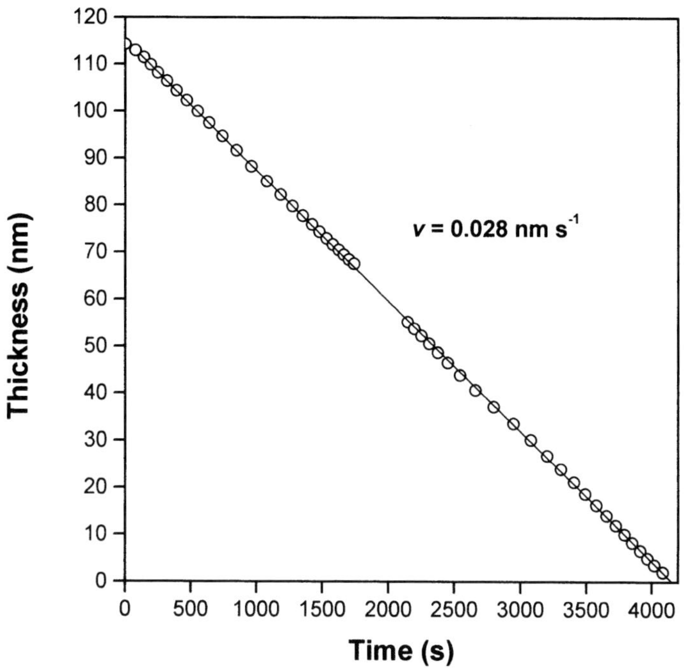

Fig. 4 *Dependence of thickness with time during chemical dissolution in a 0.001M H_2SO_4 + 0.999M Na_2SO_4 solution.*

On the contrary, for the second group (Figs 5 and 6) the Ψ–Δ changes during dissolution show a noticeable hysteresis indicating that the dissolution occurs involving, at least, a second process that takes place simultaneously with the film thinning. The break observed in 1M H_2SO_4 solutions (Fig. 5) would indicate the existence of a duplex oxide film, while the curve-crossing obtained in 1M Na_2SO_4 (Fig. 6) can be associated with development of metal surface roughness. Anyway, the optical data of the initial portion of the Ψ–Δ curve where no appreciable departure is obtained can be used to make an estimation of the initial dissolution rate, while at longer times, more elaborated models of the complex interphase are necessary for calculations.

4. Conclusions

The optical properties of anisotropic WO_3 layers either under anodising conditions or at zero-field, can be obtained through optical analysis varying six parameters in the Simplex algorithm with the only assumption that at increasing thickness the film is homogeneous and of constant refractive index and extinction coefficient.

When the field is switched off, the refractive index in both transverse and in the direction of the field increases, while the corresponding values of the extinction coefficients remain practically constant, showing that anodic WO_3 films grown in

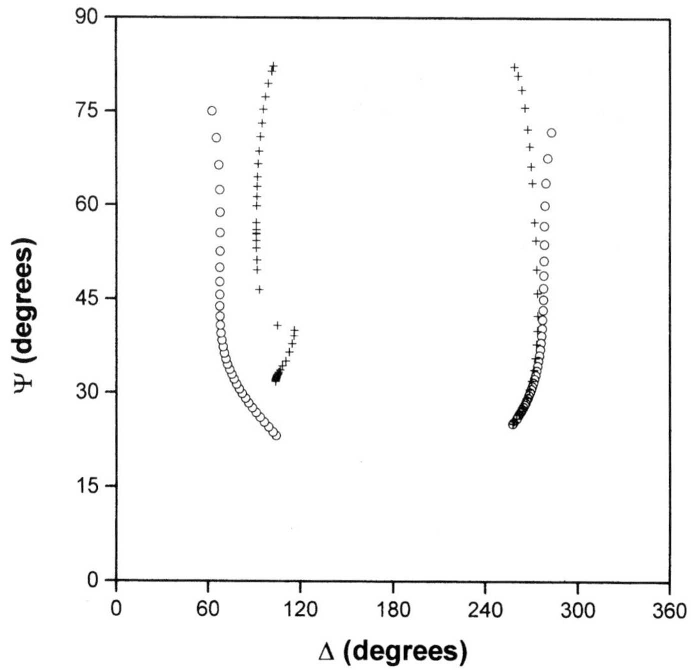

Fig. 5 *Ψ–Δ response for WO_3 anodic growth (o) and chemical dissolution (+) in a 1M H_2SO_4 solution.* j = 2 *mA cm*$^{-2}$.

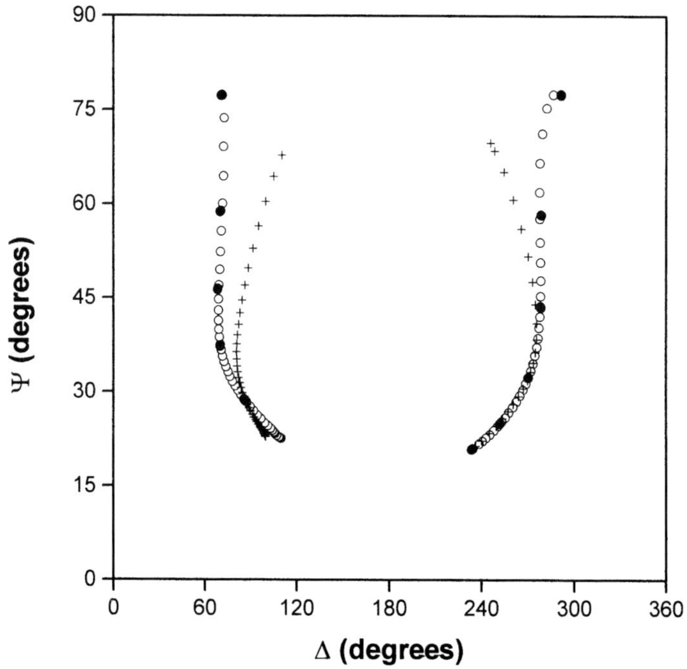

Fig. 6 *Ψ–Δ response for WO_3 galvanostatic growth (o), with current interruption (•) and chemical dissolution at open circuit (+). 1M Na_2SO_4.* j = 2 *mA cm*$^{-2}$.

sulfuric acid aqueous solutions are anisotropic in nature and that the anisotropy remains even in absence of the field.

The optical data allow the conditions to be obtained in which either the chemical dissolution behaves as a single process (film thinning) or other processes such as surface roughening, generation of a hydrated layer, etc., take place simultaneously with the film thinnning. For the first case, the evolution of thickness with time during the dissolution is linear in agreement with a zero-order rate law, i.e. the dissolution rate in the given medium is constant.

Ellipsometry is a powerful technique to provide a direct evidence to establish the conditions in which chemical dissolution processes are a single phenomenon.

5. Acknowledgements

This research has been financially supported by CONICET, CONICOR and SECYT (UNC).

References

1. T. C. Arnoldussen, *J. Electrochem. Soc.*, 1981, **128**, (1), 117–123.
2. J. W. Diggle, T. C. Downie and C. W. Goulding, *Electrochim. Acta*, 1970, **15**, 1079–1093.
3. F. El-Taib Heakal, A. S. Mogoda, A. A. Mazhar and M.S. El-Basiouny, *Corros. Sci.*, 1987, **27**, (5), 453–462.
4. M. M. Hefny, A. S. Mogoda, S. A. Salih and H.E. El-Fiky, *Thin Solid Films*, 1991, **204**, 193–202.
5. M. M. Hefny, M.S. El-Basiouny, A. G. Gad Allah and S. A. Salih, *Electrochim. Acta*, 1983, **28**, (12), 1811–1815.
6. S. M. Abd El-Motaal, N. H. Hilal and W. A. Badawy, *Electrochim. Acta*, 1994, **39**, (17), 2611–2617.
7. A. G. Gad Allah, H. A. Abd El-Rahman and M. M. Abou-Romia, *J. Appl. Electrochem.*, 1988, **18**, (4), 532–537.
8. M. S. El-Basiouny, S. A. Hassan and M. M. Hefny, *Corros. Sci.*, 1980, **20**, 909–917.
9. A. S. Mogoda, M. M. Hefny and G. A. El-Mahdy, *Corrosion*, 1990, **46**, (3), 210–214.
10. J. L. Ord and D. J. De Smet, *J. Electrochem. Soc.*, 1992, **139**, (2), 359–363.
11. J. L. Ord, *J. Electrochem. Soc.*, 1982, **129**, (4), 767–771.
12. D. J. Beckstead, G. M. Pepin and J. L. Ord, *J. Electrochem. Soc.*, 1989, **136**, (2), 362–368.
13. J. L. Ord and D. J. De Smet, *J. Electrochem. Soc.*, 1992, **139**, (3), 728–732.
14. D. J. De Smet and J. L. Ord, *J. Electrochem. Soc.*, 1989, **136**, (10), 2841–2845.
15. M. A. Pérez and M. López Teijelo, Optical anisotropy of WO_3 films, in *Proc. Symp. on Passivity and Its Breakdown*, (1997, Paris, France), (P. M. Natishan, H. S. Isaacs, M. Janik-Czachor, V. A. Macagno, P. Marcus and M. Seo, eds). The Electrochemical Society, New York, USA, 1997, pp.404–410.

16
Composite Coatings with Zirconium Oxide as Dispersed Phase and their Anti-Corrosion Properties

L. BENEA, O. DOSSENBACH*, F. WENGER†, P. PONTHIAUX† and J. GALLAND†

Department of Metallurgy and Materials Science, "Dunarea de Jos" University, 47 Domneasca St., RO-6200 GALATI, Romania

*Labor. Fur Technische Chemie, Swiss Federal Institute of Technology, ETH - Zentrum, CH - 8092 Zurich, Switzerland

†Ecole Centrale Paris, Laboratoire "Corrosion, Fragilisation, Hydrogène", Grand Voie des Vignes, 92295 Chatenay-Malabry cedex, France

ABSTRACT

Electrochemical co-deposition is one of the most important methods of providing composites with non-metallic compounds as a second phase inside a metallic or non-metallic matrix. Composite coatings can be prepared by electrodeposition from an electrolytic solution containing a suspension of insoluble ZrO_2 particles. Interesting applications have been developed for purposes of wear resistance, dry lubrication, anticorrosion, dispersion hardening and electrocatalysis. Electrochemical Impedance Spectroscopy (EIS), Scanning Electron Microscopy and microhardness measurements were used to compare the properties of a zirconia–nickel matrix composite coating with pure nickel coating. The results show that the electrochemical and corrosion properties of the zirconia composite coatings differ from those of pure nickel. Mechanical properties such as microhardness and wear resistance are also changed.

1. Introduction

Several methods have been developed to produce metal matrix composites. The most widely used technique is a mechanical mixing followed by sintering of metallic and non-metallic powders such as oxides or carbides. Another widely used method is the mixing of ceramic powders directly with the fused metals or alloys.

Electrodeposition from a metal plating bath containing a dispersion of particles (such as oxides, carbides, metal powders and plastics) involves the incorporation of these particles into the metal deposit forming a composite coating (CC). Such a coating features the properties of both the metal and the dispersed particles.

These coatings can be considered to be metal–matrix composites (MMC), obtained through electroplating. In recent years special attention has been given to composite coatings obtained by metal co-deposition of various dispersed phases during electrocrystallisation [1–8]. Metal matrix composites are attractive materials because they combine the properties of the metals and dispersed phases: high temperature strength, good resistance to thermal fatigue, good shock resistance, high thermal conductivity and low coefficients of thermal expansion. These properties lead to a range of current and potential uses in the aerospace, automotive, nuclear and electronic industries, in corrosive environments and even as active electrodes. Particle-reinforced metal matrix composites have attracted considerable attention in recent years also because of their relatively low cost and facility for manufacture by co-deposition in an electroplating bath.

This study is focused on the composite system having a nickel matrix with zirconium oxide as dispersed particles, in the average diameter range of 6 μm. The aim of the present work is to describe the effects of dispersed particles on the surface morphology and to correlate these with some properties of the obtained composite coatings.

The literature referring to composite coatings obtained through electrodeposition is characterised in terms of their mechanical properties. There is however a lack of comparative data with respect to the corrosion properties and the influence of different dispersed phases on the metallic matrix electrocrystallisation.

In this work the electrochemical properties like corrosion resistance and porosity are evaluated by linear polarisation and electrochemical impedance spectroscopy methods. The surface morphology and surface composition were studied with a scanning electron microscope and an X-ray analyser (EDAX). The Vickers microhardness was measured on the surface coatings.

2. Experimental Details

2.1. Preparation of Nickel and Nickel Composite Coatings

The nickel coating and nickel matrix composite coatings for study were obtained by electrochemical co-deposition from a sulfate–chloride nickel-plating bath. To ensure an even current distribution during the electrodeposition process, a 300 mL^3 cylindrical electrochemical cell was used having the cylindrically-shaped anode at 20 mm from the cathode surface. The cell provides a good uniformity to the metallic and composite coating, which is necessary for a better study of the co-deposition process. The dispersed particles have uniform access to the cathode surface and so effects of the differing current densities and therefore different coating compositions at the edges are eliminated. The particles were kept in suspension by mechanical or magnetic stirring at a rotation rate from 100–1500 rev/min. Electrodeposition was made on carbon steel, stainless steel, copper and nickel as supports (cathodes) of cylindrical shape and an effective area of about 10 cm^2. Thicknesses of metal and composite deposits were 150 μm for corrosion resistance and microhardness studies and 15 μm for comparative porosity investigations.

2.2. Structural and Chemical Analyses

The amount of dispersed phase in the composite coating was determined by weighing on a micro-balance and stripping the deposit. The mass percentage of the particles in the composite deposition was calculated using the formula:

$$p\% = \frac{m_p}{m_s} 100 \qquad (1)$$

where $p\%$ represents the mass percent of the particles in the composite coating; m_p is the particle mass (g); and m_s is the dissolved composite coating mass (g).
The uniformity of dispersed phase in the composite coatings was examined by light microscopy in cross-section. Scanning electron microscopy (SEM) revealed the comparative surface morphology of coatings and the uniformity of dispersed particles in the composite deposit. The surface composition was evaluated by energy dispersive X-ray analysis (EDAX).

2.3. Electrochemical Studies

Corrosion investigations were conducted using the electrochemical impedance spectroscopy method in the frequency range from 104 Hz to 0.003 Hz with an amplitude of 5 mV. After impedance measurements potentiodynamic curves were plotted with a scan rate of 0.5 mVs^{-1} between the potentials: –600 mV and +600 mV (referred to Ag/AgCl electrode). Corrosion current densities, polarisation resistance, Tafel slops and corrosion potential were determined in naturally aerated 0.5M NaCl solution and 0.5M H_2SO_4 deoxygenated by nitrogen bubbling in the solution. Correlations between impedance diagrams plotted in sulfuric acid solution with the porosity of coatings were examined.

3. Results and Discussion

3.1. Structural Aspects of Nickel matrix Composite Coatings

The dispersed phases change the structure of the metal matrix by co-deposition with the metal. Figures 1–3 show the surface structures of pure nickel electrodeposits and of the composite coatings with different contents of dispersed particles.

The oxide acts as a catalyst for metal reduction resulting in a larger number of active nucleation sites. The modification of the crystallinity in the dispersed zirconium composite coatings compared to that of pure nickel electrodeposited under the same conditions suggests that modification of their properties will also occur. The microstructural study of nickel composite coatings with ZrO_2 contents showed a uniform distribution of particles in the deposit and correlated with a good and homogeneous suspension of particles in the plating bath.
The changes in the surface morphology will have different influences on the reactivity and those other properties of composite and metal surfaces that are directly related to their surface tension, the corrosive environment and/or electrochemical uses.

3. 2. Microhardness of Composite Coatings

The possibility of incorporating ZrO_2 particles in a nickel matrix by electrodeposition leads to a new generation of coatings having interesting mechanical and/or

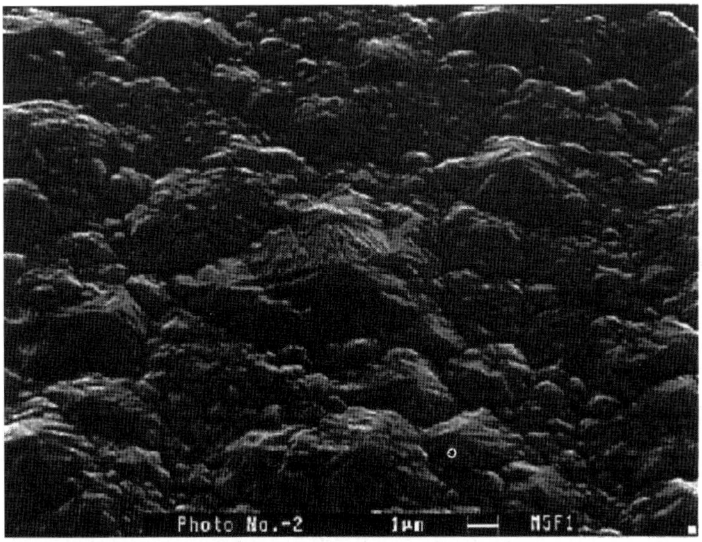

Fig. 1 SEM surface morphology of pure Ni deposit.

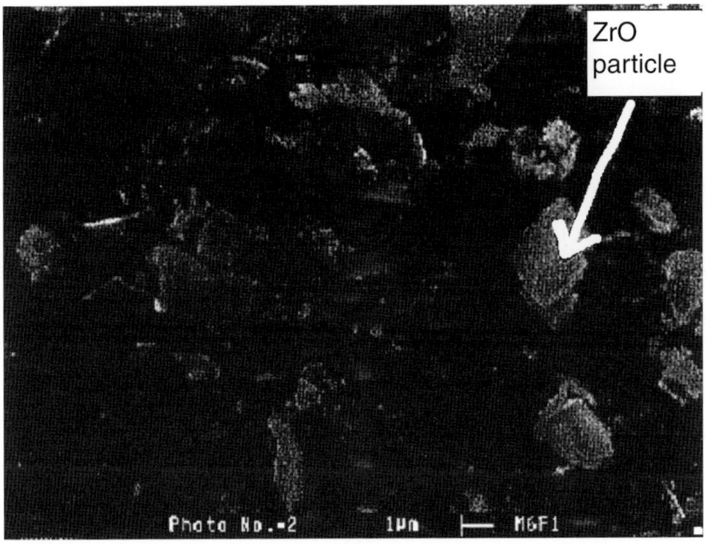

Fig. 2 SEM surface morphology of $Ni+ZrO_2$ coating (18 mass% of ZrO_2).

electrochemical properties. In our work we have studied the Vickers microhardness of composite coatings and compared these to that of pure nickel electroplating. The presence of dispersed particles in the coating increases the microhardness and can be correlated to the amount of dispersed phase in the deposit (see Table 1).

Comparative studies of corrosion wear behaviour (to determine the coefficient of friction) for pure and composite surfaces are still in progress.

Fig. 3 SEM surface morphology of Ni+ZrO_2 composite coating (46 mass% of ZrO_2).

Table 1

Type of coating	Load (g)	D_1 (µm)	D_2 (µm)	HV
Pure Ni	200	42.3	41.5	211
	100	24.9	32.4	226
	200	41.3	41.4	217
Ni + ZrO_2 20% mass	300	25.4	27.1	808
	300	22.5	31.5	763
	300	26.9	26.7	775
Ni + ZrO_2 46% mass	300	20.5	29.0	909
	300	20.3	25.5	1059
	300	24.5	20.3	1110

3.3. Corrosion and Porosity Results in Sulfuric Acid Solution

A protective coating on steel can be anodic or cathodic with respect to the substrate potential. Nickel versus steel is a cathodic coating on anodic substrate. For this reason the corrosion protection afforded by electroplated nickel to steel depends on the complete absence of any coating porosity. In the case of coating porosity corrosion of the substrate (steel) through the pores will occur. The rate of substrate dissolution will depend on the cathodic reaction controlling the corrosion process.

The effect of zirconium oxide particles co-deposited with nickel on the coating porosity was studied with a 15 μm coating thickness of pure nickel and composite on steel surfaces in the same conditions. The samples were tested for corrosion in deoxygenated 0.5M sulfuric acid solution. The impedance data are presented in Figs 4 and 5.

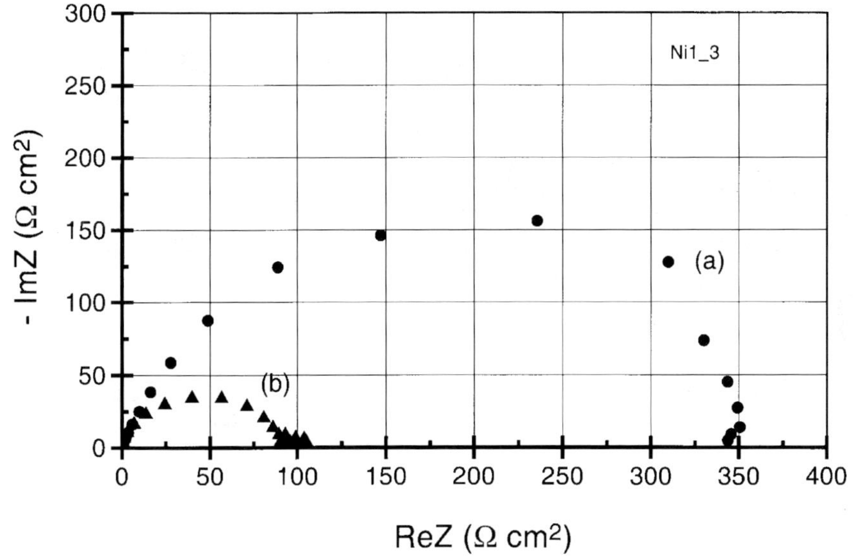

Fig. 4 *Nyquist plot of impedance data of (a) nickel zirconia composite coating, and (b) pure nickel coating in 0.5M sulfuric acid solution.*

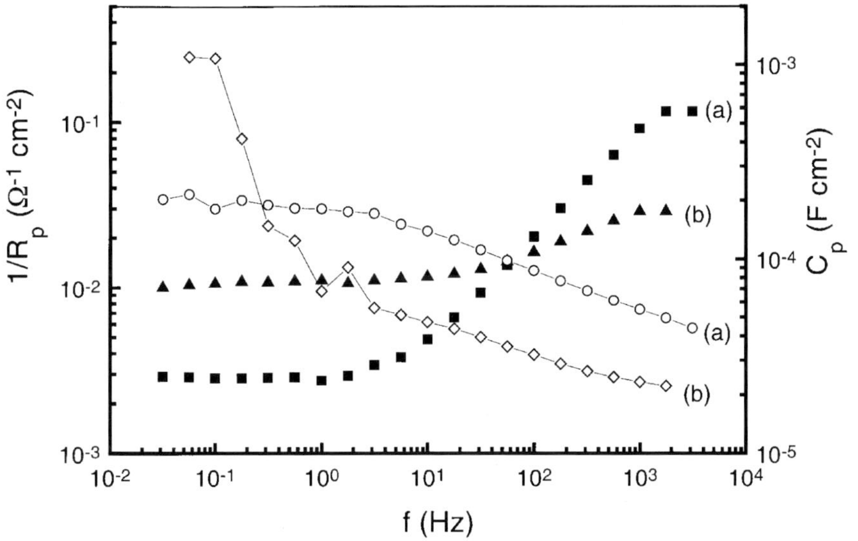

Fig. 5 *(a) Impedance data for the composite coating electrodeposition: (solid square) $1/R_p$ against frequency; (open circle) C_p against frequency, (b) Impedance data for the pure Ni electrodeposition: (solid up triangle) $1/R_p$ against frequency; (open diamond) C_p against frequency.*

The polarisation resistance of the pure nickel coating on steel was $R_p = 100\ \Omega cm^2$ with a corresponding corrosion current density of $i_{corr} = 746\ \mu Acm^{-2}$. The Tafel slopes were: $b_a = 123$ mV/decade for the anodic reaction and $b_c = 53$ mV/decade for the cathodic reaction. The large value of anodic Tafel slope and the high corrosion current density show that the corrosion process occurs on the substrate through coating pores and that the reaction rate is diffusion controlled (diffusion of oxygen through the coating pores). The presence of pores in the pure nickel coating was confirmed by visual inspection of the coating which revealed small red points of iron oxide. The increase in capacitance values at low frequency for the pure nickel coating can be explained in the same way.

The same thickness of zirconia nickel composite coating resulted in freedom from pores. The polarisation resistance of the composite coating was $R_p = 350\ \Omega cm^2$ and the corrosion current density, $i_{corr} = 198\ \mu Acm^{-2}$. The Tafel slopes differed from those of the pure nickel coating: $b_a = 31.8$ mV/decade for the anodic reaction and $b_c = 42.2$ mV/decade for the cathodic reaction. Surface examination of the sample after these measurements revealed no porosity.

We can conclude that the oxide particles co-deposited with nickel improve the protection properties in sulfuric acid solution by increasing the polarisation resistance and also by eliminating possibility of porosity in the thin layer coating.

3.4. Corrosion Behaviour in Sodium Chloride Solution

The results of corrosion testing of pure nickel and zirconia nickel composite coatings in aerated 0.5M NaCl solution, pH = 7, are presented in Figs 6 and 7. The thickness of the coatings was 150 μm in order to eliminate any influence of the substrate.

The polarisation resistance of the composite coating (see Fig. 6(A)), was $R_p = 400\ k\Omega cm^2$. From the potentiodynamic curve, (B) in Fig. 6, the calculated corrosion current density was 73 $nAcm^{-2}$ at the corrosion potential of –213 mV (Ag/AgCl). The corresponding Tafel slopes were: $b_a = 179$ mV/decade and $b_c = 130$ mV/decade for the anodic and cathodic reactions, respectively.

In the case of the pure nickel coating the polarisation resistance value was estimated at $R_p = 200\ k\Omega\ cm^2$ according to the impedance diagram (Fig.7 overlay (A)).

From the potentiodynamic diagram, overlay (B) in Fig. 7, the calculated corrosion current value is $i_{corr} = 178\ nAcm^{-2}$ with a corrosion potential situated at $E_{corr} = -236$ mV (Ag/AgCl). The corrosion potential is shifted about 23 mV to more negative values than in the case of the pure nickel coating. The corresponding Tafel slopes were the following: $b_a = 380$ mV/decade and $b_c = 133$ mV/decade for the anodic and cathodic reactions respectively. The higher value of the anodic Tafel slope for the pure nickel coating, compared with the composite coating, can be due to pitting corrosion. Pitting on the pure nickel surface was observed after corrosion testing although the aim of the experiment was only to compare general corrosion.

We can conclude that zirconium oxide particles embedded during nickel electroplating beneficially affect the corrosion resistance properties compared with pure nickel coatings. The polarisation resistance is about doubled and the corrosion current density halved in the case of the composite coating. A good effect in eliminating pitting corrosion of nickel plating in presence of chloride ions was also found. This conclusion must be proved by additional experiments.

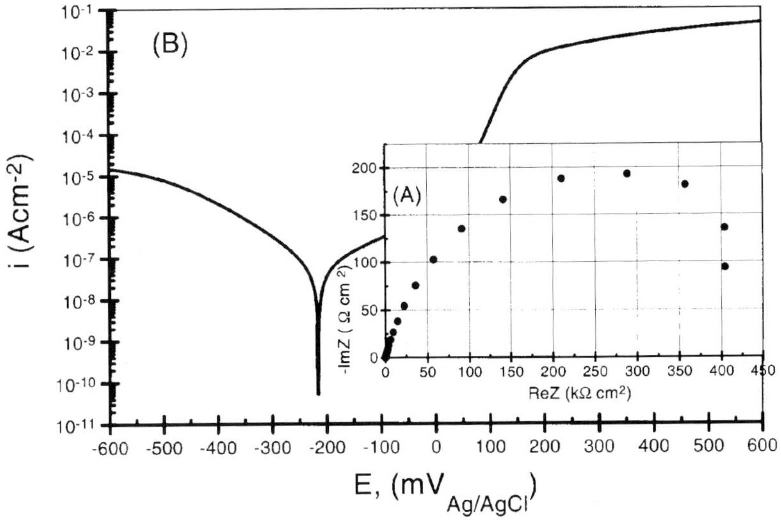

Fig. 6 *Results of corrosion measurements of zirconia nickel composite coating in aerated 0.5*M *NaCl solution: Overlay (A) Nyquist representation of impedance data; (B) Potentiodynamic curve performed at a scan sweep of 0.5 mVs^{-1}.*

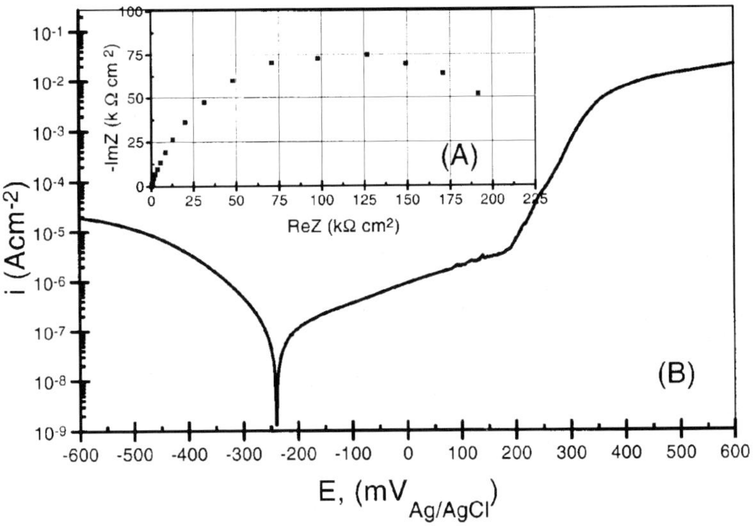

Fig. 7 *Results of corrosion measurements of pure nickel coating in aerated 0.5*M *NaCl solution: Overlay (A) Nyquist representation of impedance data; (B) Potentiodynamic curve performed at a scan sweep of 0.5 mVs^{-1}.*

4. Conclusions

- The experimental work has shown the possibility of improving the properties of nickel electroplating by incorporating zirconium oxide particles in the coating during the electrodeposition process.

- The composite coating with zirconium oxide particles represents a qualitatively new material coating compared with pure deposited nickel.
- The zirconium oxide particles dispersed in the nickel plating bath induce structural modifications in the nickel matrix during co-deposition.
- Modification of crystallinity in the composite coatings as compared with pure nickel plating affect the properties of the composite surface.
- The presence of dispersed particles in the nickel coating influences the microhardness i.e. the higher the amount of dispersed phase in the deposit the greater the microhardness.
- Zirconium oxide particles co-deposited with nickel improve the corrosion protection properties in sulfuric acid solution by increasing the polarisation resistance and also by eliminating any possible porosity in the thin layer coating.
- The ZrO_2 dispersed particles embedded during nickel electroplating have a good effect on the corrosion properties. Thus, compared with pure nickel coatings, the polarisation resistance increases by a factor of two and the corrosion current density is about halved. The possibility of pitting corrosion of nickel plating in presence of chloride ions seems to be eliminated on the composite surface.

References

1. L. Benea, Composite Electrodeposition – Theory and Practice, (PORTO FRANCO, ed.). Romania, ISBN 973 557 490 X, (1998).
2. S. W. Watson, *J. Electrochem Soc.*, 1993, **140**, 2235.
3. G. Maurin and A. Lavanant, *J. Appl. Electrochem.*, 1995, **25**, 1113–1121.
4. J. Fransaer, J. P. Celis and J. R. Roos, *J. Electrochem. Soc.*, 1992, **139**, 413–425.
5. J. Fransaer, J. P. Celis and J. R. Roos, *Met. Finish.*, 1993, **91**, 97–100.
6. L. Benea and G. Carac, *Metall. New Mater. Res.*, 1997, **2**, 1–19.
7. L. Benea, Proceedings volume *Passivity and Its Breakdown* (P. M. Natishan, H. S. Isaacs, M. Janik–Czachor, V. A. Macagno, P. Marcus and M. Seo, eds). Sept. 1997, ISBN 1-56677-179-X.
8. L. Benea, *Mater. Manuf. Process.*, 1999, **14**, (2), 231–242.

Part 4

Passivity

17
Effect of Potential and Sulfate Ions on the Growth of Cuprous Oxide Films on Copper in Neutral Solutions

J. O. ZERBINO, C. TAPIA* and R. O. LEZNA
Instituto de Investigaciones Fisicoquímicas Teóricas y Aplicadas (INIFTA), (UNLP. CONICET. CIC), C. C. 16, Suc. 4, (1900) La Plata (Argentina)
*Inst. Química, Univ. Cat. Valparaíso, Chile

ABSTRACT
The passive behaviour of copper in borax solutions of pH 7.4 has been analysed between $E = -0.32$ V and $E = 0.73$ V (RHE) with the aid of voltammetry, photocurrent, ellipsometry and *in situ* reflectance techniques. Oxide formation can be explained as a sequence of Cu_2O growth (inner layer) and dissolution precipitation of Cu(II), (outer hydrated oxide layer). Significant variations in the amplitude of the photocurrent were detected as a function of the anodic potential limit, the oxide growing time and the addition of 8 mM sulfate to the electrolyte. These effects are related to changes in the compactness and defect or excess of cations accumulated in the inner layer.

1. Introduction

In previous work the passive behaviour of copper has been studied in aqueous solutions at pH 9.2 where the metal shows the most effective passivity. For potentials cathodic with respect to the bulk formation of CuO, the passive oxide layer consists of a duplex structure: a dehydrated inner part of the passive layer (i.p.p.l.) in contact with the metal and an outer hydrated copper(II) oxide layer (o.p.p.l.). The kinetics and growth of the i.p.p.l. are dependent on the anodic potential limit and the growth time [1,2]. This effect is related to variations in the number of cation defects in the Cu_2O layer.

Several works have been reported on the passive films formed on copper in different media. Artificial patinas are commonly applied to protect the metal from further corrosion. The presence of different ions in the electrolyte and the effect of ultraviolet light were found to modify the rate of growth and structure of the patina. Patinas are prepared in acid media and become consolidated with the exposure time by raising the pH from 2.5 to values near neutrality. Sulfate concentration higher than 6 mM seems to promote the formation of cuprite [3,4]. X-ray photoelectron and Auger electron spectroscopies and other techniques are usually applied for patina examination. Most of these methods are capable of producing various kinds of chemical and mineralogical data rather than information about macro-physical properties of the patina [5].

In this work, voltammetry complemented by ellipsometric, reflectance and photocurrent techniques were used to characterise *in situ* the growth of the oxide layers on Cu in the –0.32 to 0.70 V potential region vs (RHE). The influence of the addition of 8 mM sulfate to the electrolyte has been analysed.

2. Experimental

Experiments were made at room temperature in borax buffer solutions of pH 7.4 ($Na_2B_4O_7$ 0.005M, H_3BO_3 0.18M), solution *a*, and in the same solution with the addition of 8 mM Na_2SO_4, solution *b*. Experimental conditions were similar to those described in a previous work [1,2,6,7]. The copper electrode was made by axially fitting a polycrystalline copper rod (99.99% purity) into a Teflon sheath. It was polished to a mirror finish with alumina of 0.3 and 0.05 μm. The sampled ellipsometric area of the electrode, vertically placed in the cell, was about 1 mm^2 and the electrode area 0.4 cm^2. All experiments were performed with bubbling of nitrogen and the potentials were referred to a reversible hydrogen electrode (RHE) in the same solution.

The photocurrent was measured using a lock-in analyser EG&G, PAR 5210. The whole area of the electrode was illuminated (through a quartz window placed in front of the electrode at a distance of 13 mm) with a mercury lamp 100 W, 600 B General Electric H100A 4-T. A filter (Corning Glass Works 7.54) selected the ultraviolet light in the region 240 nm < λ < 400 nm, $\bar{\lambda}$ = 320 nm. The frequency of the chopper was 20 Hz.

Ellipsometry was used illuminating the electrode with monochromatic light in the visible wavelength range (400 nm < λ < 700 nm) with an incident angle of 70°. The ellipsometric parameters Δ (Δ being the phase shift between the parallel and perpendicular components of the electrical vector (E_s, E_p)), and tangent Ψ, the change in amplitudes ratio of the components (E_s, E_p), were obtained upon reflection. Wavelengths were selected by interposing appropriate filters at 405, 450, 492, 546 and 580 nm.

Ultraviolet-visible (UV-VIS) spectra were obtained at an angle of incidence of 45°, using a computerised optical multichannel analyser (EG&G PAR OMAIII) fitted with a cooled Si diode array [8]. The detector is unintensified and made up of 1024 channels. This rapid scan spectrometer, with a 14-bit resolution, was employed to obtain integral spectra resulting from the co-addition in computer memory of a variable number of exposures (depending on signal/noise), each averaged 0.03 s on the diode array chip. Diffraction orders higher than unity were sorted out by appropriate filters.

A freshly polished Cu electrode in the cell attains an open circuit potential between 0.670 and 0.690 V. The experimental procedures were as follows:

Procedure A: The ellipsometric parameters were obtained after 5 min of holding the potential at E_c = – 0.312 V. The reference electrode was a Pt/H_2 cell in the same electrolyte connected to the system through a Luggin-Haber capillary tip. The ellipsometric parameters Δ and Ψ, and the amplitude, δi, of the photocurrent were recorded as a function of potential, E, and the time, τ. The potential was scanned from E_c up to E_a at a sweep rate ν = 0.5 mV s^{-1} followed by a potential holding at E_a during a time τ. A cathodic scan from E_a up to E_c was subsequently performed.
Procedure B: Reflectance spectra were taken during a potential program similar to

that of procedure A between E_c and E_a at a sweep rate $\nu = 5$ mV s^{-1} or 1 mV s^{-1}. The potential was measured against a saturated calomel electrode, connected through a ground glass stopcock. In the text all potentials are referred to vs RHE. In all cases spectra were calculated as $(R-R_{ref})/R_{ref}$ with the reference spectra, R_{ref}, taken at −0.320 V. The system optical response is plotted as $\Delta R/R$, where R is the surface reflectance, as a function of wavelength, λ.

3. Results and Discussion

Figure 1 shows the simultaneous detection of the photocurrent, δi, and the ellipsometric parameter Δ during the formation of the Cu_2O layer; δi is measured following procedure A by illuminating the electrode with λ = 320 nm (Fig. 1a). The current/potential voltammogram is very similar to that reported in previous experiments [1,2,5]. The parameter Δ measured with incident light of λ = 580 nm is plotted in Fig. 1(b) as a function of the potential E. The potential is held at $E_a = 0.685$ V for a time τ = 10 min. A higher change in δi and in Δ is observed in solution *b* compared to solution *a*. The Δ/Ψ values measured after oxide reduction are close to the initial values corresponding to the bare surface.

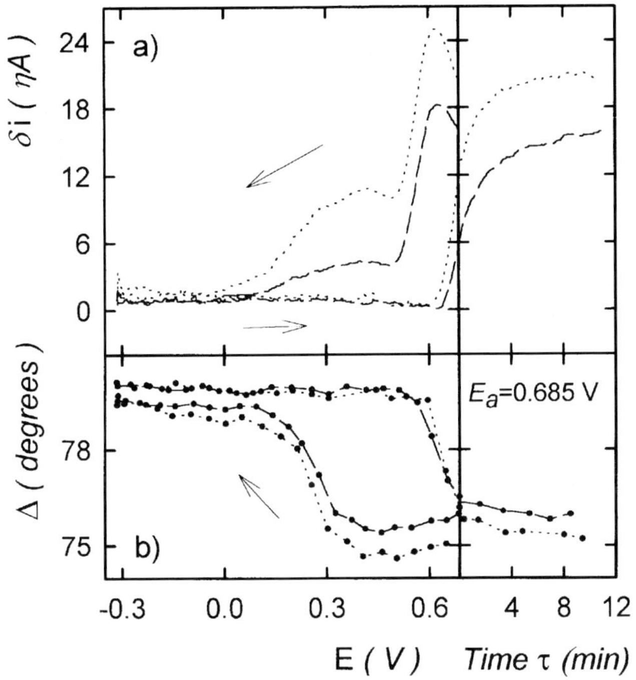

Fig. 1 Photocurrent δi and ellipsometric phase Δ during the first potential scan between $E_c = -0.320$ *V and the anodic limit* $E_a = 0.685$ *V at* $\nu = 0.5$ *mV s*$^{-1}$. *At* E_a *the potential was held for a time τ = 11 min followed by the start of the negative-going scan. - - - - - : solution* a, *borax pH 7.4; .... : solution* b, *borax pH 7.4 with addition of 8 mM* Na_2SO_4.

During the anodic scan (Fig. 1), Δ decreases whereas Ψ remains constant (not shown). A comparison between ellipsometric experimental results and calculations considering the growth of the i.p.p.l. and the o.p.p.l. layers allows these processes to be identified in different potential regions [1]. The decrease in Δ (at quasi constant Ψ) may be considered as a measure of the increase in thickness of the i.p.p.l. for these interfaces. The experiment shown in Fig. 1 verifies the reversible (no kinetics implied) formation and reduction of the i.p.p.l. [1,2,7,9].

The $\delta i/E$ plot shows two well defined peaks during the reduction scan. The photocurrent peak observed during the cathodic scan between 0.50 V and 0.15 V is three times higher in the presence of sulfate than in borax. The growth of the i.p.p.l. gives rise to a photocathodic current. The Cu_2O layer is a *p*-type semiconductor due to a deficiency of copper [7,10]. The Δ/E profiles observed in Fig. 1(b), in the presence and in the absence of sulfate in solution are very similar. The δi maximum is about 20% greater in solution *b* than in solution *a*.

For thin films the simplest modelling is to assume the photocurrent to be proportional to the absorption coefficient, *k*, and to the thickness *d*. However, accumulation of charge at boundaries and traps may demand a more complex relationship [11–14]. The $\delta i/E$ profile of Fig. 1(a), may be related to changes in the space charge and bias potential during the growth and reduction of the i.p.p.l. layer.

The evolution of δi under procedure A with different anodic limits E_a = 0.645 V and E_a = 0.685 V is shown in Fig. 2. Higher values for δi are observed in solution *b*

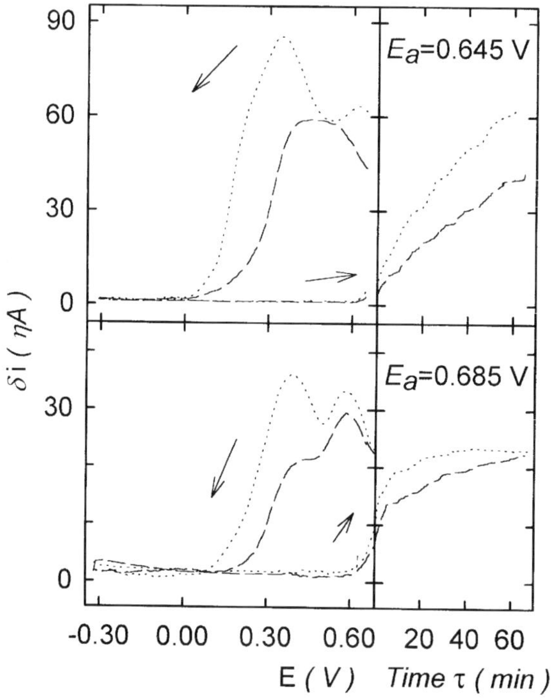

Fig. 2 *Evolution of* δi *as a function of potential and time during scans at 0.5 mV* s^{-1} *between* E_c *and* E_a *= 0.645 V and* E_a *= 0.685 V. - - - - - : borax solution* a*, ... : sulfate solution* b.

than in solution *a* during the entire growth and reduction processes. For $E_a = 0.645$ V δi increases continuously with the time τ while for $E_a = 0.685$ V it attains a plateau for $\delta i = 23$ nA. The higher increase of δi at $E_a = 0.645$ V may be related to an increase in thickness or optical absorption developed during the potential holding at E_a [12,13]. In Fig. 3 are plotted the ellipsometric data measured at the constant potential $E_a =$ 0.645 V and $E_a = 0.685$ V corresponding to the experiments of Fig. 2. The changes in Δ, $\delta\Delta$, are higher in the case of solution *b* than for solution *a*. The maximum value for $\delta\Delta$, reached at the more anodic potential 0.685 V, is smaller than that at 0.645 V. This behaviour indicates a thinner or less compact i.p.p.l. at the higher potential. In the same way the higher $\delta\Delta$ observed in the presence of sulfate at both potentials indicates a thicker and/or a more dense i.p.p.l. in the presence of sulfate. The Δ/Ψ values measured at $E_c = -0.312$ V after keeping the potential at E_a for a time $\tau = 65$ min are different from those corresponding to the fresh polished electrode. Figure 3(a) shows an increase in Ψ (about 0.5 degree) and values of Δ similar to those of the bare surface. Theoretical simulations show that this effect does not indicate a change in roughness. Instead it agrees with the growing of an external layer of hydrated oxide o.p.p.l. [1]. The different Δ/Ψ slopes related to the Δ/Ψ values corresponding to the bare electrode depend on the compactness and optical density of the o.p.p.l. [15]. This increase in Ψ is not observed in experiments with times τ equal or lower than 10 min (Fig. 1). The Δ/Ψ values measured after reduction in Fig. 3(b), in the presence of sulfate indicate a small increase in roughness [1]. The formation of the o.p.p.l. and the changes in

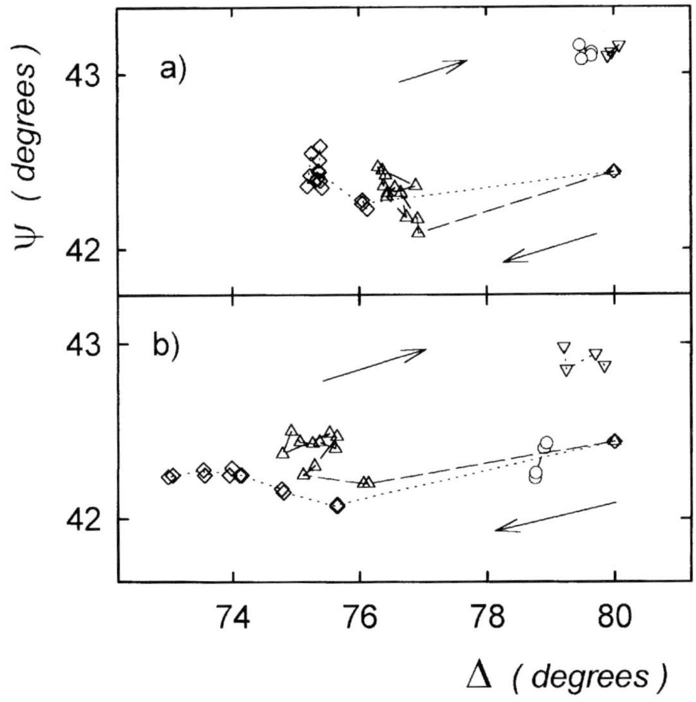

***Fig. 3** Δ/Ψ plot corresponding to Fig. 2. (a) $E_a = 0.685$ V, (b) $E_a = 0.645$ V. (△) solution* a *at E_a; (◇) solution* b *at E_a; (▽) solution* a *at E_c; (o) solution* b *at E_c.*

roughness are irreversible processes. The changes in the Δ/Ψ values measured after reduction in Fig. 3 are small in comparison with the total change in Δ/Ψ observed at E_a. Note the higher sensitivity in the Ψ scale in relation to that of Δ.

The Δ/Ψ values measured at E_a correspond to the formation of the i.p.p.l. Changes in the optical response attributed to the growth of the o.p.p.l. and increases in roughness are not detected for $\tau = 10$ min. and are still small for $\tau = 65$ min. On the other hand, previous work has shown that an important contribution to the o.p.p.l. dissolution–precipitation process takes place during the reduction scan of the i.p.p.l. in addition to that occurring during the potential hold at E_a [1]. A single layer model was assumed to evaluate the structure of the i.p.p.l. during the potential detention at E_a. Figure 4 shows the predicted values of Δ/Ψ for different optical constants n and k, $1.41 < n < 1.83$, $0.04 < k < 0.10$, for data measured on illuminating with $\lambda = 580$ nm. The dashed line corresponds to 3 nm increases in d. The dotted line in Figs 4(a) and 4(b) corresponds to changes in n for the same d values and in Fig. 4(c) to each 0.02

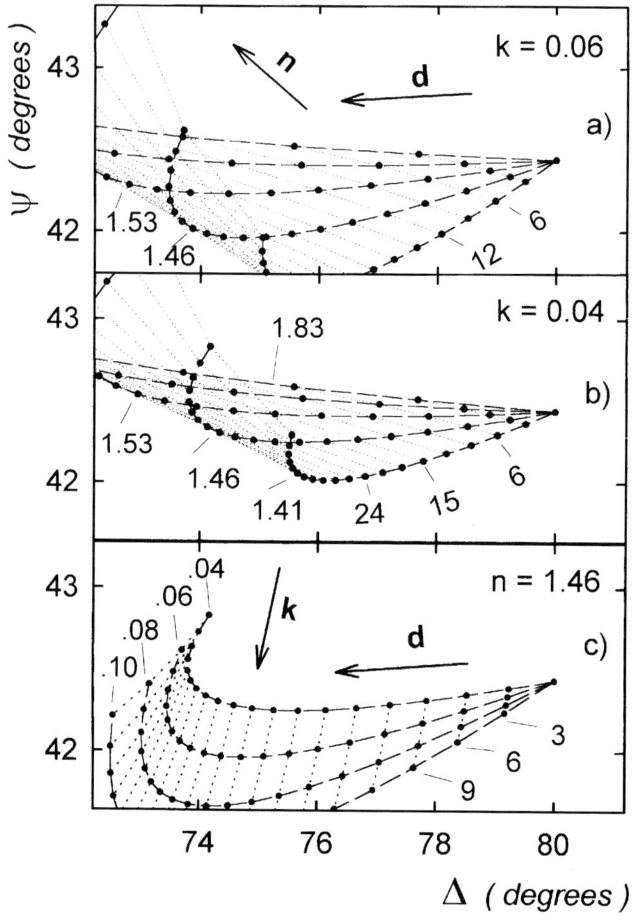

Fig. 4 Calculated values of Δ/Ψ for a single film. The filled circles correspond to d *increases each 3 nm. (a)* n = *1.41, 1.46, 1.53, 1.63, 1.83 and* k = *0.06. (b)* n = *1.41, 1.46, 1.53, 1.63, 1.83 and* k = *0.04. (c)* n = *1.46,* k = *0.04, 0.06, 0.08, 0.10.*

increase in k, at constant d. The comparison of Figs 3 and 4 shows that the changes in the Δ/Ψ measured at E_a may be related to either changes in d or increases in n of the i.p.p.l.

Assuming a single homogeneous film, the fit of the Δ/Ψ optical measurements taken at different wavelengths allows the unequivocal determination of d, n and k [16,17]. This simple model evaluates the thickness (d), compactness (n) and optical absorption (k) of the i.p.p.l. and estimates the changes of these parameters during the time τ. Although the i.p.p.l. may present a complex structure and a small contribution of the o.p.p.l. cannot be disregarded, this single film model leads to real average d, n and k values of the passive layer growth at E_a. The optical indices corresponding to solutions with and without the addition of sulfate are plotted in Figs 5 and 6. The n and k values corresponding to solutions *b* are higher than those of solution *a*. This indicates a more compact i.p.p.l. in the presence of sulfate [15]. The values of n and k corresponding to $E_a = 0.645$ V are higher than those of $E_a = 0.685$ V. This indicates a less compact film or a more porous i.p.p.l. structure at more anodic potentials. The stepwise increase of d, assuming n and k to remain constant over four thicknesses, leads to unequivocal solutions of the ellipsometric equations [16,17]. This method allows the dependence of the optical indices with τ to be estimated. Figure 5 shows a small increase of d with τ. In the case of the experiment of the Fig. 6 in solution *b* with $E_a = 0.645$ V a net increase of n and k with time is observed attended by values of d of about 15 nm. This indicates a d value independent of τ. In the case of $E_a = 0.685$ V the fitted n and k are smaller than those at $E_a = 0.645$ V and a decrease in n and k attended by a small increase in the d values for longer durations can be observed. This indicates a less compact i.p.p.l., which increases the porosity and thickness during the potential holding.

A limiting film thickness was observed at pH 9.2 for potentials positive to 0.650 V. This potential corresponds to the transition between region II and region III. Region II is characterised by a parabolic growth of the i.p.p.l. The cathodic limit of region II corresponds to the detection of the first protrusions assigned to the growth of oxide nuclei [1,9]. In region III a limiting thickness of the i.p.p.l. was detected at potentials positive to 0.650 V while above 0.700 V a sharp increase of Ψ indicates the growth of the o.p.p.l. related to a dissolution–precipitation process [1,2].

The photocurrent peak feature shown in Fig. 2 for $E_a = 0.685$ V in solution *a* is similar to that found in a previously reported result obtained at pH 9.2 [7] and consistent with those obtained after reduction of bulk CuO layers [13]. At 0.600 V a noticeable increase of δi with respect to $\delta\Delta$ was observed. This potential is close to the zero current and to the flat band potentials. The photocurrent peak at 0.600 V on the back scan indicates that other processes are being detected in the photocurrent response which are not reflected in the ellipsometric parameters measured at 580 nm.

The higher δi observed at $E_a = 0.645$ V in comparison with that at $E_a = 0.685$ V may be related to a more compact i.p.p.l.. The potential located between the two $\delta i/E$ peaks defined in Fig. 1(a) is in agreement with the more cathodic potential of the dehydrated Cu/CuO redox couple 0.570 V related to the hydrated Cu/CuO couple 0.609 V [1]. These effects may be attributed to different compactness. Brochantite ($Cu_4SO_4(OH)_6$) and cuprite (Cu_2O) patches were detected in thicker patinas grown at lower pH [3,4]. The presence of sulfate increases the development of the double

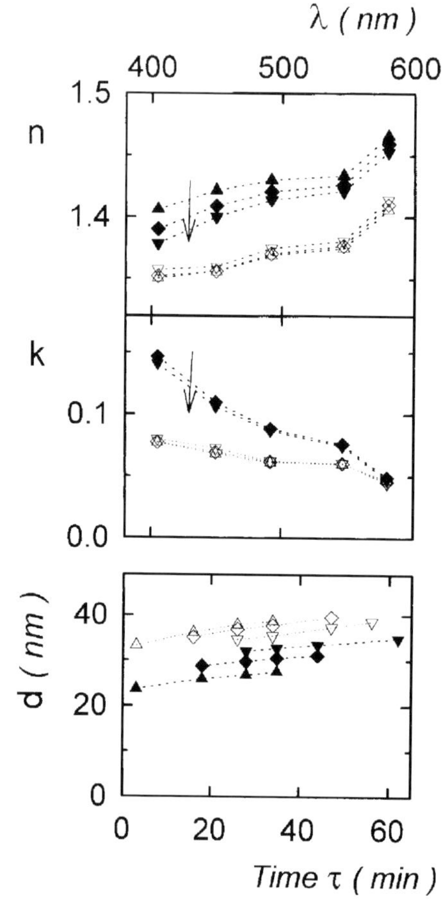

***Fig. 5** Values of* n *and* k *fitted for* $E_a = 0.685$ *V. Solution a open symbols. Solution* b *filled symbols. Lower thickness range : △▲, higher thickness range: ▽▼.*

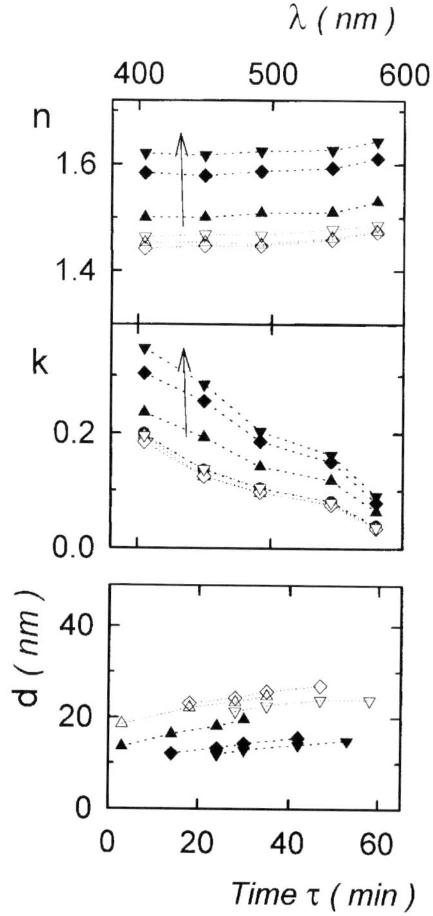

***Fig. 6** Values of* n *and* k *fitted for* $E_a = 0.645$ *V. Solution* a *open symbols. Solution* b *full symbols. The arrows indicate the fit for sets of data corresponding to increasing time* τ.

peak shown at pH 7.4 which may be attributed to regions in the i.p.p.l. of different porosity or compactness (Figs 1 and 2).

Figure 7 shows the changes in relative reflectance against wavelength taken each 100 mV between –0.312 V and 0.705 V (Procedure B). A single peak is observed at ~ 566 nm at the more negative potentials. As the potential is made more positive another maximum starts developing at *ca.* 420 nm that grows in intensity faster than the signal at 566 nm therefore becoming rapidly dominant. Bands at 460 nm and at 520 nm have been reported by Wieder and Czanderna [18] for Cu_2O, the absorption in the blue region being less intense than in the green. The higher relative absorption at *ca.* 420 nm detected in our samples as the potential is being made more positive may probably arise from the development of the $Cu_{0.67}O$ phase [18] that presents a strong absorption in the 400 nm region . The $Cu_{0.67}O$ absorption coefficient, *k*, has

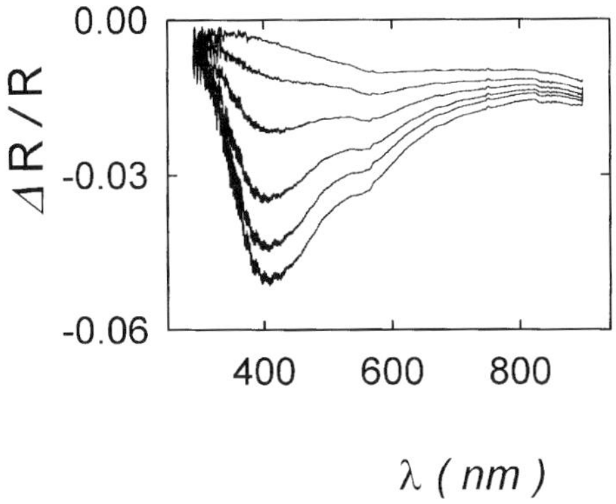

Fig. 7 $[(R - R_{-0.312\ V})/R_{-0.312\ V}]$ *during a potentiodynamic scan at* $v = 5\ mV\ s^{-1}$ *between* $E_c = -0.312$, $E_a = 0.705$ *V.*

been measured to be about ten times as large as that of Cu_2O practically throughout the ultraviolet-visible region.

The optical responses during the growing and reduction of cuprous oxide species are plotted in Figs 8 and 9 as a function of time. The potential/time programme applied to the electrode is shown in the upper part of the diagram. The relative reflectance at 421 nm and 566 nm, $\Delta R/R$, presents a clear dip at the potential where the surface starts getting oxidised. The optical response is reversed on the back scan of the linear potential sweep in an almost reversible way, that is, the surface reflectance is seen to go back to the initial value. The spectra are very similar in solutions *a* and *b*. However the change with the potential of $\Delta R/R$ in solution *b* is about 50% greater than that in solution *a* (Fig. 8). This indicates a more dense i.p.p.l. in the presence of sulfate.

Figure 9 exhibits a potential perturbation that includes a time holding τ of 35 min at 0.720 V. The lower frame has two traces corresponding to the 421 nm and 566 nm reflectances. The absorption for the two photon energies exhibit an abrupt increase at the onset of the oxide formation followed by a much smaller growing slope. The crossing of the two reflectance lines at a given potential is likely to stem either from the presence of two species in the film or from a change of the oxide stoichiometry with potential. The start of reduction of the oxide layer brings about an interesting phenomenon. The absorbance, that is supposed to decrease with the oxide layer thickness, in fact *increases* during the first stages of reduction, the expected steep fall in absorbance is observed sometime later when the oxide layer is stripped off the surface. The initial increase in absorbance upon reduction may be due to the generation of copper ion vacancies in the lattice. It has been shown [19] that increments in absorbance are directly related to the amount of excess oxygen in nickel oxide structures. A carrier hopping process involving metal vacancies has been proposed to account for the high-frequency conductivity of that material [19].

The features of the *n* and *k* spectra corresponding to Figs 5 and 6 are in agreement with that reported for Cu_2O [1,2,18,20]. However, the *n* and *k* values are lower than those reported for Cu_2O thereby indicating porosity. The increase of the absorption at $\lambda = 420$ nm in comparison to that at $\lambda = 570$ nm shown in Fig. 8 indicates the presence of the $CuO_{0.67}$ phase. This phase has a structure similar to that of Cu_2O

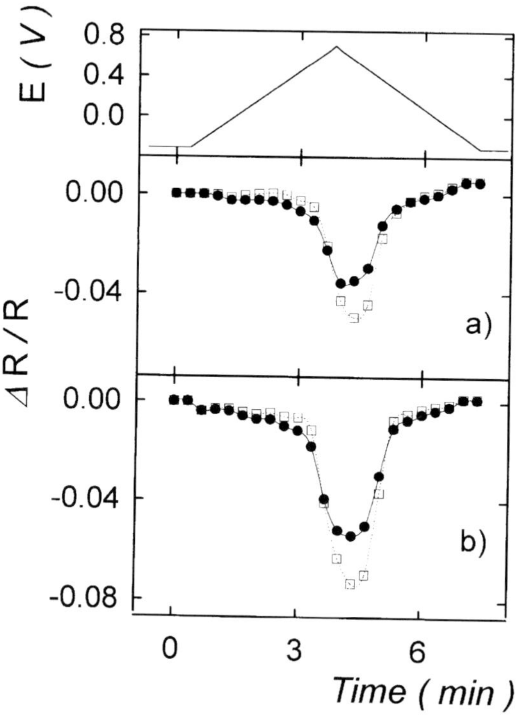

Fig. 8 ΔR/R *intensities as a function of time/potential.* $v = 5\ mV\ s^{-1}$. *(a) solution* a, *(b) solution* b. *(□)* $\lambda = 566$ *nm and (●)* $\lambda = 421$ *nm. Potential/time programme as in upper frame.*

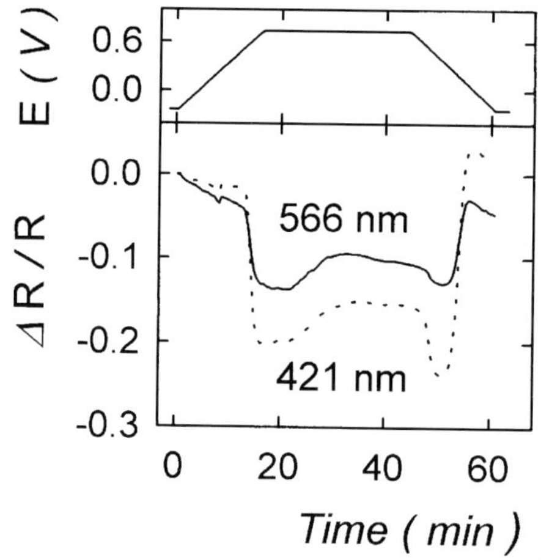

Fig. 9 ΔR/R *intensities at* $\lambda = 566$ *and* $\lambda = 421$ *nm as a function of time/potential.* $v = 1\ mV\ s^{-1}$ *Solution* a. *Potential/time programme as in upper frame.*

with defects, which have the same X-ray diffraction and electron diffraction lines as Cu_2O and shows at $\lambda \approx 400$ nm an absorption coefficient at least four times higher than that of Cu_2O [21–23].

The presence of sulfate in solution leads to a larger photocurrent, reflectance and ellipsometric response which probably indicates a larger i.p.p.l. density. This effect is only observed when the oxide is formed at a slow sweep rate. On the other hand, at potentials positive to 0.65 V an increase in the formal oxidation state (that is a higher excess of Cu(II) in the i.p.p.l. or a larger deficiency in copper) was observed through photocurrents and reflectance measurements.

The high anodic potential and the dissolution process [1,7] promote the deficiency of copper. On the other hand, the presence of sulfate in solution leads to the co-precipitation of sulfate basic salts that gives rise to a barrier hindering the copper dissolution [3,4,24]. This effect decreases the deficiency of copper in the i.p.p.l. by promoting the formation of Cu_2O which, in turn, leads to a larger i.p.p.l. density.

4. Conclusions

Photocurrent–potential profiles in borax buffer solutions depend on the anodic limit E_a in a potential scan and on the presence of sulfate in solution. The appearance of double peaks in the $\delta i/E$ negative-going sweep is probably related to patches of different compactness. The oxide formed at slow sweep rates in the presence of sulfate ($5 > \nu > 0.5$ mV s^{-1}) shows an increase of about three times in the photocurrent during the oxide reduction as compared with the sulfate-free solution. The complex structure of the photocurrent peaks depend on the thickness, the optical absorption and the compactness of the barrier oxide layer. The compactness of the inner part of the passive layer (i.p.p.l.) decreases with increasing potentials. The higher δi in sulfate-containing solutions in comparison with sulfate-free solutions is related to an increase in optical absorption and a higher compactness of the i.p.p.l.

5. Acknowledgements

The authors are grateful to the Organización de Estados Americanos (OEA), the Comisión de Investigaciones Científicas de la Prov. de Bs. As. (CIC), and the Consejo Nacional de investigaciones Científicas y Técnicas (CONICET) for grants in support this work. J. O. Z. is member of the Research Career of CIC and R. O. L. of the Research Career of CONICET.

References

1. J. O. Zerbino and M. F. L. de Mele, *J. Appl. Electrochem.*, 1997, **27**, 335.
2. M. F. L. de Mele, M. R. Viera and J. O. Zerbino, *J. Appl. Electrochem.*, 1997, **27**, 396.
3. R. Vera, G. Layan and J. I. Gardiazabal, *Bol. Soc. Chil. Quim.*, 1995, **40**, 149– 156.
4. B. Rosales, R. Vera and G. Moriena, *Corros. Sci.*, 1999, **41** (4), 625– 651.
5. A. O. Salnick and W. Faubel, *Appl. Spectroscopy*, 1995, **49** (10), 1516.
6. R. O. Lezna, S. Juanto and J. H. Zagal, *J. Electroanal. Chem.*, 1998, **452**, 221.
7. J. O. Zerbino, *Electrochim. Acta* (Special Issue on Electrified Interfaces) 1999, **45**, 819–825.
8. K. Rajeshwar, R. O. Lezna and N. R. De Tacconi, *Anal. Chem.*, 1992, **64**(7), 429A.

9. V. Maurice, H. H. Strehblow and P. Marcus, *J. Electrochem. Soc.*, 1999, **146** (2), 524.
10. G. D. Zhou, H. Shao and B. H. Loo, J. Electroanal. Chem., 1997, **421**, 129.
11. A. Aruchamy and A. Fujishima, *J. Electroanal. Chem.*, 1989, **272**, 125.
12. H. Gerischer, *Corros. Sci.*, 1989, **29**, 257.
13. L. M. Peter, *Ber. Bunsenges Phys. Chem.*, 1987, **91**, 419.
14. W. Paatsch, *Ber. Bunsenges. Phys. Chem.*, 1977, **81**, 645.
15. O. A. Albani, L. M. Gassa, J. O. Zerbino, J. R. Vilche and A. J. Arvia, *Electrochim. Acta*, 1990, **35** (9), 1437.
16. A. E. Bolzan, J. O. Zerbino, E. Macchi and A. J. Arvia, *Thin Solid Films*, 1993, **233**, 82.
17. J. O. Zerbino, W. J. Plieth and G. Kossmehl, *J. Appl. Electochem.*, 1991, **21**, 935.
18. H. Wieder and A. W. Czanderna, *J. Appl. Phys.*, 1966, **37**, 184.
19. D. P. Snowden and H. Saltsburg, *Phys. Rev. Lett.*, 1965, **14**, 407.
20. B. Karlsson, C. G. Ribbing, A. Ross, E. Valkonnen and T. Karlsson, *Phys. Scr.*, 1982, **25**, 826.
21. R. L. Deutscher and R. Woods, *J. Appl. Electrochem.*, 1986, **16**, 413.
22. R. G. Greenler, R. R. Rahn and J. P. Schwartz, *J. Catalysis*, 1971, **23**, 42.
23. J. Y. Malvault, J., Lopitaux, D. Delahaye and M. Lenglet, *J. Appl. Electrochem.*, 1995, **25**, 841.
24. D. Tromans and T. Ahmed, *J. Electrochem. Soc.*, 1998, **145** (2), 601.

18
Passive Salt Film Formation on Silver Electrodes

V. BRUNETTI, H. M. VILLULLAS and M. LÓPEZ TEIJELO
INFIQC, Depto de Fisicoquímica, Facultad de Ciencias Químicas, Universidad Nacional de Córdoba, Pabellón Argentina, Ala 1, Ciudad Universitaria. 5000 Córdoba, Argentina

ABSTRACT

The formation of Ag_2CrO_4 films on silver electrodes takes place through a dissolution–precipitation mechanism. The early stages of growth of the salt film involve a three-dimensional nucleation process with growth controlled by diffusion of chromate ions. After nucleation, the growth is controlled by the resistance of the electrolyte in the pores of the layer until the surface is almost completely covered. The passive film formed under ohmic control is porous and further growth occurs under diffusion control of chromate ions into the pores. The properties of passivating layers depend on the formation conditions. Changes of the film electric properties with time are associated with variations in the layer structure.

1. Introduction

The electroformation of anodic films has been widely studied and it is well known that the kinetics of growth and the nature of the passive state are determined by the chemical, electrical and structural properties of the new phase. The kinetically dominant process for the growth of anodic oxides on valve metals is the migration of ions through the oxide film where resistance increases with thickness. In contrast, on active metals like silver and copper, the formation of passive films involves either nucleation of a new phase on the surface by a solid state mechanism or results from the saturation of the interfacial region by a dissolution–precipitation mechanism. The kinetics of the passivation process are determined, in both cases, by surface coverage and film properties.

Salt film formation during anodic polarisation of active metals is related to electrochemical processes of economic importance such as pitting corrosion, electrochemical machining and electrolytic polishing and has received considerable attention. It has been shown that, in general, the active–passive transition is caused by the growth of an insulating or poorly conducting layer. Among the numerous examples where passivation is produced by the formation of an insoluble salt that can be found in the literature, it is worth mentioning the oxidation of Cd to form CdS [1], the formation of $PbSO_4$ on Pb in sulfuric acid [2] and the passivation of Cu in solutions containing chloride [3] or Ag in electrolytes containing sulfide, halides, chromate, etc. [4–11]. In particular, the electroformation and electroreduction of insoluble silver salts is interesting due to its relation to galvanic cells, sea water batteries, second class reference electrodes and corrosion and passivation of silver in different media.

In this work, the active–passive transition of silver caused by the formation of a poorly conducting film of Ag_2CrO_4 is described. It is shown that the properties of passive layers are dependent on the conditions of formation and time.

2. Experimental

The electrochemical measurements were performed in a three-compartment electrolysis cell using a gold sheet as counter electrode and a $Hg/Hg_2SO_4/Na_2SO_4$ (1M) as reference electrode. Nevertheless, all potentials are referred to the standard hydrogen electrode (SHE). The working electrode consisted of a polycrystalline silver rod (Koch–Light, 99.999% purity) of 2 mm dia. mounted in a Teflon holder which exposes a circular area. Before the experiments the electrode surface was abraded with emery paper and then polished mechanically with alumina (1, 0.3 and 0.05 μm) on a polishing cloth (Microcloth, Buehler). The electrode was cleaned repeatedly with purified water, immersed in the working solution and cathodised at 0.115 V for 10 min. Potentiodynamic curves were obtained by applying single or repetitive triangular potential sweeps at different scan rates (ν). The EIS measurements were carried out at the open circuit potential employing an Impedance Spectrum Analyzer Zahner IM5D. The amplitude of the perturbation signal was 10 mV and the frequency range was 0.1 Hz–200 kHz.

Ellipsometric data were obtained by using a Rudolph Research rotating-analyser automatic ellipsometer (vertical type, 2000 FT model) equipped with a 75 W tungsten lamp as light source and a filter of 632.8 nm. The working electrode was mounted in a Teflon holder and placed horizontally in the cell compartment provided with two plane glass windows adequate for optical measurements. The light reflected by the surface of the electrode passed through a rotating analyser to a photomultiplier tube for detection. The sinusoidal output of the photomultiplier was digitised by a HP216 microcomputer and the values of the ellipsometric angles Ψ and Δ (defined in terms of the ratio of the electric-field amplitudes and phase of the perpendicular and parallel components of the reflected polarised light) were calculated by the computer from the three Fourier coefficients of the photomultiplier signal. This signal was averaged twice and data were collected in continuous mode during the experiments (*ca.* 1 s between points). All measurements were performed at an incidence angle of 70.00 degrees.The Ψ–Δ plots were analysed according to different models.

Solutions of Na_2CrO_4 were prepared from AR chemicals and purified water (Milli Ro-Milli Q system). Borax was added to the solutions (pH 9.3) to ensure that the change in CrO_4^{2-} concentration by formation of $Cr_2O_7^{2-}$ can be neglected and to avoid interference of the silver oxide formation. Measurements were performed at 25°C under nitrogen gas saturation.

The morphology of the surface layer was characterised by scanning electron microscopy (SEM).

3. Results and Discussion

The general voltammetric behaviour of a polycrystalline silver electrode in a solution containing chromate is shown in Fig. 1. The positive scan exhibits an anodic current peak that corresponds to the formation of a silver chromate film followed by a more or less sharp decrease in current due to the fact that the silver salt layer is insulating or poorly conducting and produces electrode passivation. In the negative scan, a cathodic current peak which is attributed to the reduction of the electroformed layer is observed.

Although the potentiodynamic *j/E* profile is apparently simple, the electroformation of the passivating salt layer is a rather complex process which occurs through a sequence of different processes starting with the active dissolution of silver in potential region I of the figure:

$$Ag \rightarrow Ag^{+}\,(aq) + e^{-} \tag{1}$$

Active dissolution of silver produces a continuous increase in the concentration of silver ions near the electrode surface until a critical value is reached and the precipitation of the salt layer that passivates the electrode takes place. The early stages of growth of the Ag_2CrO_4 films also involve a nucleation process. From the analysis of the rising parts of current transient experiments [12], it was shown that a

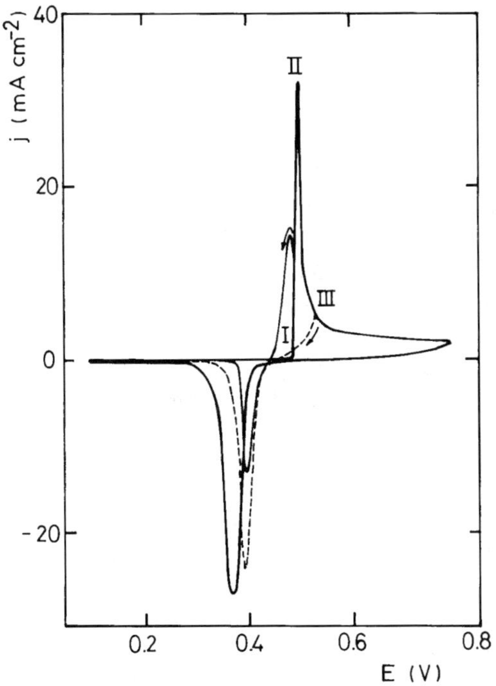

Fig. 1 *Potentiodynamic j/E profile for polycrystalline silver in 0.1M Na_2CrO_4 + 0.1M borax (pH 9.3) at 0.05 Vs^{-1} for different positive switching potentials.*

three-dimensional progressive nucleation process with diffusion controlled growth takes place.

As the salt film grows, the surface coverage increases, the passivating film acts as a mechanical barrier to the current flow and the rate is controlled by the resistance of the electrolyte in the pores (region II). The experimental data follow closely the behaviour predicted for the layer-pore resistance model indicating that the rate of the overall process in this potential region is controlled by the resistance of the electrolyte in the pores of the passivating layer [14]. According to the layer–pore resistance model, a solid insulating film starts to grow at certain points and spreads over the electrode surface at a rate controlled by the resistance of the porous layer.

After the drastic drop in current produced by the almost complete coverage of the surface by the salt layer, the current measured in region III, which is also a function of sweep rate, decreases with $t^{-1/2}$. This dependence would indicate a process that takes place under diffusion control. However, in experiments carried out using rotating disc electrodes it was found that current is independent of rotation rate. Thus, the kinetically dominant process during further growth of the passive film seems to be the diffusion of chromate ions into the pores of the film [14].

Figure 2 shows a SEM micrograph for a passivated electrode. It can be seen that, as a result of the progressive nucleation process, grains of different sizes are observed and, as would be expected for a compound formed by a dissolution–precipitation mechanism, they have irregular shape. The salt layer that covers completely the surface exhibits a rather complex morphology and is porous.

Since the rate of growth of the porous layer is controlled by ohmic resistance until the surface is almost completely covered and then, further growth takes place under diffusion control of chromate ions within the pores, the formation of a duplex layer as proposed by Druskovich *et al.* [11] could be expected. In addition, appearance of a duplex layer has also been predicted from the simulation of dissolution–

Fig. 2 SEM micrograph of the surface electrode after passivation.

precipitation processes treated as generalized random processes [15]. Figure 3 shows the *j* vs *t* profile obtained for the reduction of the Ag_2CrO_4 porous layer by a potential step. The response observed would indicate that different kinetic effects could be involved in the reduction process or that different portions of the passive layer have different structure and/or stability.

The optical response of the system during film growth was obtained by *in situ* ellipsometry. Salt films were grown potentiodynamically until potential values prior to oxygen evolution to avoid optical interference of bubbles. The Ψ–Δ plot for the initial portion of the anodisation, involving the active dissolution and nucleation processes, is shown in Fig. 4 for a potentiodynamic experiment carried out at 0.005 V s^{-1}. The variation of optical parameters for the first stages of film formation is different from that observed during growth of the passive salt layer and could not be reproduced using simple models.

In order to follow the thickening of the passive layer the potential was swept up to a value within region III and subsequently held at that limit. For this type of experiment, the simple approach of comparing Ψ–Δ curves simulated using a single isotropic layer model with the experimental data showed that they could not be adequately described. Thus, to take into account the complex nature of the passive film uniaxial anisotropy was considered. The Ψ–Δ plot of Fig. 5 shows the experimental data compared with two curves calculated for a single layer with uniaxial anisotropy. It can be seen that the experimental data tend to coincide with each one of the calculated curves in different ranges of thickness indicating that the optical properties of the poorly conducting phase change slightly but continuously during growth. Moreover, it can be seen that as film thickness increases the experimental data shift towards the curve simulated with higher values of refractive indices indicating that during film thickening a decrease in the average water content of the passive layer occurs.

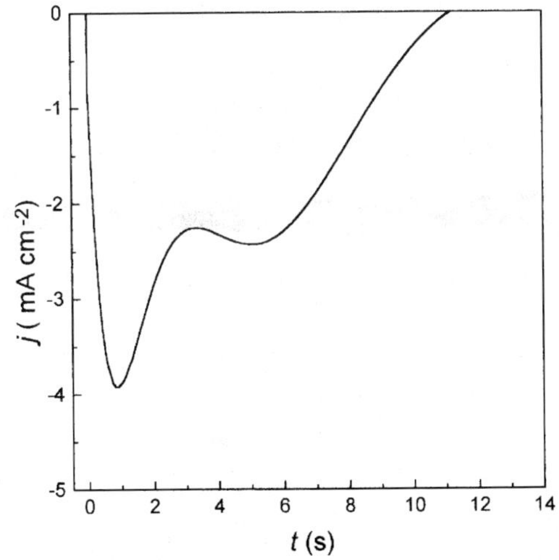

Fig. 3 *Potentiostatic current transient for the reduction of the silver chromate layer at 0.435 V. 0.1*M *Na_2CrO_4 + 0.1*M *borax (pH~9.3).*

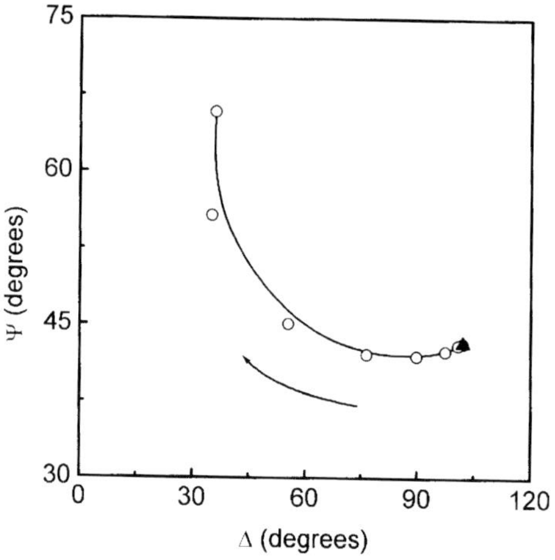

Fig. 4 *Ψ–Δ plot for the first stages of anodisation of polycrystalline silver at 0.005 Vs^{-1} in 0.1M Na_2CrO_4 + 0.1M borax (pH 9.3); (λ = 632.8 nm. (▲) bare substrate.*

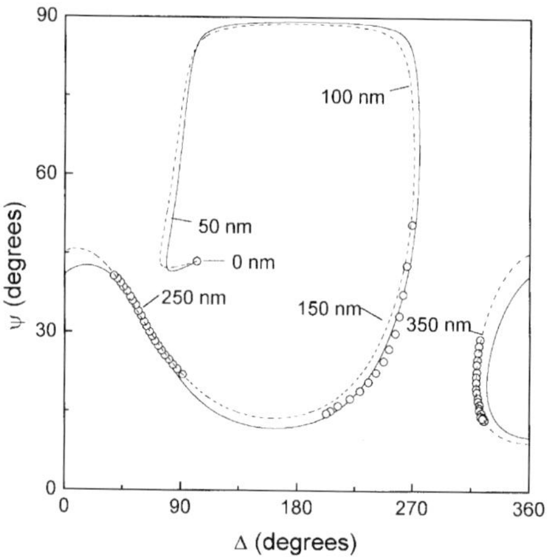

Fig. 5 *Ψ–Δ plot for anodisation of polycrystalline silver at 0.005 Vs^{-1} in 0.1M Na_2CrO_4 + 0.1M borax (pH 9.3); λ = 632.8 nm. Experimental data for film growth (○) and calculated curves for a single layer with uniaxial anisotropy: (—) n_{ord} = 1.795, k_{ord} = 0.188, n_{ext} = 1.670, k_{ext} = 0.138; (----) n_{ord} = 1.885, k_{ord} = 0.194, n_{ext} = 1.928, k_{ext} = 0.172.*

Aiming to gain a better insight on the electrical properties and stability of the passive layers, impedance measurements were also carried out. The Ag_2CrO_4 films were grown potentiodynamically up to a given potential and then kept at the open circuit potential. Despite the complex nature of the passive film, it behaves somewhat as a single porous layer as can be seen in Fig. 6, where the capacitive behaviour observed at high frequencies is associated with the capacity of the salt film and the impedance obtained at intermediate frequencies is attributed to the resistance of the electrolyte in the pores of the layer [16]. It is clear that the response changes with time while the electrode is kept at the open circuit potential. It is also clear that the main change is

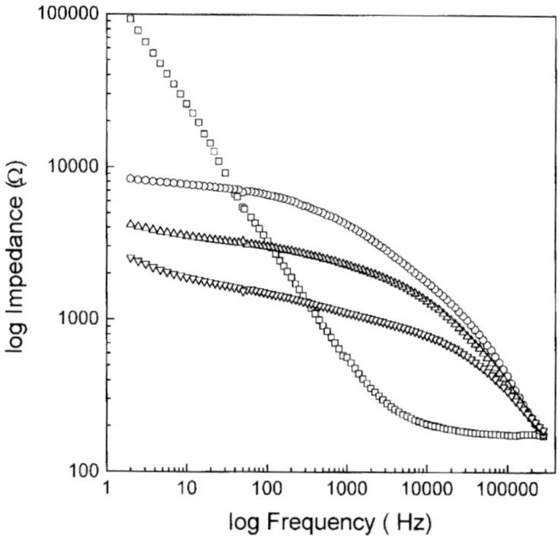

Fig. 6 Effect of time on the Bode diagrams for a silver chromate film: 1 (□), 30 (△) and 60 (▽) min. (○) Bode plot for polycrystalline silver before film formation. 0.1M Na_2CrO_4 + 0.1M borax (pH 9.3).

observed in the range of intermediate frequencies and is associated with the value of the resistance of the electrolyte in the pores which decreases considerably (for example, at ~ 1 kHz it drops from ~ 4300 to ~900 Ω) while the passive film capacity at high frequencies appears to remain approximately constant (~5 nF). This variation can be interpreted taking into account that the resistance of the electrolyte in the pores $R(\theta)$ which is dependent on surface coverage, θ, can be expressed in terms of the film thickness d and the specific conductivity of the electrolyte κ as [13]:

$$R(\theta) = \frac{d\ \theta^{1/2}}{\kappa A\ (1-\theta)} \tag{2}$$

Thus, the change in the value of the resistance with time as shown in Fig. 6 should be due to a decrease in the average thickness of the passive layer, associated probably with changes in structure and/or dehydration while coverage remains approximately constant.

On the other hand, for passive layers grown at different sweep rates, the impedance response showed distinct values for the resistance of the electrolyte in the pores that exceed greatly the differences in the anodic charge, indicating that the surface coverage reaches different values. Therefore, films formed at low sweep rates have a lower porosity.

It is important to note that although both ellipsometric and impedance data indicate changes in structure that appear to be accompanied by a variation of hydration degree of the salt layer, they do not provide the same information. The ellipsometric results are the response of the interphase as a whole and, consequently, include the electrolyte in the pores as well as hydration of the passive film. In contrast, the impedance values in the intermediate range of frequencies correspond only to the resistance of the electrolyte in the pores of the layer.

4. Conclusions

The active–passive transition of silver electrodes in solutions containing CrO_4^{2-} occurs by a dissolution–precipitation mechanism. The structure of the passive layer seems to be rather complex. A slight but continuous variation of optical properties during growth was observed. The variation of the electrical properties of the salt film with time can be associated with processes of ageing or re-structuring of the porous layer (dehydration).

5. Acknowledgements

This research has been financially supported by CONICET, CONICOR and SECYT(U.N.C.). V.B. thanks the fellowship granted by CONICET.

References

1. L. M. Gassa, S. B. Saidman, J. R. Vilche and A. J. Arvia, *J. Appl. Electrochem.*, 1991, **21**, (9), 816–823.
2. F. E. Varela, J. R.Vilche and A. J.Arvia, *Electrochim. Acta*, 1994, **39**, (3), 401–406.
3. A. J. Calandra, N. R. de Tacconi, R. Pereiro and A. J. Arvia, *Electrochim. Acta*, 1974, **19**, (12), 901–906.
4. T. Kattan, S. Szpak and D. N. Bennion, *J. Electrochem. Soc.*, 1974, **121**, (6), 757–764.
5. M. F. L. de Mele, R. C. Salvarezza, M. D. Vazquez Moll, H. A. Videla and A. J. Arvia, *J. Electrochem. Soc.*, 1986, **133**, (4), 746–752.
6. V. I. Birss and C. K. Smith, *Electrochim. Acta*, 1987, **32**, (2), 259–268.
7. V. I. Birss and G. A. Wright, *Electrochim. Acta*, 1982, **27**, (10), 1429–1438.
8. V. I. Birss and G. A. Wright, *Electrochim. Acta*, 1982, **27**, (10), 1439–1444.
9. V. I. Birss and G. A. Wright, *Electrochim. Acta*, 1981, **26**, (12), 1809–1818.
10. V. I. Birss and G. A. Wright, *Electrochim. Acta*, 1982, **27**, (1), 1–8.
11. M. Druskovich, I. M. Ritchie, P. Singh and Z. Hong-guang, *Electrochim. Acta*, 1989, **34**, (3), 409–414.
12. V. Brunetti, H. M. Villullas and M. López Teijelo, *Electrochim. Acta*, 1999, **44**, (17), 2843–2852.
13. D. Devilliers, F. Lantelme and M. Chemla, *Electrochim. Acta*, 1986, **31**, (10), 1235–1246.
14. V. Brunetti, H. M. Villullas and M. López Teijelo, *Electrochim. Acta*, 1999, **44**, (26), 4693–4700.
15. B. Aurian-Blajeni, M. Kramer and M. Tomkiewicz, *J. Phys. Chem.*, 1987, **91**, (3), 600–605.
16. V. Brunetti, H. M. Villullas and M. López Teijelo, EIS Study of Silver Salt Films, in *Proc. Symp. on Passivity and Its Breakdown*, 31 August – 5 September, 1997, Paris, France, pp.973–980, (P. M. Natishan, H. S. Isaacs, M. Janik-Czachor, V. A. Macagno, P. Marcus and M. Seo eds.). The Electrochemical Society, New York, USA, 1997.

19
Effect of High Temperature Ageing on the Electrochemical Behaviour of C-22 Alloy

R. B. REBAK, N. E. KOON and P. CROOK
Haynes International Inc., 1020 West Park Ave., Kokomo, IN, 46904-9013, USA

ABSTRACT

When Ni–Cr–Mo alloys are exposed to temperatures above 873 K precipitation of second phases can occur. As a result, these alloys lose ductility and resistance to corrosion in some aggressive acidic environments. The aim of this research work was to characterise the corrosion resistance and the anodic behaviour of C–22 alloy (Ni–22Cr–13Mo–3W) as a function of ageing temperature and ageing time. Ageing was performed at 922 K, 978 K and 1033 K for times between 1 h and 1000 h. Results show that, in acidic solutions, the corrosion resistance of C–22 alloy decreases as the ageing temperature and the ageing time increases, both for oxidising and reducing environments. This decrease in corrosion resistance is caused by a localised depletion of either Cr or Mo. At a constant ageing temperature, the depletion of Mo occurs faster than the depletion of Cr.

1. Introduction

The current design concept for the high level nuclear waste containers in the USA is based on a metallic multibarrier system. One of the designs of the containers calls for a 10 cm thick external barrier of carbon steel and a 2 cm thick internal barrier of a nickel based C–22 alloy [1–3]. Another more recent design specifies a 2 cm thick external layer of C–22 alloy and a 5 cm thick internal layer of type 316 stainless steel. The C–22 alloy (Ni–22Cr–13Mo–3W) was selected for its excellent resistance to localised corrosion (pitting corrosion, crevice corrosion and stress corrosion cracking) in chloride-containing environments [4–12].

Due to the heat generated by the radioactive decay of the waste, the containers, which will be buried at the Yucca Mountain site, might experience temperatures as high as 523 K during their first 1000 years of emplacement. The lifetime design of the containers is 10 000 years and their maximum allowed temperature is 623 K. It would be useful to know how this long term exposure to temperatures as high as 623 K would affect the corrosion resistance of C–22 alloy; however, thermal stability studies of C–22 alloy at 623 K for such long times are impractical in the laboratory. Previous studies have shown that the mechanical and corrosion properties of C–22 alloy did not change when it was aged for up to 40 000 h at 700 K [7,9]. However, it is known that precipitation of detrimental second phases occurs when Ni–Cr–Mo alloys are aged at temperatures above 873 K [13–18]. The precipitation of these second phases, especially the intermetallic phases, can affect the corrosion resistance and mechanical properties of the aged Ni–Cr–Mo alloys [13–15,18–20]. Phase stability studies based on corrosion testing of C–22 alloy are limited [20].

In order to quantify the nature and the volume of the second phases that form during the ageing of C–22 alloy and to determine their influence on its corrosion behaviour, systematic studies on the kinetics of second phase nucleation and growth are required. It is not certain what are the lower temperature boundaries for the formation of the detrimental second phases in Ni–Cr–Mo alloys that are found to develop at temperatures higher than 873 K. Moreover, it is not clear what are the phases, precipitation volumes, location of the precipitation, etc. that would affect considerably the corrosion rate of C–22 alloy in conditions where the containers will be buried. One first approach would be to study the precipitation kinetics at the higher temperature shorter time conditions and to identify the relationship between precipitation and changes in the corrosion susceptibility in aggressive standard electrolyte solutions. If it can be assumed that the ageing mechanism remains the same at lower temperatures, these high temperature data could also be used to predict the possibility that these detrimental phases would form — in a volume that will affect the corrosion resistance of C–22 alloy — at temperatures in the order of 623 K for times as long as 10 000 years (the design life of the containers).

Isothermal ageing studies of C–22 alloy in the temperature range between 533 K and 1073 K for times between 30 min and 20 years are being currently carried out at Haynes International Inc. However, the purpose of the work presented in this paper was to determine the changes in the corrosion resistance and electrochemical behaviour of HASTELLOY® C–22® alloy* when it is aged only in the temperature range 922 K to 1033 K for times only between 1 h and 1000 h.

2. Experimental Technique

Samples of C–22 alloy were aged for 1, 3, 6, 10, 30, 60, 100 and 1000 h at 922, 978 and 1033 K. Ageing was performed in air and the samples were allowed to cool down in air under normal convection conditions. Samples were prepared from a 0.25 in. (6.35 mm) thick plate (Heat 2277–6–3181).

Changes in the corrosion resistance and anodic behaviour of C–22 alloy were determined using standard immersion tests and electrochemical techniques [21]. For the immersion tests, the samples had a finished grinding of abrasive alumina paper number 120 and were degreased in acetone. Immersion tests included ASTM G 28 A (boiling solution of 50% H_2SO_4 + 42 gL^{-1} of $Fe_2(SO_4)_3$) and ASTM G 28 B (boiling solution of 23% H_2SO_4 + 1.2% HCl + 1% $FeCl_3$ + 1% $CuCl_2$) [21]. Both solutions are oxidising and the second solution also contains 23500 ppm chloride ions which could induce localised attack in many alloys. Each ASTM G 28 A and G 28 B corrosion test lasted 24 h. Immersion corrosion tests were also performed using a boiling 2.5% HCl solution. Each of these tests lasted 96 h (the electrolyte solution was changed every 24 h). Electrochemical tests included polarisation resistance (ASTM G 59) and

*HASTELLOY® and C–22® are registered trademarks of Haynes International, Inc.

potentiodynamic polarisation (ASTM G 61) [21]. These tests were carried out in a one-litre, three-electrode cell (ASTM G 5) [21] provided with a condenser to avoid changes in the concentration of the solution during the testing time. The potential of the working electrodes was controlled using a pair of equidistant and interconnected platinum electrodes. The electrochemical tests were carried out in deaerated solutions of 1M HCl pH ~ 0 at 339 K. The potential scan was started 150 mV below the open circuit potential and the scan rate was 0.5 mVs^{-1}. The samples (working electrodes) had a finished grinding of abrasive alumina paper number 400 and were degreased in acetone.

3. Experimental Results

3.1. Immersion Tests

Figures 1 and 2 show the general corrosion rate (mass loss) of C–22 alloy as a function of ageing time at the three different ageing temperatures in the ASTM G 28 A and ASTM G 28 B solutions, respectively. Figure 1 shows that, at each ageing temperature, the corrosion rate gradually increases as the ageing time increases. At 922 K, a minimum ageing time of 10 h is needed before the corrosion rate starts to increase. At an ageing time of 1000 h, the aged samples corroded at the same rate independently of the ageing temperature.

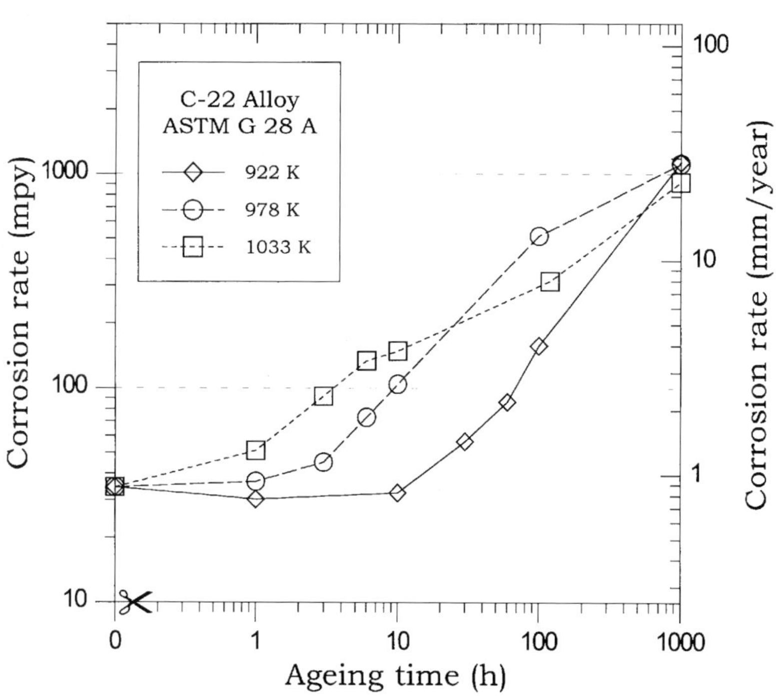

***Fig. 1** Corrosion rate of the aged C-22 alloys in the ASTM G 28 A Test.*

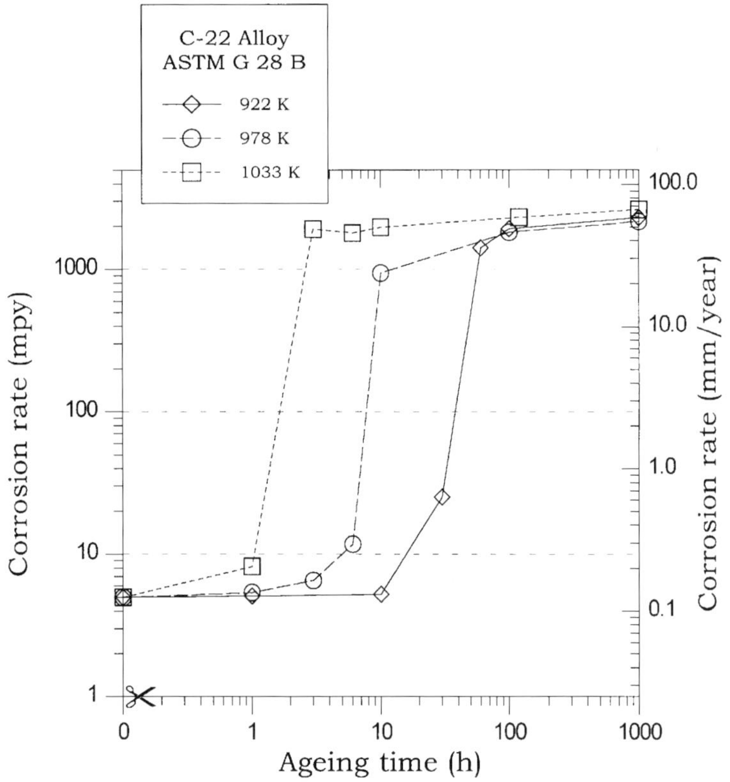

Fig. 2 Corrosion rate of the aged C-22 alloys in the ASTM G 28 B Test.

Observation of the tested coupons in the ASTM G 28 A solution showed that the corrosion rate increased as the ageing time increased because of the formation of corrosion pits at triple grain boundary junctions. At the higher ageing times a higher density of corrosion pits was observed. For an ageing time of 1000 h, at the three ageing temperatures, the corrosion pattern was generalised leaving behind a crystallographic microstructure, i.e. some crystallographic planes corroded preferentially. Metallographic cross-section of the corroded coupons showed that the depth of intergranular attack for coupons aged for 10 h at 1033 K was comparable to coupons aged for 1000 h at 922 K. In both cases the intergranular attack penetration was approximately 350 μm.

Figure 2 shows that, in the ASTM G 28 B solution, at each ageing temperature the corrosion rate increased abruptly by more than two orders of magnitude after a certain threshold ageing time. The higher the temperature the lower the threshold time. After an ageing time greater than 100 h all the aged samples corroded at the same rate independently of the ageing temperature.

Observation of the tested coupons showed that, below the threshold time, the aged samples had only a few isolated shallow crystallographic pits; however, above the threshold time a high density of deep crystallographic pits was observed. As the ageing time increased above the threshold time the number of pits increased but their depth decreased, therefore the overall corrosion rate remained approximately the same. Metallographic cross section of the corroded coupons showed that, as in the case of the ASTM G 28 A solution, the depth of pitting corrosion for coupons aged for 10 h at 1033 K was comparable to coupons aged for 1000 h at 922 K. In both cases the pitting corrosion penetration was approximately 650 µm.

Figure 3 shows the general corrosion rate (mass loss) of C–22 alloy as a function of ageing time at the three different ageing temperatures in a boiling 2.5% HCl solution. The corrosion rate increased with the ageing temperature and the ageing time. Examination of the corroded coupons showed that, even in the mill annealed condition, a slight preferential attack at some grain boundaries occurred. The intergranular attack increased with ageing temperature and ageing time. For example, at the ageing temperature of 922 K, as the ageing time increased the intergranular attack became wider and deeper. At an ageing time of 1000 h, preferential corrosion at twins inside the grains was observed.

3.2. Electrochemical Tests

Figure 4 shows the corrosion rate of C–22 alloy measured by the polarisation resistance technique in deaerated 1M HCl at 339 K. The corrosion rates in Fig. 3 are much higher than those in Fig. 4 because of the higher temperature in Fig. 3. Figure 4 shows that, at the lower temperature of 339 K, an ageing time higher than 10 h is required to detect changes in the corrosion rate of the aged alloys. The relationship between the

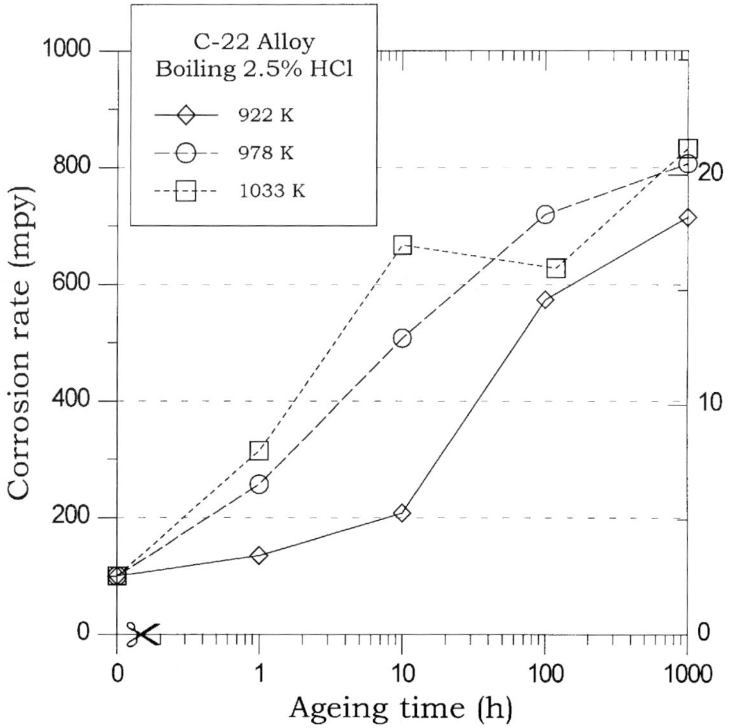

***Fig. 3** Corrosion rate of the aged C-22 alloys in 2.5% HCl at the boiling temperature.*

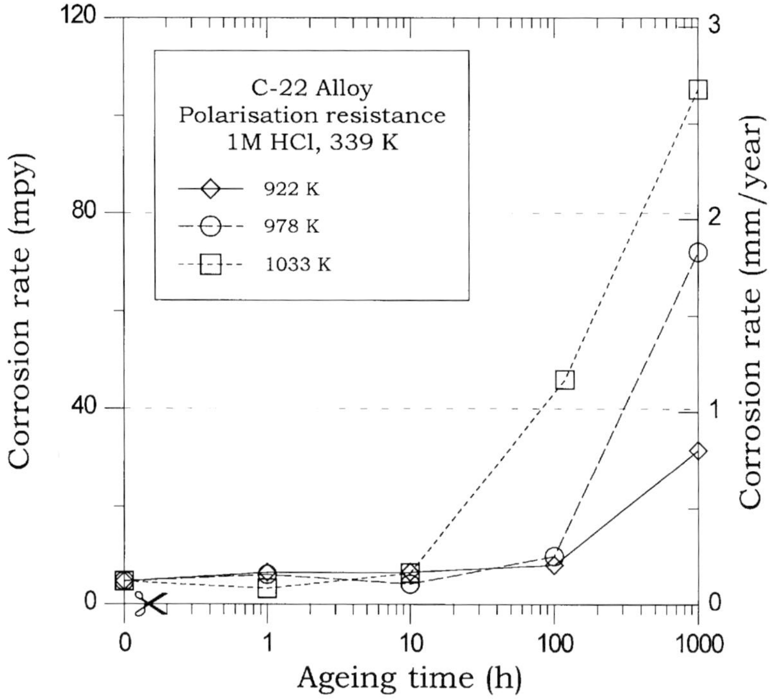

***Fig. 4** Corrosion rate of the aged C-22 alloys in 1M HCl at 339 K.*

critical current for passivation and the ageing temperature and ageing time is the same as that of the general corrosion rate measured by polarisation resistance (Fig. 4).

3.3. Potentiodynamic Polarisation Tests

Figure 5 shows typical potentiodynamic polarisation curves for C–22 alloy in deaerated 1M HCl solution at 339 K. The cathodic portion of the curve is the same for the mill annealed (MA) as for the 1000 h aged materials. The critical current for passivation and the passive current density of the mill annealed material are the lowest and they increase as the ageing temperature increases. Figure 6 shows the effect of the ageing time on the anodic behaviour of C–22 alloy. Figure 6 shows that, at 1033 K, an ageing time higher than 100 h is necessary before there is a significant increase in the critical current for passivation and in the anodic passive current.

4. Discussion

When Ni–Cr–Mo alloys such as C–22 alloy are exposed to temperatures above 873 K, several solid state reactions can occur. Below near 873 K, the presence of long range ordering (LRO) reactions has been reported, [9,17–19] and at higher temperatures

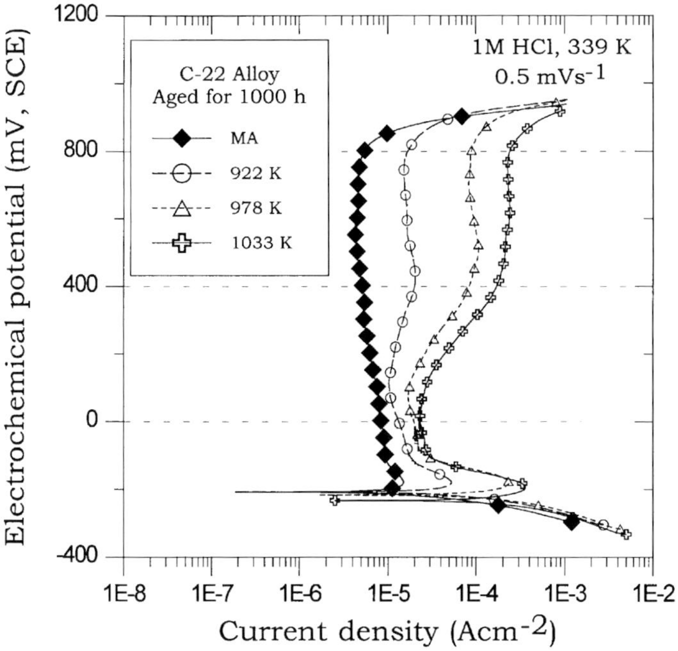

Fig. 5 *Effect of ageing temperature on the anodic behaviour of aged and MA C-22 alloy in 1*M *HCl at 339 K.*

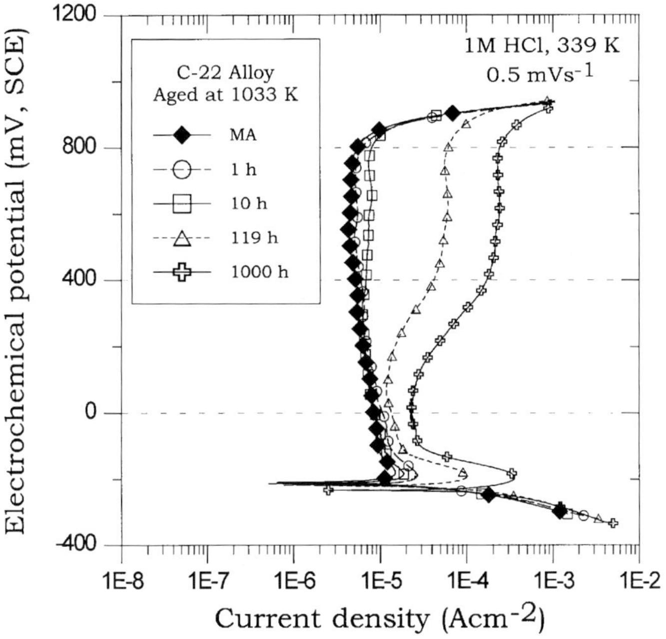

Fig. 6 *Effect of ageing time on the anodic behaviour of aged and MA C-22 alloy in 1*M *HCl at 339 K.*

the presence of secondary carbides were observed [14,19]. However, the largest degradation in the corrosion resistance and in mechanical properties is generally attributed to the formation of two intermetallic phases known as μ (mu) and *P* phases [13,14,16–20]. These phases are similar both in their chemical composition and their topologically-close-packed (TCP) crystal structure. They tend to be rich either in chromium or molybdenum, therefore depleting the immediate matrix in these two elements that give C–22 alloy its resistance to corrosion. The resistance to general corrosion under reducing conditions (boiling 2.5% HCl) is given by Mo and under oxidising conditions (ASTM G 28 A) by Cr. The resistance to localised corrosion promoted by chlorides (ASTM G 28 B) is given both by Cr and Mo. As an alloying element, W behaves similarly to Mo. Cieslak *et al.* [22] measured the composition of particles found in welded structures. They reported that the composition of the μ and *P* phases in C–22 alloy could contain up to 38.7% Mo and 6.3% W compared to the composition of the base γ (gamma) phase of 13% Mo and 3% W. The amount of Cr in the intermetallic phases was similar to that of the base gamma phase [22].

Figures 7 and 8 show that, in the initial stages of ageing at 922 K and 978 K, the corrosion rate increased faster in a reducing solution than in an oxidising solution. After 10 h of ageing the corrosion rate in the oxidising solution increased faster than in the reducing solution. Results from Figs 7 and 8 seem to suggest that, in the earlier

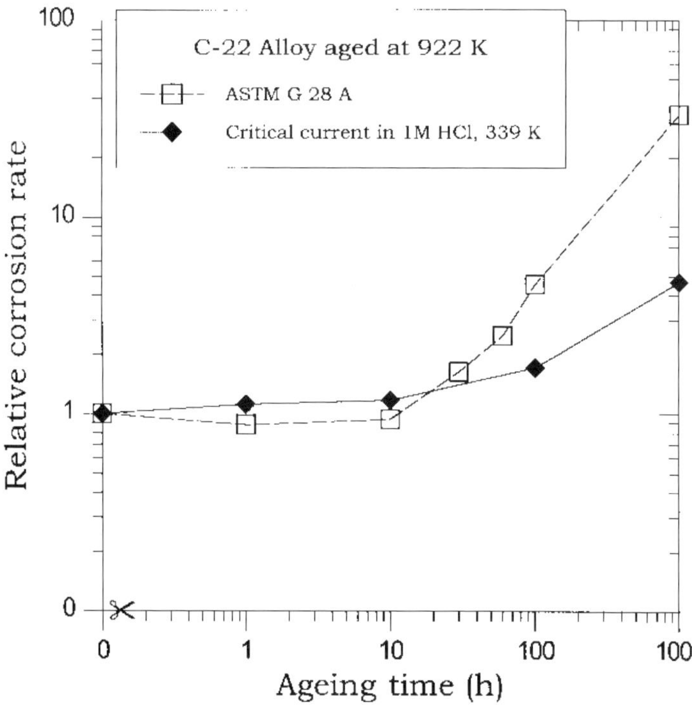

Fig. 7 *Effect of ageing time on the relative corrosion behaviour of C-22 alloy in 1*M *HCl at 339 K and in the ASTM G 28 A solution.*

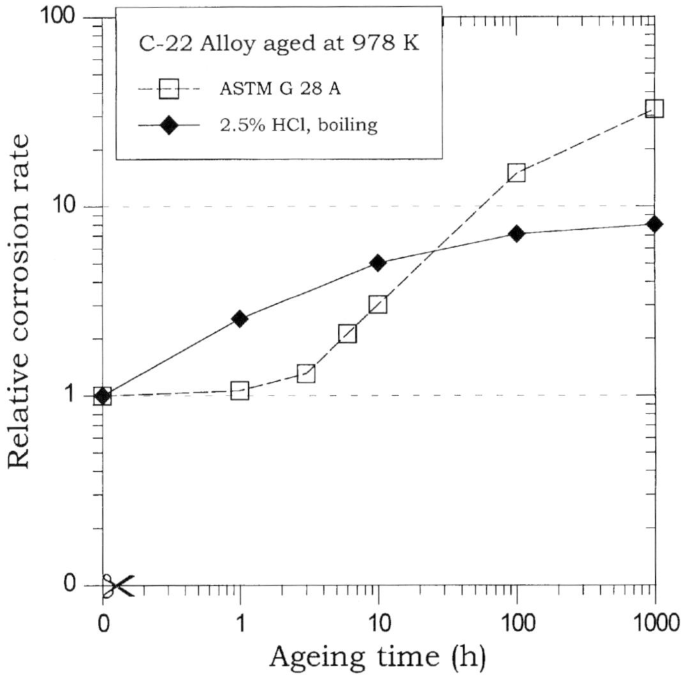

Fig. 8 *Effect of ageing time on the relative corrosion behaviour of C-22 alloy in boiling 2.5% HCl solution and in the ASTM G 28 A solution.*

stages of ageing (less than 10 h), the depletion of Mo is faster than the depletion of Cr; however, for times higher than 10 h, the depletion of Cr starts to be important. Data in Fig. 7 shows that under oxidising conditions, the corrosion of C–22 alloy seems to be lower for the 1 h and 10 h ageing than for the MA condition. This could be a consequence of the local depletion of Mo which caused an enrichment of Cr in the Ni matrix. This effect is not observed for the samples aged at 978 K (Fig. 8).

A similar effect of ageing in reducing and oxidising conditions is also shown in Fig. 9 for the ASTM G 28 B solution. However, Fig. 9 (overleaf) shows that the increase in the corrosion rate in the ASTM G 28 B solution exceeds the boiling 2.5% HCl curve at an earlier ageing time than that for the ASTM G 28 A solution (Fig. 8). As mentioned above, the depletion of elements around the second phase precipitates seems to initiate with Mo and W and then progress with Cr, as detected by corrosion tests in oxidising and reducing conditions (Figs 7 and 8). On the other hand, the corrosion rate in the ASTM G 28 B solution would increase sharply when the concentration of both Cr and Mo (and also W) in regions adjacent to the precipitates falls below the critical concentration for protection against pitting corrosion.

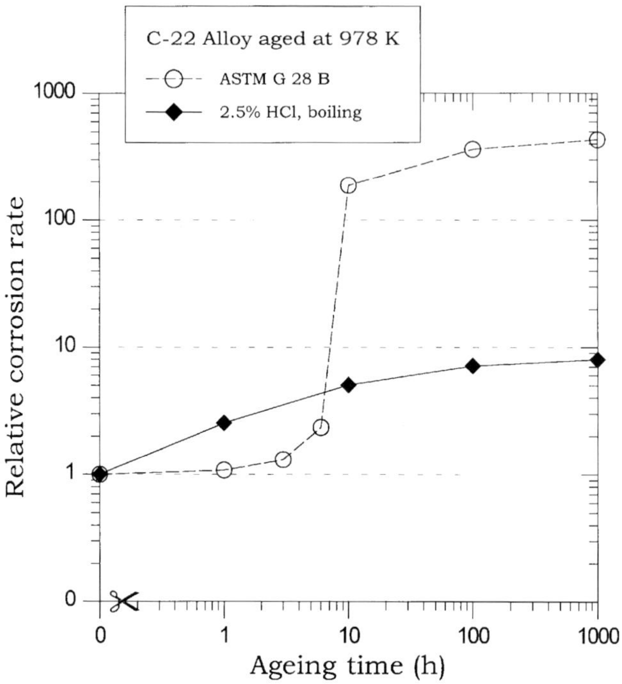

Fig. 9 Effect of ageing time on the relative corrosion behaviour of C-22 alloy in boiling 2.5% HCl solution and in the ASTM G 28 B solution.

5. Conclusions

1. Ageing of C–22 alloy in the temperature range 922–1033 K causes a decrease in its corrosion resistance both under oxidising and reducing conditions.

2. In general, in both types of environments, the corrosion rate increases with the ageing temperature and the ageing time.

3. At a constant ageing temperature, the localised depletion adjacent to the second phase precipitates seems faster for Mo than for Cr.

4. The early stages of second phase formation are detected better by corrosion testing under reducing conditions.

References

1. U. S. Nuclear Waste Technical Review Board, 1997 Findings and Recommendations, Report to The U. S. Congress and The Secretary of Energy, Arlington, Virginia (April 1998).
2. D. McCright and W. L. Clarke, *Corrosion'98*, Paper 159, NACE International, Houston, Tx (1998).

3. A. A. Sagüés, Paper QQ14.1 in *Proc. Symp. on Scientific Basis for Nuclear Waste Management XXII* at the Fall Meeting of the Materials Research Society, Boston, 30 Nov.–4 Dec., 1998, Vol. 556 (In Print).
4. P. E. Manning and J. D. Schöbel, *Werkst. und Korros.*, 1986, **37**, 137–145.
5. A. I. Asphahani, *Arab. J. Sci. Eng.*, 1989, **14**, (2), 317–335.
6. S. J. Lukezich, The Corrosion Behavior of Ni-Base High Performance Alloys in Simulated Repository Environments, MS Thesis, The Ohio State University, 1989.
7. R. B. Rebak, in *Proc. Symp. on Passivity and Its Breakdown*, The Electrochemical Society, PV 97–26, pp.1001–1012 (1998).
8. A. K. Roy, D. L. Fleming and B. Y. Lum, *Mater. Perform.*, 1998, (3), 54.
9. R. B. Rebak and N. E. Koon, *Corrosion'98*, Paper 153, NACE International, Houston, Tx (1998).
10. K. A. Gruss, G. A. Cragnolino, D. S. Dunn and N. Sridhar, *Corrosion'98*, Paper 149, NACE International, Houston, Tx (1998).
11. R. B. Rebak and P. Crook, in *Proc. Symp. on Critical Factors in Localized Corrosion III*, Fall Meeting of the Electrochemical Society, Boston, 1–6 Nov., 1998, PV 98–17, pp.289–302 (1999).
12. D. S. Dunn, G. A. Cragnolino and N. Sridhar, Paper QQ14.6 in *Proc. Symp. on Scientific Basis for Nuclear Waste Management XXII*, at the Fall Meeting of the Materials Research Society, Boston, 30 Nov.–4 Dec., 1998, Vol. 556, pp.879–886.
13. R. B. Leonard, *Corrosion*, 1969, **25**, 222.
14. F. G. Hodge, *Corrosion*, 1973, **29**, 375.
15. S. J. Matthews, in *Proc. 3rd Int. Conf. on Superalloys*, Claitor's Publishing Division, Baton Rouge, 1976, p.215.
16. M. Raghavan, B. J. Berkowitz and J. C. Scanlon, *Met. Trans.*, 1982, **13A**, 979.
17. H. M. Tawancy, *J. Mater. Sci.*, 1996, **31**, 3929.
18. T. S. Edgecumbe Summers, M. A. Wall, M. Kumar, S. J. Matthews and R. B. Rebak, Paper QQ9.71 in *Proc. Symp. on Scientific Basis for Nuclear Waste Management XXII* at the Fall Meeting of the Materials Research Society, Boston, Nov. 30–4 Dec., 1998, Vol. 556, pp.919–926.
19. H. M. Tawancy, R. B. Herchenroeder and A. L. Asphahani, *J. Met.*, 1983, (6), 37.
20 U. L. Heubner, E. Altpeter, M. B. Rockel and E. Wallis, *Corrosion*, 1989, **45**, 249.
21. Annual Book of ASTM Standards, Volume 03.02, American Society for Testing and Materials, West Conshohocken, PA, 1998.
22. M. J. Cieslak, T. J. Headley and A. D. Romig Jr., *Metall. Trans.*, 1986, **17A**, 2035.

20
Electrochemical Study of FINEMET–Co Amorphous Alloys

A. ALTUBE, A. R. PIERNA, F. F. MARZO, M. SISTIAGA and A. LORENZO
Chemical Engineering and Environment Dept., University of the Basque Country, P. O. Box 1379, 20080 San Sebastian, Spain

ABSTRACT
Cyclic voltammetry and Electrochemical Impedance Spectroscopy were used to study the electrochemical behaviour of various cobalt-containing FINEMET alloys in aqueous KOH solutions. Both as-quenched and thermally relaxed samples have been analysed. The results show that the electrochemical behaviour of FINEMET–Co amorphous alloys is related to the effect of cobalt on the solubility of iron oxides in potassium hydroxide. When samples are submitted to thermal relaxation processes, the oxidation charge decreases due to the topological short range order, changes produced in the samples. The extent of these changes depends on the cobalt content.

1. Introduction

FINEMET type amorphous alloys [1,2] have special electrochemical, magnetic and mechanical characteristics that make them very interesting from a technological point of view. The cobalt-containing alloys [3,4] have soft magnetic properties which make them very attractive for the production of magnetic sensors and some other electronic devices.

However, the heating caused during the use of, for example, an electrical transformer with a magnetic core made in this kind of alloy, may lead to an eventual variation of its physico–chemical properties. The effect of this heating on the structure of the material will depend on the temperature reached [5] and on the alloy composition [6]. One way to evaluate the effect of the change of the atomic structure on the properties of the alloy is by studying its behaviour from the point of view of electrochemistry.

In the present work, some electrochemical properties such as corrosion resistance are evaluated by cyclic voltammetry and electrochemical impedance spectroscopy (EIS). Although it is possible to obtain important mechanistic information [7,8] from EIS data no attempt has been made to do so in this presentation. The aim of this work is to describe the effect of different amounts of cobalt from 0 to 10 at.% on the electrochemical behaviour of FINEMET type alloys in highly alkaline media. The basic composition of FINEMET alloys may be written as $Fe_{73.5}Cu_1Nb_3Si_{13.5}B_9$ (F) but for sake of simplicity, alloys will be referred to as F–Co_X (X = 0 to 10). The alloys were also isothermally treated at a temperature of 673 K for 30 min to study the influence of the relaxation processes (and maybe structural changes) on their electrochemical properties.

2. Experimental details

Crystalline ingots of various F–Co_X ($X = 0, 2, 4, 6, 8, 10$) were prepared by means of argon arc melting of pure metals and binary alloys (FeSi and FeB) from Goodfellow. Amorphous ribbons were obtained by the plannar flow casting method as described in previous work [9]. The actual composition of the alloys was verified by ICP (Inductively Coupled Plasma) and the amorphicity of the structure checked by X-ray diffraction. Figure 1 shows the spectrum obtained for a FINEMET alloy; the absence of any sharp peak on the plot confirms the amorphicity of the sample.

The cyclic voltammetry measurements were carried out with a Princeton Applied Research Model 273 (PARC 273) potentiostat–galvanostat under PARC-M270 software control. The EIS experiments were performed with a Solartron 1250 frequency response analyser attached to the potentiostat. Frequency sweeps were carried out potentiostatically from 50 000 Hz to 1 mHz superimposing a 5 mV a. c. amplitude over a bias potential of –400 mV. The EIS results obtained were analysed using Zview and EQUIVCRT software packages. The latter has been developed to fit equivalent circuits to the experimental results.

The cyclic voltammetry experiments were performed in a conventional three electrode cell. Additionally, in EIS experiments, a platinum sheet was connected in parallel with the reference electrode (Ag/AgCl) by a capacitor in order to achieve a reference system of low resistance which is important for avoiding phase shifts at high frequencies. A 0.5N KOH solution was used as electrolyte. The working electrode

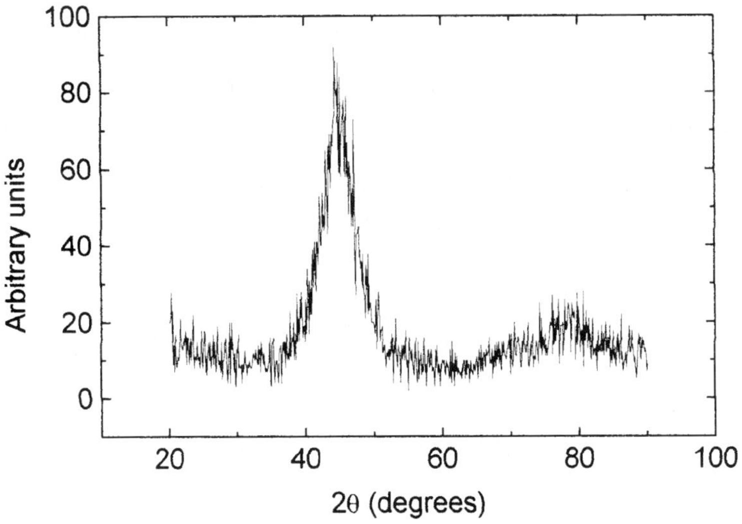

Fig. 1 *XRD pattern of a FINEMET amorphous alloy.*

(WE), consisted of a sample of the alloy with 3 cm^2 of active surface and the circuit was completed with a platinum sheet of 3 cm^2 as counter electrode (CE).

Two types of pretreatments were carried out before the impedance measurements: the first, called treatment A, involved the stabilisation of the alloy at a constant potential of –400 mV vs Ag/AgCl. The second consisted of the electrochemical growth by cyclic voltammetry of an oxide layer before the stabilisation of the electrode at –400 mV vs Ag/AgCl and is referred to as treatment B.

After each experiment the KOH solution was analysed in order to measure the content of dissolved metals (cobalt and iron). The analyses were performed by UV-visible spectrophotometry using a Shimadzu UV–160 spectrophotometer. Iron was determined by the o-phenanthroline method and the nitroso-R salt method was used to measure the cobalt content as described by Onishi [10].

The thermal stability of the alloys was studied by Differential Scanning Calorimetry (DSC) with a TA Instruments MDSC-2920 calorimeter, under helium stream at a scan rate of 10 K min^{-1}. The results obtained are shown in Fig. 2. The plots show an exothermic peak at 819 K (onset temperature = 800 K) for FINEMET amorphous alloy corresponding to the nanocrystallisation of the sample. The addition of cobalt to the alloy does not significantly shift the onset temperature of this peak, that is, it does not affect the thermal stability of the amorphous phase.

In this work, alloy samples were isothermally treated at 673 K for 30 min. The annealing of an amorphous sample below the temperature of nanocrystallisation

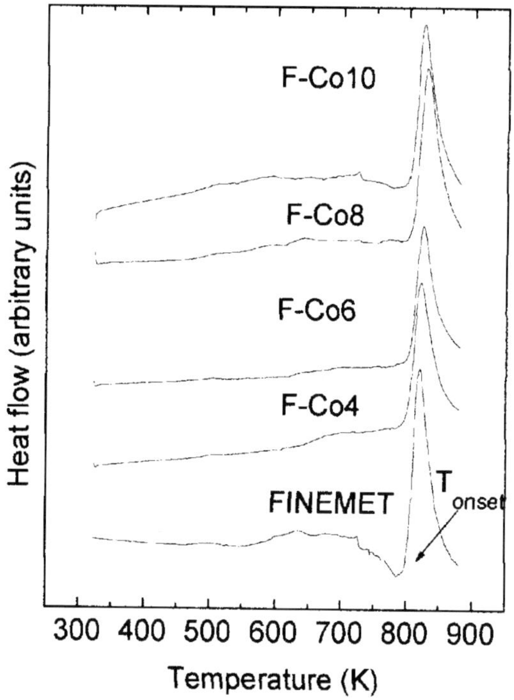

Fig. 2 DSC curves of F-Co$_x$ alloys obtained under helium stream. Scan rate = 10 K min^{-1}.

(819 K for F–Co_X alloys) modifies the short range order. This modification can be reversible (CSRO or chemical short range ordering) when it is related to the chemical changes in the region nearest to the atom, or irreversible (TSRO or topological short range ordering) if the changes reflect a decrease in the free volume created in the alloy during the production process. Filipecki *et al.* [11] reported that after thermal annealing at 673 K for 30 min irreversible topological short range ordering occurs in Fe–Co based metallic glasses due to a gradual decrease in the free volume created during the production process. On the other hand, Serebryakov *et al.* [12] reported that as-quenched and 673 K annealed CoFeSiNbB alloys lead to similar XRD (X-ray diffraction) spectra which correspond to an amorphous material, and to some small changes in Mössbauer data, related to the structural evolution of the alloy due to association of atoms. These changes are very small when samples are heated at 673 K.

3. Experimental Results and Discussion

3.1. As-quenched State

Figure 3 compares the first and tenth cycle of a typical cyclic voltammogram of an amorphous FINEMET alloy (a) with that of an F–Co10 alloy (b). The sweeps were carried out from $E_c = -1300$ mV to $E_a = +550$ mV at a scan rate of 20 mV s^{-1}. The plot of the first cycle of the pure FINEMET alloy (Fig. 3a) shows an anodic peak at –715 mV vs Ag/AgCl (peak II) and two cathodic peaks (peak I′ at –970 mV and peak II′ at –1070 mV). The repetitive potential cycling results, after ten sweeps, in an increase of the voltammetric charge. This increase, which is related to the peaks II and II′, indicates the accumulation of oxidised species forming a surface layer. These peaks (II and II′) are related to the formation of FeOOH and its conjugate and peak I′ to the electroreduction of the inner layer of the oxide film [13] .

The addition of cobalt to the alloy composition (Fig. 3b) leads to some changes on the plot. Taking into account the results obtained for pure cobalt [14], the response of the first oxidation reaction of cobalt should be located at –820 mV for the $Co/Co(OH)_2$ process and at –800 mV for the Co/CoO, but for the first cycle, it may be seen that the first anodic peak on the voltammogram (peak III) is found at $E = -115$ mV. This contribution is, usually, related to the $Co(OH)_2 \rightarrow Co_3O_4$ transition which may result from the reaction of surface oxides present on the alloy. The plot related to the tenth cycle shows again peak II. This peak has a similar height to that of the FINEMET alloy but it is broader than the former and may be considered as a sum of the contribution of various oxidation processes of iron and cobalt. Peak III is seen as a shoulder on the right side of peak II. Finally, there is a new contribution just before the upper limit of the potential scans (peak IV) which is, usually, related to the formation of CoOOH [14,15].

As explained before, for F–Co_X alloys, peak II should be considered as the sum of contributions of iron and cobalt oxidation reactions. It was not possible to make a deconvolution, so the total area corresponding to the anodic contributions of the voltammogram was measured. The results are listed in Table 1. In every case the anodic charge measured decreased when cobalt was added to the alloy composition.

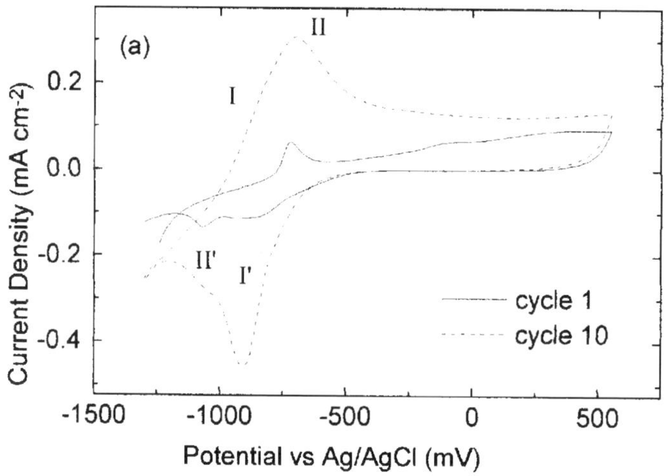

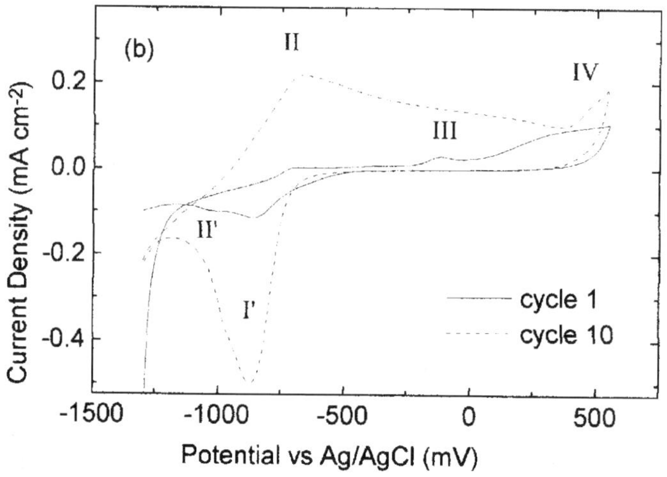

Fig. 3 *Voltammograms of the first cycle (solid line) and 10th cycle (dotted line) of (a) FINEMET and (b) F–Co10 alloys in 0.5N KOH.* $E_c = -1300$ *mV,* $E_a = 550$ *mV, scan rate = 20 mV* s^{-1}.

Table 1. *Anodic charge measured on the voltammograms after 10 cycles treatment for F–*Co_X *alloys and Fe content on the alloy*

	Fe content (at.%)	Anodic Charge (mCm^{-2})
FINEMET	73.5	11.8
F-Co2	71.5	11.2
F-Co4	69.5	10.4
F-Co6	67.5	9.8
F-Co8	65.5	9.4
F-Co10	63.5	9.0

Taking into account that the anodic charge is proportional to the amount of oxidised metal, it would seem that cobalt prevents the oxidation of iron.

The analysis of the electrolytic solutions showed that both, iron and cobalt, had dissolved in KOH during the experiments after treatment A, that is, after keeping the sample at a constant potential of –400 mV for 28000 s. Furthermore, the measured amounts of iron increased with cobalt content for as-quenched (AQ) alloy samples as listed in Table 2. Cobalt cannot passivate easily in highly alkaline media as iron does so it cannot form a compact oxide layer to prevent the oxidation and dissolution of the alloy. The growth of the oxide layer (treatment B) results in the decrease of the ion concentration into the solution. No measurable amounts of cobalt were found but, on the other hand, the iron concentration increased again with cobalt content. The results obtained suggest that the oxide film growth over the surface forms a protective layer against corrosion but taking into account the concentration increase with the cobalt content, this metal does not prevent the oxidation of the alloy. Considering the results in Tables 1 and 2, the decrease of the anodic charge should be due to the increase of the dissolution process, which makes the growth of the oxide layer difficult.

The electrochemical characteristics of the double layer formed on the alloy surface were also studied by EIS. Figure 4 (a) depicts the Nyquist diagrams for FINEMET and F–Co10 alloys after treatment A. The Nyquist plot of a FINEMET alloy is characterised by one simple capacitive loop at high frequencies corresponding to the double layer relaxation [16], and a constant phase element (CPE) at low frequencies. This type of constant phase element also called a Warburg element, is usually related to the resistance of the mass transfer and the overall shape of the plot corresponds to a mass diffusion controlled dissolution process. When cobalt is added to the alloy composition the charge transfer resistance (R_T) increases with cobalt (Table 3) as this element is nobler than iron.

Figure 4 (b) shows the results obtained after treatment B. As explained before, the potentiodynamic cycling which was carried out just before the stabilisation leads to the formation of oxide layers which are mainly composed of Fe_3O_4 which may form a compact film over the alloy surface. Two simple capacitive loops are present in the plot corresponding to the FINEMET alloy. The measured R_T is about four times higher than in the non-cycled case. This response indicates that the iron oxide layer is protective against corrosion. The addition of cobalt results in a decrease of the R_T

Table 2. *Measured amounts of metallic ion found in the solutions for some as-quenched and thermally treated F–Co alloys after treatments A (Stabilisation at –400 mV vs Ag/AgCl) and B (25 cycles between* $E_c = -1300$ *and* $E_a = +550$ *mV at 100 mVs*$^{-1}$ *and stabilisation at –400 mV vs Ag/AgCl)*

	Treatment A		Treatment B	
	Fe (μg cm^{-2})	Co (μg cm^{-2})	Fe(μg cm^{-2})	Co (μg cm^{-2})
FINEMET	16.4	—	8.5	—
F-Co2	16.8	0.19	11.4	—
F-Co6	17.3	0.28	13.2	—
F-Co10	25.8	0.35	14.6	—

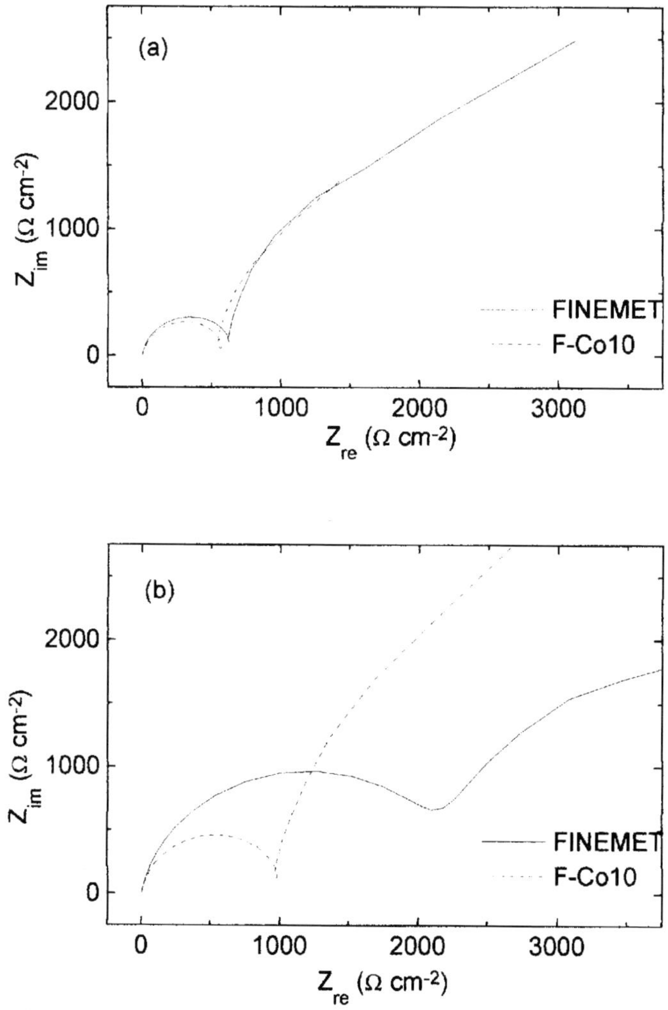

Fig. 4 *Impedance diagrams of as-quenched and FINEMET and F–Co10 alloys in 0.5*N *KOH at potential* E = *–400 mV after (a) treatment A and (b) treatment B. Conditions: frequency sweep from 50000 Hz to 1 mHz, a.c. amplitude: 5 mV.*

and also in the insertion of a new Warburg element in the range of frequencies measured. Taking into account that the oxide layers should be of iron oxide, the substitution of a part of this by an element, cobalt, which does not passivate in highly alkaline media, gives a oxide layer that is less protective than that obtained with the FINEMET alloy. As a result metal ions can diffuse through the layer and dissolve in the KOH solution.

3.2. Thermally Treated Samples

Figure 5 shows the voltammograms obtained for the thermally treated FINEMET and F–Co6 alloys. The CV treatments consisted in 25 cycles of triangular sweeping

Table 3. *Calculated parameters of the impedance elements of high frequency loops for some F–Co alloys under treatment A and B.* Q_{dl} *was used instead of* C_{dl} *in order to avoid data dispersion due to the surface roughness.* R_t *= charge transfer resistance,* Q_{dl} *= constant phase element,* n *= exponent of the impedance of the capacitor* ($Z_{cap} = Q_{dl}(j\omega)^{-n}$)

	Treatment A			Treatment B		
	R_T (Ω cm^{-2})	Q_{dl} (F cm^{-2})	n	R_T (Ω cm^{-2})	Q_{dl} (F cm^{-2})	n
F-Co2	358	8.7 10^{-4}	0.92	2536	3.4 10^{-4}	0.90
F-Co4	367	5.8 10^{-4}	0.91	2168	3.9 10^{-4}	0.90
F-Co6	383	5.9 10^{-4}	0.90	1343	4.7 10^{-4}	0.92
F-Co8	458	3.5 10^{-4}	0.91	1344	5.0 10^{-4}	0.91
F-Co10	585	3.4 10^{-4}	0.90	1264	5.2 10^{-4}	0.91

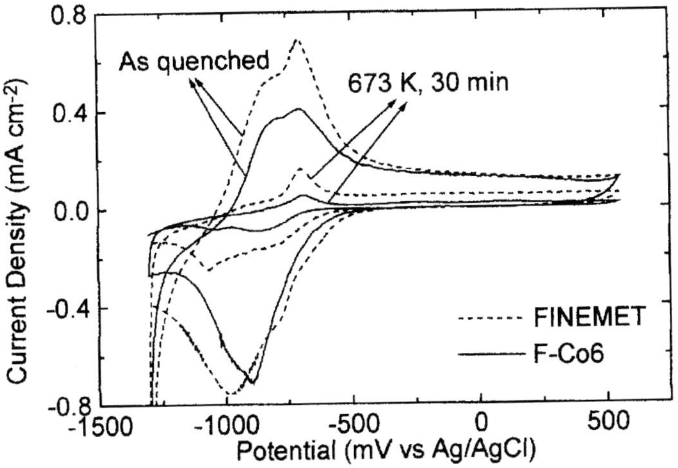

Fig. 5 *Comparison of the voltammograms of FINEMET and F–Co6 alloys in the as-quenched and relaxed states in 0.5*N *KOH after 25 cycles. Scans from 1300 to +550 mV vs Ag/AgCl at 100 mV* s^{-1}*, recorded at 20 mV* s^{-1}*.*

from $E_c = -1300$ mV to $E_a = +550$ mV at a scan rate of 100 mV s^{-1}. The plots were then recorded at a scan rate of 20 mV s^{-1}. There are no important changes in the peak potentials after the thermal treatments but they are much smaller than those obtained for as-quenched samples. The decrease of the anodic charge in the plots indicates that the oxidation reactions take place to a lower extent, that is, relaxation processes improve the electrochemical characteristics of the alloys.

The analysis of the electrolytic solutions after treatment A (Table 4) reveals similar amounts of iron for all cobalt containing alloys. There were no measurable amounts of cobalt into the KOH solutions. It has been reported that cobalt has a higher affinity than iron for silicon [6], thus, during the TSRO it should tend to form Co–Si pairs. On the other hand, the decrease of the Fe concentration found in the solution after treatment B compared to treatment A indicates the protective behaviour of the oxide layers.

Table 4. *Measured ion amounts found in KOH solutions for some thermally treated F–Co$_X$ alloys*

	Treatment A Fe (μg cm^{-2})	Treatment B Fe (μg cm^{-2})
FINEMET	20.5	----
F-Co2	15.5	2.6
F-Co6	15.0	2.5
F-Co10	15.4	2.1

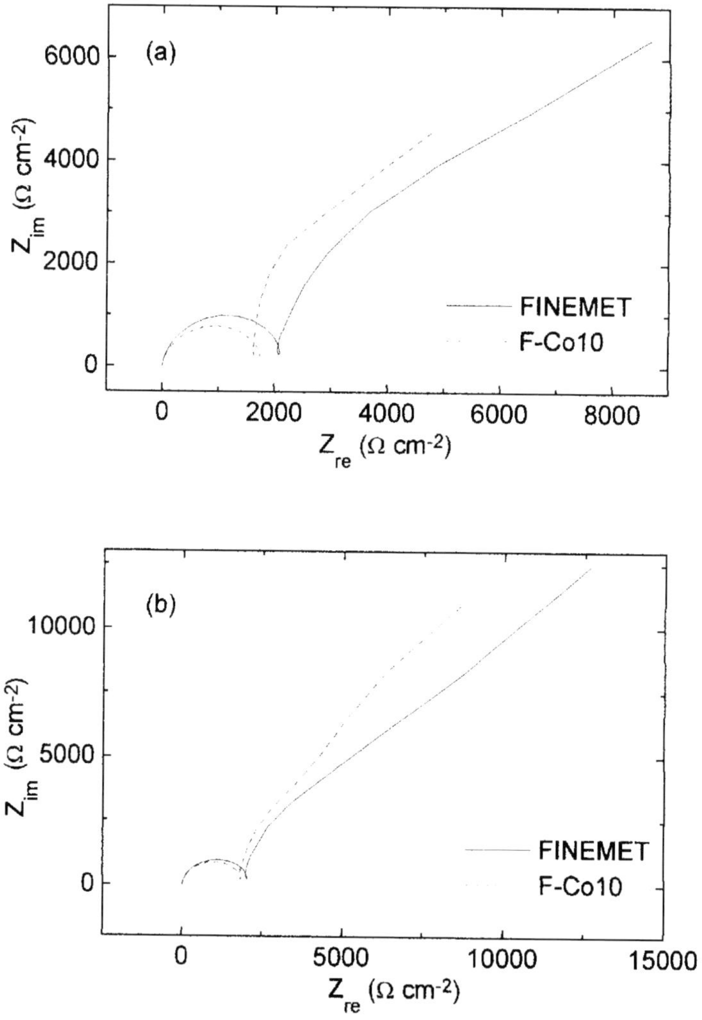

Fig. 6 *Impedance diagrams of thermally treated FINEMET and F–Co10 alloys in KOH 0.5*N *at potential* E = *–400 mV after (a) treatment A and (b) treatment B. Conditions: frequency sweep from 50000 Hz to 1 mHz, a.c. amplitude: 5 mV.*

Figure 6 (a) shows the impedance plots obtained for the treated alloys after both treatment A and B. It may be seen that the shape of the response obtained for the relaxed FINEMET alloy is very similar to that obtained for the as-quenched sample but the charge transfer resistance increases as expected for a more stable state. When very small amounts of cobalt (2 at.%) are present in the alloy composition the R_T value decreases, but when more than 4% of atoms is added, the charge transfer resistance increases again and the values measured are only slightly lower than those obtained for the as-quenched state (Table 5). The addition of cobalt deestabilises the oxide layer but, taking into account that only iron dissolves when it is substituted by cobalt, the resistance tends to increase again.

Figure 6 (b) shows the plots obtained when the annealed samples are submitted to treatment B. The effect of the cycling is to form a protective oxide layer. The performance of this layer deteriorates again at high contents of cobalt. For both treatments, the overall shape of the plots corresponds to a diffusion controlled process.

The results obtained suggest that the relaxation phenomena oppose the activation effect produced by addition of cobalt to FINEMET alloys, and result in the increase of resistance of the treated alloys compared with the as-quenched samples.

4. Conclusions

The presented results reveal the following conclusions:

- The addition of cobalt to the FINEMET alloys decreases the corrosion resistance of the alloys by enhancing the dissolution processes.
- The oxide layer formed during potentiodynamic cycling prevents the further oxidation of the alloys. The performance of this layer decreases with increase of cobalt content.
- Thermal annealing results in an improvement of the charge transfer resistance of the cobalt containing alloys due to the topological short range order changes in the alloy.

Table 5. *Calculated parameters of the impedance elements of high frequency loops for some thermally treated F–Co alloys with treatment A and B.* Q_{dl} *was used instead of* C_{dl} *in order to avoid data dispersion due to surface roughness.* R_t = *charge transfer resistence,* Q_{dl} = *constant phase element,* n = *exponent of the impedance of the capacitor* ($Z_{cap} = Q_{dl}(j\omega)^{-n}$)

	Treatment A			Treatment B		
	R_T (Ω cm^{-2})	Q_{dl} (F cm^{-2})	n	R_T (Ω cm^{-2})	Q_{dl} (F cm^{-2})	n
F-Co2	2064	4.4 10^{-5}	0.92	3497	1.2 10^{-4}	0.89
F-Co4	944	3.4 10^{-5}	0.92	2893	2.9 10^{-5}	0.93
F-Co6	987	1.1 10^{-4}	0.92	2381	7.6 10^{-5}	0.91
F-Co8	1492	1.0 10^{-4}	0.91	2190	6.8 10^{-5}	0.90
F-Co10	1856	3.4 10^{-5}	0.87	1965	1.2 10^{-4}	0.91

5. Acknowledgements

The authors acknowledge the financial support of the Comisión Interministerial de ciencia y Tecnología (CICYT) (project MAT95-0961-C02-02) and the Ministerio de Educacíon y Cultura (MEC) (project PB97-1119-C02-02) for this work.

References

1. K. Hono, K. Hiraga, Q. Wang, A. Inoue and T. Sakurai, *Acta Metall. Mater.*, 1997, **40**, 2137.
2. F. F. Marzo, A. R. Pierna, A. Altube, M. Sistiaga, A. Lorenzo, A. Salaverria and A. Fernández, *Mater. Sci. Forum*, 1998, **289–292**, 1047.
3. J. Füzer, P. Kollár, A. Zorkovská, P. Sovak, P. Matta and M. Konc, *Mater. Sci. Eng.*, 1997, **A226–228**, 199.
4. L. Kraus, V. Haslar, P. Duhaj, P. Svec and V. Studnicka, *Mater. Sci. Eng.*, 1997, **A226–228**, 626.
5. E. Jakubzyk, Z. Mandecky and J. Filipecki, *J. Non-crystalline Solids*, 1995, **192–193**, 509.
6. A. Altube, A. R. Pierna, F. F. Marzo, M. Sistiaga and A. Lorenzo, in *Non-Crystalline and Nanoscale Materials*. Ed. World Scientific, 1998, 341.
7. J. R. MacDonald, *Impedance Spectroscopy*. Ed. Wiley and Sons, 1987.
8. D. D. MacDonald, *Electrochim. Acta*, 1990, **35**, 1509.
9. F. F. Marzo, A. R. Pierna, A. Altube, M. Sistiaga, A. Lorenzo and X. Artola, *Mater. Sci. Forum*, 1998, **289–292**, 1041.
10. H. Onishi (Ed.), *Photometric Determination of Traces of Metals*. Wiley, New York, 1986.
11. J. Filipecki, Z. Mandecky, C. F. Conde and A. Conde, *J. Mater. Sci.*, 1998, **33**, 2171.
12. A. Serebryakov, V. Sedykh, N. Novokhatskaya and V. Stelmukh, *Nanost. Mater.*, 1996, **7**, 461.
13. A. R. Pierna, F. Fernández and A. Lorenzo, *J. Braz. Chem. Soc.*, 1995, **6**, 55.
14. W. K. Behl and J. E. Toni, *J. Electroanal. Chem.*, 1971, **31**, 63.
15. H. Gómez Meier, J. R. Vilche and A. J. Arvia, *J. Electroanal. Chem.*, 1982, **134**, 251 and 1982, **138**, 367.
16. O. E. Barcia and O. R. Mattos, *Electrochim. Acta*, 1990, **35**, 1601.

21
Influence of Thermal Relaxation and Nanocrystallisation on the Corrosion Behaviour of $Fe_{(73.5-X)}Cu_1Nb_3Si_{13.5}B_9Co_{X(1,2,3,4,5)}$ Metallic Amorphous Alloys

F. F. MARZO, A. R. PIERNA, A. LORENZO, A. ALTUBE, M. SISTIAGA and J. ARANOA

Chemical Engineering and Environment Dept., University of the Basque Country, P. O. Box 1379, 20080 San Sebastian, Spain

ABSTRACT

The electrochemical behaviour of $Fe_{(73.5-x)}Cu_1Nb_3Si_{13.5}B_9Co_{X(1,2,3,4,5)}$ amorphous, isothermally annealed, and nanocrystallised FINEMET type alloys has been investigated in 0.5N KOH and in 2N HCl solutions using cyclic voltammetry and potentiodynamic polarisation techniques respectively. The anodic voltammetric charges obtained in the growth of oxide layers in 0.5N KOH solution, are not sensibly modified by the addition of small quantities of cobalt. The thermal treatment improves its resistance to corrosion in alkaline media as the Co at.% content is increased. The additon of Co does not sensibly modify the polarisation curves of amorphous alloys in acid media. The most interesting aspect in the polarisation curves of the nanocrystalline alloys, obtained by isothermal annealing at 600°C for 30 min, is the appearance of two passivation processes during the anodic sweeping of potential suggesting the dissolution of two different phases present in the alloy. An attempt has been made to model the anodic dissolution of these alloys as a material with two phases of different activity.

1. Introduction

Since its development in 1988 by Yoshizawa *et al.* [1], the $Fe_{73.5}Cu_1Nb_3Si_{13.5}B_9$ alloy with nanoscale crystallites, also known as FINEMET type alloy, has received much attention. As a result of the excellent soft magnetic properties of these materials, they have attracted particular attention for structural and magnetic properties studies.

While the magnetic properties, crystallisation kinetics and structural changes of FINEMET type alloys have been studied extensively, there is little information available on the electrochemical properties of these alloys [2,3]. As previously reported, the soft magnetic properties of the Fe-based amorphous alloys could be significantly modified by the addition of transition metal elements. The addition of chromium to FINEMET type alloys appears to be especially interesting with respect to the combination of soft magnetic properties and corrosion resistance. The influence of Cr addition on the soft magnetic properties of FINEMET alloys has been discussed recently [4,5] and the influence of nanocrystallisation on the corrosion behaviour of $Fe_{(73.5-X)}Cu_1Nb_3Si_{13.5}B_9Cr_{X(0-5)}$ FINEMET type alloys has been studied [6]. The addition of Co [7,8] and Mo [9] to nanocrystalline FINEMET type alloys has also been investigated.

Moreover, the corrosion behaviour of amorphous alloys is significantly sensitive to the presence of microcrystallites in the amorphous matrix [10,11]. The ultrafine grains of nanocrystalline alloys, obtained by crystallisation from the amorphous state, change the electrochemical behaviour of these alloys. The passive films of the nanocrystalline FINEMET alloys and FINEMET alloys modified with chromium are formed as a result of preferential dissolution of iron from the nanocrystalline phase [6]. Therefore, it may be supposed that the rate of dissolution of the nanocrystals is much higher than the rate of dissolution of the amorphous matrix. The present study was made to determine the influence of the microstructural changes and alloy composition on the electrochemical behaviour of $Fe_{(73.5-X)}Cu_1Nb_3Si_{13.5}B_9Cr_{X(1-5)}$ alloys in alkaline media (designated F–CoX(1–5) alloys; where F = $Fe_{(73.5-X)}Cu_1Nb_3Si_{13.5}B_9$). On the basis of potentiodynamic polarisation curves, an attempt has been made to model the anodic dissolution of the nanometer-sized α–Fe(Si) crystallites embedded in a residual glassy metallic matrix.

2. Experimental

Alloy ingots were prepared by arc melting in a water cooled copper crucible with He. From these ingots, $Fe_{(73.5-X)}Cu_1Nb_3Si_{13.5}B_9Cr_{X(1-5)}$ amorphous alloy ribbons of about 5 mm wide and 20–30 μm thick were prepared by the Planar Flow Casting method. The amorphous structure of the samples was verified by X-ray diffraction.

The electrochemical measurements were made in a typical three-electrode cell. Working electrodes consisting of amorphous alloy ribbons were used. Electrode potentials were measured against an Ag/AgCl reference electrode with a built-in electrolyte bridge to prevent chloride ion diffusion into the electrolyte solution. A platinum sheet was used as a counter electrode. Before the electrochemical measurements, the samples were degreased in acetone and rinsed with distilled water. The experiments were made at room temperature. The potentials were controlled with a Princeton Applied Research Model 273 potentiostat.

Surface oxide layers were obtained on the F–CoX(1–5) amorphous and heat treated alloys when these were subjected to repetitive triangular potential sweep (RTPS) treatment between different limits and with different sweep rates in 0.5N KOH alkaline solution.The thermal behaviour of the alloys was studied by Differential Scanning Calorimetry (DSC). The isothermal annealing was carried out in a quartz tube under a helium atmosphere at 400 and 600°C. Moreover, from potentiodynamic polarisation curves various kinetic parameters namely, corrosion potential, critical passivity current density and Tafel slopes, were obtained to determine the electrochemical corrosion of these alloys.

3. Results and Discussion

3.1. Alkaline Media

A standard procedure was adopted in obtaining the optimal electrochemical variables for oxide growth on $Fe_{(73.5-X)}Cu_1Nb_3Si_{13.5}B_9Cr_{X(1-5)}$ alloys in 0.5N KOH solution. Thus, the potential was first swept during 200 cycles between a +0.2 V fixed anodic limit and a variable cathodic limit between –1.2 V and –3.4 V, at a sweep rate of 0.2 Vs^{-1}. Cyclic voltammograms, from which the anodic charge capacity of the oxide layers was measured, were recorded after each treatment between potential limits of –1.2 V and +0.2 V at an analytical scan rate of 0.2 Vs^{-1} (control voltammograms). Table 1 displays the optimal cathodic limits, L_{copt}, obtained from this procedure.

As seen in Fig. 1, for samples with Co<3 at.%, it is necessary to reach high cathodic limits to obtain an appreciable amount of anodic voltammetric charge during the cyclic treatment of potential. The same electrochemical behaviour can be found in FINEMET alloys, $Fe_{(73.5-X)}Cu_1Nb_3Si_{13.5}B_9$, and in FINEMET type alloys modified with Cr and Mo [6,9]. The voltammetric charges obtained below –2.0 V are relatively small for the five samples, and it was necessary to reach –2.8 V to obtain appreciable values of charge in the alloys of Co<3 at.%. With increasing cathodic potential, a comparable voltammetric charge is observed, reaching a maximum value of about 40 mC cm^{-2}. The L_{copt} was then fixed at –2.8 V for the five alloys, and the potential was swept between the L_{copt} and a variable anodic limit between –0.2 V and +0.6 V at a sweep rate of 0.2 V s^{-1}. When the anodic limit is displaced towards more positive values the voltammetric charge increases, reaching stable values at +0.2 V. For anodic limit values higher than +0.6 V, the voltammetric charge decreases. The results obtained for the F–Co1 alloy are given in Fig. 2. The optimal anodic limit, L_{aopt}, was fixed at +0.2 V. The same results were obtained for F–CoX(2–5) amorphous alloys. With L_{copt} and L_{aopt}, the potential was swept at a sweep rate between 0.1 and 2 V s^{-1}. While for L_{copt}, a marked dependence of Co at.% on oxide growth was observed, L_{aopt} and S_{opt} were not appreciably modified by the cobalt content in the alloy. The optimal sweep rate, S_{opt}, was 0.2 V s^{-1}. Finally, with L_{copt}, L_{aopt}, and S_{opt}, the influence of the number of cycles was studied in amorphous, annealed and nanocrystalline alloys. The results are shown in Figs 3–5.

Several models have been proposed to explain the composition of the passive films on iron [12]. In the double layer model, the film has a two-layer structure with different chemical compositions. The protective inner layer, anhydrous, would be probably formed by Fe_3O_4, while in the outer layer hydrated iron oxide would dominate. For many metals, e.g. platinum and iron [13] the cathodic limit in potential cycling experiments corresponds to extensive rupturing of the protective inner layer

Table 1. *Optimal cathodic limits obtained in the generation and growth of the oxide surface films in the F–CoX(1–5) amorphous alloys in 0.5*N *KOH solution*

F-CoX	L_{copt} **(V)**
X = 1	–2.80
X = 2	–2.80
X = 3	–2.25
X = 4	–2.25
X = 5	–2.25

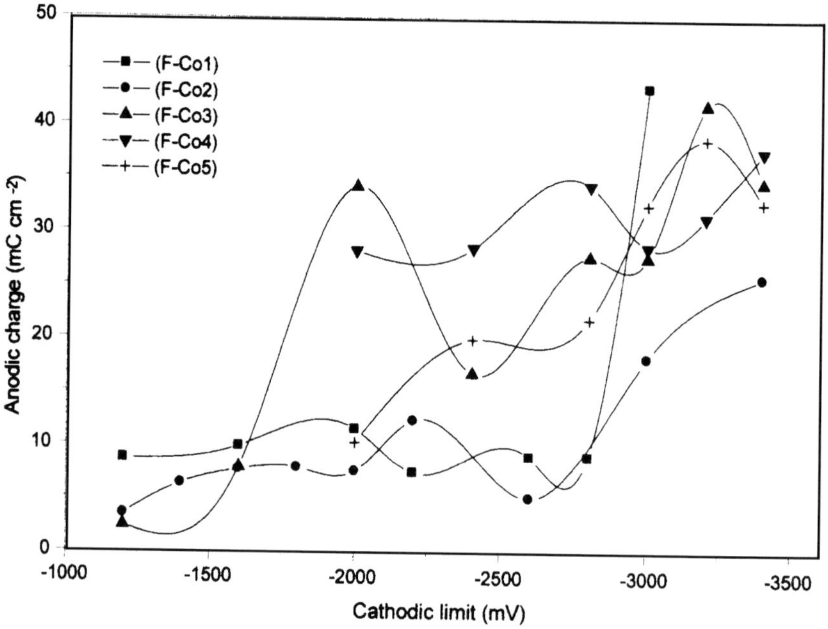

Fig. 1 Anodic voltammetric charge, Q, obtained for F–CoX(1–5) amorphous alloys associated with the anodic contribution of control voltammograms as a function of the cathodic limit after 200 cycles at a potential sweep rate of 0.2 V s^{-1} in 0.5N KOH solution. The analytical scan was registered between –1.2 V and +0.2 V at a potential sweep rate of 0.2 Vs^{-1}.

in the oxide film. In the F–CoX amorphous alloys, the voltammetric charge associated with the oxide film grown on potential cycling does not increase significantly at potentials of –2.0 V, i.e. in the hydrogen evolution region, as the cobalt content in the alloy is greater than 3 at.%. These results suggest that the presence of an adequate cobalt content in the alloy provides a high electrocatalytic activity for hydrogen evolution. As a result, the hydrogen overvoltage decreases significantly and the film growth is enhanced.

The anodic voltammetric charges of F–CoX(1–5) amorphous alloys obtained with the optimal electrochemical variables are shown in Fig. 3. This charge indicates the accumulation of anodically formed oxide surface layer during potential cycling. The addition of cobalt to the alloy does not modify significantly the anodic charge. It is only F–Co4 that shows an anomalous electrochemical behaviour in these five samples. As the alloys obtained by rapid solidification are amorphous and homogeneous in their composition, the change in their electrochemical behaviour may be the result of a non-homogeneous structure of F–Co4 amorphous alloy arising from the use of different melting temperatures during production. It has been shown that the electrochemical properties of alloys melted at different temperatures were significantly influenced by the initial melt temperature [14,15]. It can be seen in Fig. 4, that when the alloys were annealed at 400°C for 30 min, their electrochemical behaviour was quite similar. These results support the hypotesis that the F–Co4 alloy shows an irreversible modification of short range order associated with the structural evolution of the alloy during annealing treatment [16].

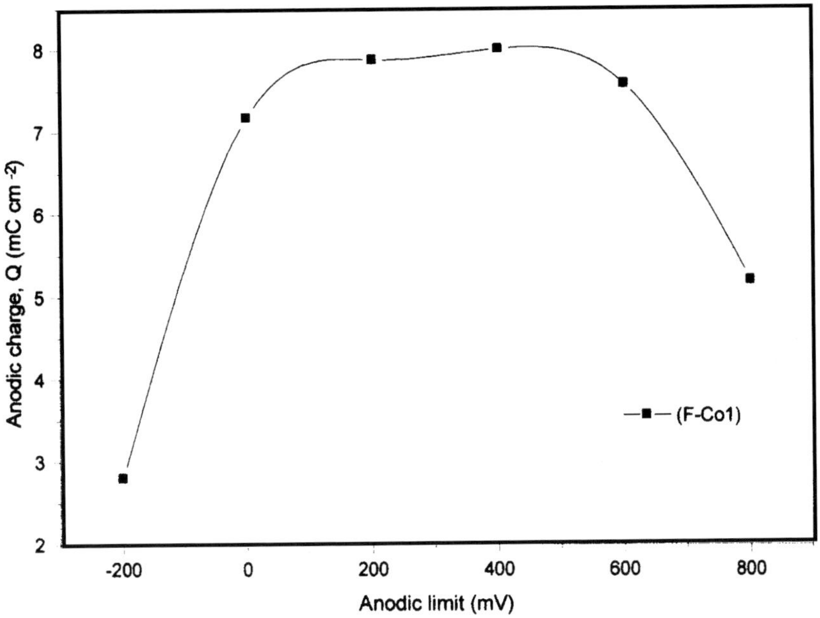

Fig. 2 *Anodic voltammetric charge, Q (mC cm^{-2}), obtained for the F–Co1 alloy associated with the anodic contribution of the control voltammograms as a function of the anodic limit after 200 cycles at a potential sweep rate of 0.2 V s^{-1} in 0.5N KOH solution. The upper limit was fixed at –2.8 V. The analytical scan was registered between –1.2 V and +0.2 V at a potential sweep rate of 0.2 Vs^{-1}.*

In Fig. 6 the cyclic voltammograms are given for FINEMET, $Fe_{40.5}Cu_1Nb_3Si_{13.5}B_9Co_{33}$ (F–Co33 alloy), and $Fe_{20}Co_{60}B_{20}$ amorphous alloys. Only the $Fe_{20}Co_{60}B_{20}$ amorphous alloy exhibits a very complex peak multiplicity. However, in the voltammogram of the F–Co33 alloy only a peak in the anodic oxidation is observed, as in the voltammogram of the FINEMET alloy. From the literature on electrochemical processes and hydrous oxide growth on the single components of amorphous alloys, such as iron [17,18] and cobalt [19], the surface oxidation/reduction processes on F–CoX(1–5) amorphous alloys can be described as follows. Peaks 1 (see Fig. 6) for FINEMET and F–Co33 alloys are related mainly to the oxidation of Fe to form $Fe(OH)_2$, and the transition $Fe(OH)_2$/FeOOH although somewhat higher oxides have been proposed in the oxidation of iron species at these potentials [20]. In the $Fe_{20}Co_{20}B_{20}$ alloy the peak 1 may be interpreted by oxidation of cobalt with formation of a cobaltous hydroxide layer, $Co(OH)_2$. At this potential the simultaneous oxidation of iron is expected from the presence of a shoulder at more positive potentials. The anodic peak 2 may be attributed to the further oxidation of cobalt hydroxide.

The behaviour observed in Fig. 6 suggests that the formation of oxides in F–Co33 is inhibited. The same electrochemical behaviour was found in F–CoX(1–5) amorphous alloys. It has been reported [8] that the oxidation of FINEMET type alloys modified with cobalt is prevented by superficial silicates on the alloy, and this seems to be responsible for the decrease in the anodic charges of voltammograms in Fig. 6. The L_{aopt} was +0.2 V (see Fig. 2). At more negative potentials than L_{aopt} the peaks in

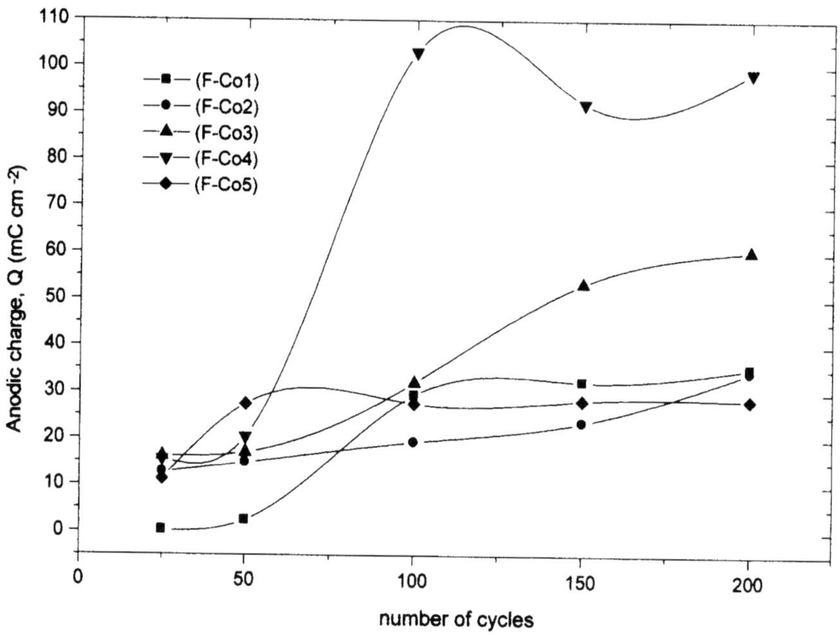

Fig. 3 Anodic voltammetric charge Q (mC cm^{-2}), obtained for the F–CoX(1–5) amorphous alloys associated with the anodic contribution of the control voltammograms as a function of the number of cycles in 0.5N KOH solution. The upper limit was the L_{opt} of each alloy. The L_{opt} and S_{opt} were fixed at +0.2 V and 0.2 V s^{-1} respectively for all the alloys. The analytical scan was registered between −1.2 V and +0.2 V at a potential sweep rate of 0.2 Vs^{-1}.

the voltammograms are related to the formation of iron and cobaltous hydroxides, although at more positive potentials than L_{aopt}, only the further oxidation of cobalt hydroxide is expected. As these oxidation proceses are inhibited, the voltammetric charge does not show anodic potential dependence. Only when the anodic limit reaches the onset of the oxygen evolution is the voltammetric charge decreased. This seems to be due to the high electrocatalytic activity for oxygen evolution at these potentials and the destruction of the oxide layer by gas evolution in the surface of the alloy.

The thermal behaviour of F–CoX(1–5) was studied by DSC. The results obtained for FINEMET and F-Co4 alloys (the other samples have similar behaviour) are shown in Fig. 7. The plot shows an exothermic peak at 546°C for FINEMET alloy, corresponding to the nanocrystallisation of the sample [21]. The cobalt addition does not significantly modify the nanocrystallisation process of the samples. Nanocrystalline alloys were obtained by isothermal annealing at 600°C for 30 min. Figure 5 shows the anodic voltammetric charges of F–CoX(1–5) nanocrystalline alloys obtained with the optimal electrochemical variables. As can be seen, nanocrystallisation provides a higher corrosion stability in alkaline media than does the amorphous state when the Co at.% is increased. The peaks of cyclic voltammograms for F–CoX(1–5) nanocrystalline alloys are decreased, compared with amorphous alloys, but no changes in the potential peaks are observed. Investigations to the corrosion stability of these nanocrystalline alloys in alkaline media are continuing.

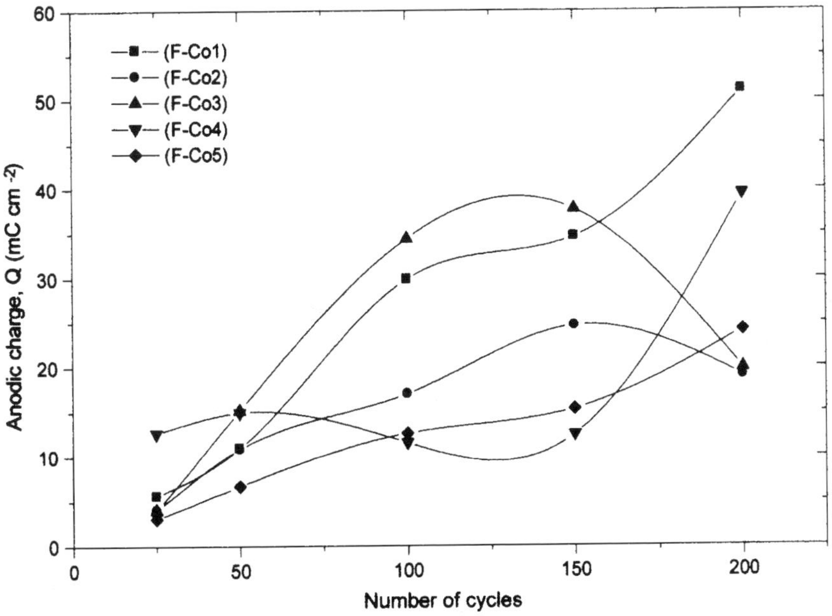

Fig. 4 *Anodic voltammetric charge, Q (mC cm^{-2}), of annealed alloys obtained from the F–CoX(1–5) amorphous alloys after isothermal annealing at 400°C for 30 min, associated with the anodic contribution of the control voltammograms as a function of the number of cycles in 0.5N KOH solution. The upper limit was the* L_{opt} *of each alloy. The* L_{opt} *and* S_{opt} *were fixed at +0.2 V and 0.2 V s^{-1} respectively for all the alloys. The analytical scan was registered between –1.2 V and +0.2 V at a potential sweep rate of 0.2 Vs^{-1}.*

3.2. Acid Media

The potentiodynamic polarisation curves of F–CoX(1–5), amorphous alloys, obtained in 2N HCl solution, are given in Fig. 8. Various kinetic parameters namely, corrosion potential (E_{corr}), critical passivity current density (i_{crit}; at peak), Tafel slopes (β_a and β_c), and polarisation resistance (R_p) were calculated from these curves. The results obtained are summarised in Table 2.

As can be seen in Fig. 8, the F–CoX(1–5) amorphous alloys show an active dissolution without any revelation of passivity in 2N HCl solution. The addition of Co to the FINEMET type alloy in the as-quenched state does not sensibly modify the polarisation curves.

The corrosion potential for the F–Co5 alloy is displaced slightly towards more noble potentials. From Table 2 there appears to be no systematic change in kinetic parameters with the increase of Co at.% in the F–CoX(1–5) amorphous alloys.

The polarisation curves of F–CoX nanocrystalline alloys, obtained by isothermal annealing under helium atmosphere at 600°C for 30 min, are given in Fig. 9. The kinetic parameters obtained for these curves are summarised in Table 3. As in the case of F–CoX(1–5) amorphous alloys, there appears to be no systematic change in kinetic parameters with the cobalt addition. The experimental results of Fig. 10 indicate, from the two well resolved peaks of the curves, that the anodic polarisation

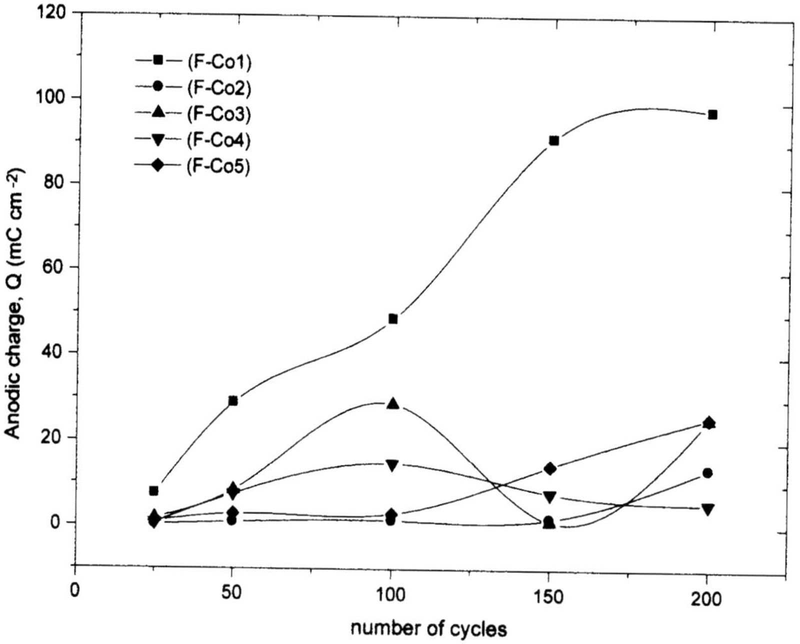

***Fig. 5** Voltammetric charge Q (mC cm^{-2}), of nanocrystalline alloys obtained from the F–CoX(1–5) amorphous alloys after isothermal annealing at 600°C for 30 min, associated with the anodic contribution of the control voltammograms as a function of the number of cycles in 0.5N KOH solution. The upper limit was the L_{opt} of each alloy. The L_{opt} and S_{opt} were fixed at +0.2 V and 0.2 V s^{-1} respectively for all the alloys. The analytical scan was registered between –1.2 V and +0.2 V at a potential sweep rate of 0.2 Vs^{-1}.*

experiment in nanocrystalline alloys proceeds by at least two steps. After the first passivation potential there is a typical drop in the anodic current at the onset of passivity. On further gradual increase of the potential, the current rises again due to the oxidation of the metal.

Nanocrystallisation leads to a shift in the onset of the first anodic peak of the alloy. A similar influence of nanocrystallisation was observed in FINEMET alloys where the nanocrystalline phase within a residual amorphous matrix is more active and is preferentially dissolved compared to the amorphous phase [6]. It has also been reported that compositional changes and grain growth change the electrochemical behaviour of amorphous Fe–B–S–Cu–Nb amorphous alloys [22].

Controlled annealing of FINEMET was used to study the structural changes occurring on crystallisation. The alloy generally demostrates multiple stages of crystallisation. It was initially felt that the result of the primary crystallisation process was a body centred cubic α–Fe(Si) phase, but there is increasing evidence now that the primary product is an ordered DO_3 phase having a stoichiometry close to $Fe_{80}Si_{20}$ [23]. The crystallisation process in these alloys involves subsequent secondary transformation of the remaining amorphous grain bondary phase to form various intermetallic compounds. In the alloys that form the ordered DO_3 phase ($Fe_{80}Si_{20}$) on crystallisation, the process can be denoted by the mass balance equation [24]:

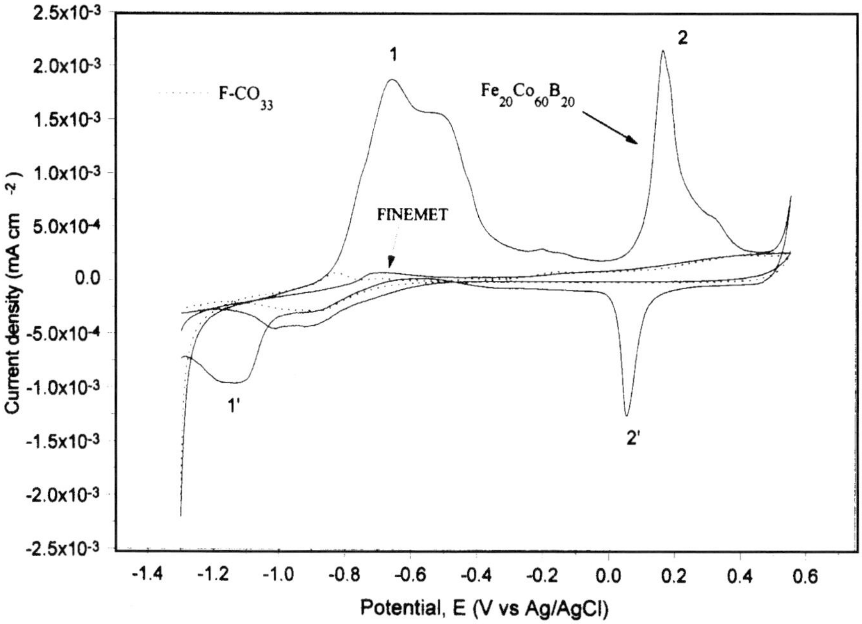

Fig. 6 *Typical voltammograms for FINEMET,* $Fe_{40.5}Cu_1Nb_3Si_{13.5}B_9Co_{33}$ *(F–Co33), and* $Fe_{20}Co_{60}B_{20}$ *amorphous alloys obtained in 0.5N KOH solution at room temperature. The sweep rate was fixed at 0.2 V* s^{-1}*.*

$$Fe_{73.5}Cu_1Nb_3Si_{13.5}B_9 \rightarrow 0.675\ Fe_{80}Si_{20}(\text{nano.}) + 0.325Fe_{60}B_{28}Nb_8\ (\text{amor.}) + Cu$$

The above mass balance equation is valid after completion of the first stage of crystallisation, and in any intermediate stage the volume fraction of $Fe_{80}Si_{20}$ phase as well as the composition of the remaining amorphous region will have different values from that indicated above. The $Fe_{78-X}Co_XSi_9B_{13}$ amorphous alloys partially crystallised show the presence of a bcc solid solution α–Fe(Co, Si) and a Fe_3B-type phase at the end of the first crystallisation stage [16]. Therefore, there are three factors that could affect the electrochemical responses of these alloys, namely, the presence or absence of different phases, the volume fraction of these phases and the composition of the residual amorphous matrix. Moreover in FINEMET alloys modified with Cr [25], it has been reported that the magnetic properties of nanophase does not depend on the chromium content of the alloy, and this suggests that Cr should be excluded from nanocrystals. Thus, passivating elements in FINEMET type alloys, like Cr or Co, differently distributed between the nanocrystalline phase and the amorphous matrix during grain growth, can be expected to change the corrosion rates of these phases.

On the basis of potentiodynamic polarisation curves a model has been proposed to characterise the anodic dissolution of amorphous alloys containing crystalline clusters. L. Péter *et al.* [26] have derived equations for the electrochemical behaviour of amorphous alloys in which the area under the peak of polarisation curves is proportional to the charge required for the dissolution of the much more active secondary phases. In F–CoX(1–5) nanocrystalline alloys, the surface of the samples

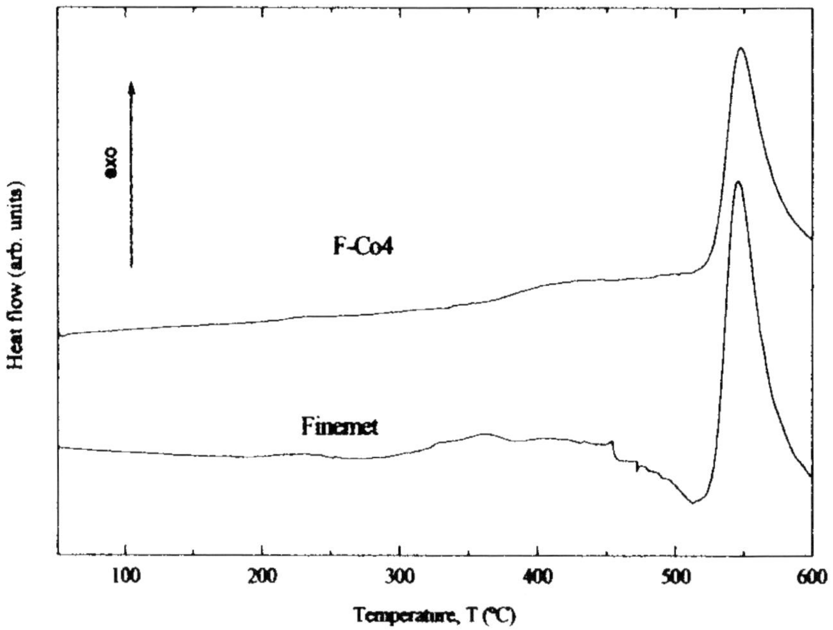

Fig. 7 DSC curves of FINEMET and F–Co4 alloys at a heating rate of 10°C min^{-1}.

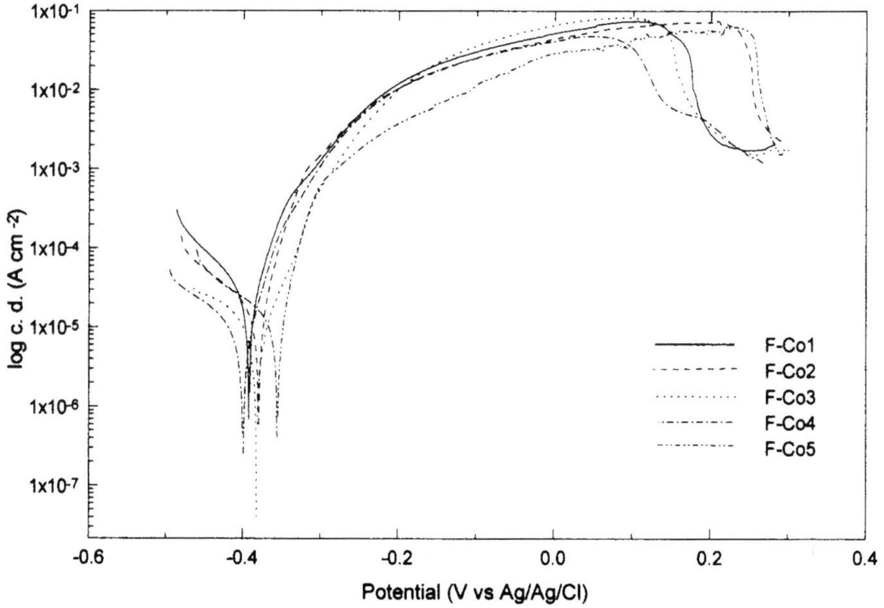

Fig. 8 Potentiodynamic polarisation curves of F–CoX(1–5) amorphous alloys obtained in 2N HCl at room temperature. The sweep rate was fixed at 20 mV min^{-1}.

is partially covered with nanocrystalline phase, and it may be supposed that the corrosion rates of nanocrystalline grains, obtained by crystallisation from the

Table 2. *Kinetic parameters obtained from polarisation data of Fig. 8 for F–CoX(1–5) amorphous alloys in 2N HCl at room temperature*

Alloy	E_{corr} mV (Ag/AgCl)	i_{cp} (mA cm^{-2})	$\beta_a \times 10^3$ (V/decade)	$\beta_c \times 10^3$ (V/decade)	I_{corr} (µA cm^{-2})	R_p (Ω cm^{-2})
F-Co1	–389	73.8	43.92	115.5	30.5	40.5
F-C02	–379	71	33.65	147.5	18.1	49.0
F-Co3	–382	81.9	51.89	123.9	9.8	143.9
F-Co4	–379	83	39.12	139.3	9.0	124.9
F-Co5	–357	64	27.03	143.1	11.5	58.8

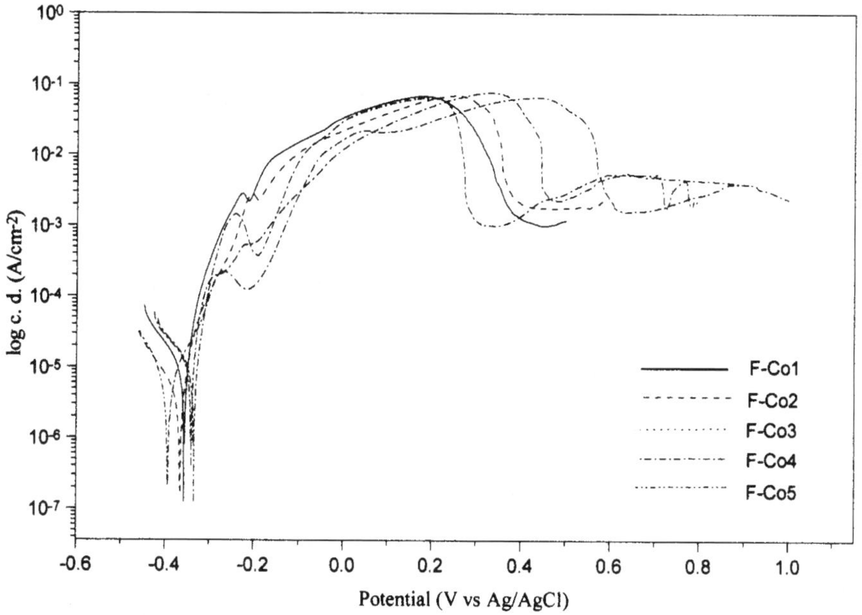

Fig. 9 *Potentiodynamic polarisation curves of F–CoX(1–5) nanocrystalline alloys obtained in 2N HCl at room temperature. The sweep rate was fixed at 20 mV min^{-1}.*

amorphous state, are much higher than the corrosion rate of the amorphous matrix. On the basis of this electrochemical behaviour an attempt has been made to model the anodic dissolution of the nanometre-sized α–Fe(Si) crystallites embedded in a residual glassy metallic matrix. The equation that applies to this model is [26]:

$$j_t = k_k C_k \exp(-b_k E) + k_{am} \exp(b_{am} E) + k_n \Theta_0 \exp\left[\frac{k_n}{\nu \delta b_n} \exp(b_n Eo)\right] \exp\left[b_n E - \frac{k_n}{\nu \delta b_n} \exp(b_n E)\right] \quad (1)$$

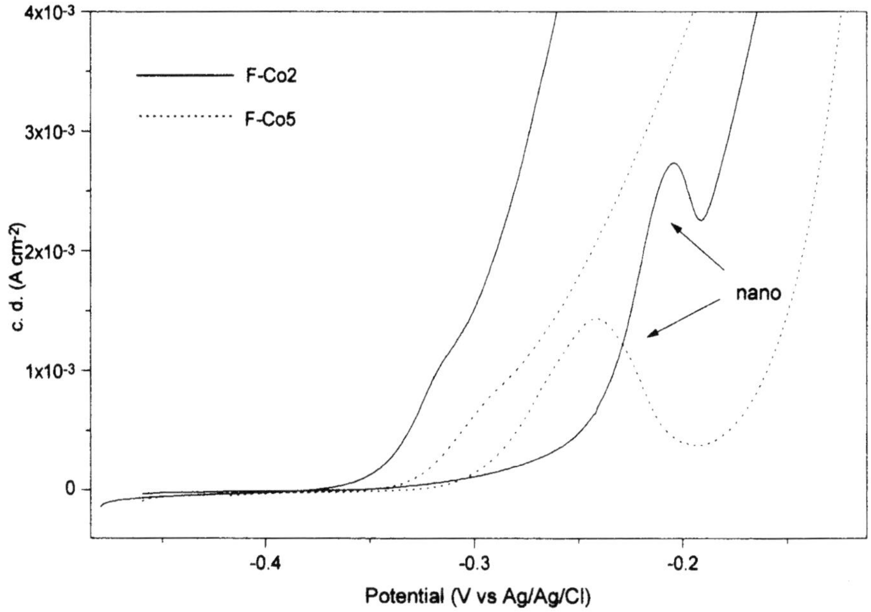

Fig. 10 *Potentiodynamic polarisation curves of F–Co2 and F–Co5 amorphous and nanocrystalline alloys obtained in 2N HCl at room temperature. The sweep rate was fixed at 0.02 V min*$^{-1}$.

The total current density, j_t, is assumed to be the sum of the anodic current densities originating from the dissolution of the nanocrystaline phase and the amorphous matrix, with characteristic rate constants K_n, and K_{am} respectively (independent of the electrode potential); b_n and b_{am} are the Tafel constants characteristic of the anodic dissolution of the nanocrystalline phase and the amorphous matrix respectively, E is the potential referring to a given reference electrode, ν the sweep rate, E_0 the starting potential, Θ_0 the fraction of the sample surface covered with the nanocrystalline phase

Table 3. *Kinetic parameters obtained from polarisation data of Fig. 9 for F–CoX(1–5) nanocrystalline alloys in 2N HCl at room temperature*

Alloy	E_{corr} **mV (Ag/AgCl)**	i_{cp} **(mA cm^{-2})**	$\beta_a\times10^3$ **(V/decade)**	$\beta_c\times10^3$ **(V/decade)**	I_{corr} **(μA cm^{-2})**	R_p **(Ω cm^{-2})**
F-Co1	–335	68	28.94	124.0	9.67	75.06
F-C02	–365	69	34.06	114.5	1.17	208.75
F-Co3	–415	57.6	130.8	156.4	13.13	230.06
F-Co4	–368	36.1	47.14	143.3	5.47	253.06
F-Co5	–334	63.3	28.98	142.0	10.31	70.72

before anodic polarisation (at $t = 0$), the proportionality factor (related to $Q(t)$, the charge passing through unit surface during time t on account of the dissolving nanocrystalline phase and $\Theta(t) = \Theta_0 - (Q(t)/\delta)$). The terms related to the cathodic process are c_k, b_k and k_k.

In Figs 11 and 12, the computer simulation of the polarisation curves is compared with the experimental results obtained for the F–Co2 and F–Co5 nanocrystalline alloys. The results presented in these Figures are in good agreement with experimental data and enable us to describe the electrochemical behaviour of the samples. Based on the simulation results it seems that the addition of Co mainly decreases the fraction of nanocrystalline phase present on the sample surface before anodic polarisation, Θ_0, and the dissolution rate of the nanocrystalline phase, k_n.

The results obtained for the dissolution rate of the amorphous phase, k_{am} are comparable. As the dissolution rate of the nanocrystalline phase is increased the potential peak corresponding to the maximum of the curve is shifted towards anodic potentials. The source for the dissolution rate dependence of the nanocrystalline phase may be explained by the different volume fractions of the different type of phases formed after primary crystallisation. Furthermore, it seems that the presence of Co 5 at.%, remaining in the amorphous phase during primary crystallisation, was beneficial since in the F–Co5 alloy the values of critical passivity current density and passivation potential were lower than in the F–Co2 alloy.

On the other hand, the crystallisation and subsequent grain growth of Fe–Si–B–Nb–Cu alloys are influenced by Cu and Nb additions [24,27,28]. As F–CoX(1–5)

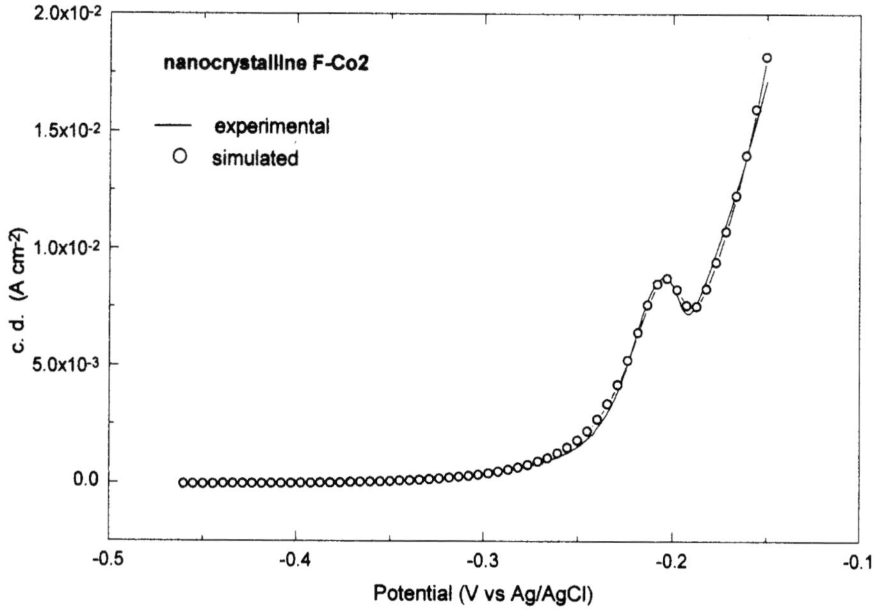

Fig. 11 *Experimental polarisation curve and simulated curve, obtained on basis of eqn (1), compared for F–Co2 nanocrystalline alloy. Data used for simulation are:* k_k: *0.5E–4* Am^{-2}; c_k: *2 mol* L^{-1}; b_k: 0.482×10^{-2} Am^{-2}; k_{am}: 769.67×10^{-3} Am^{-2}; Θ: 0.12×10^{-2}; k_n: 175.97×10^{6} Am^{-2}; v: 3.33×10^{-4} Vs^{-1}; δ: *360.342;* b_n: *80.629* V^{-1}; E_o: *-0.46 V;* b_{am}: *24.939* V^{-1}.

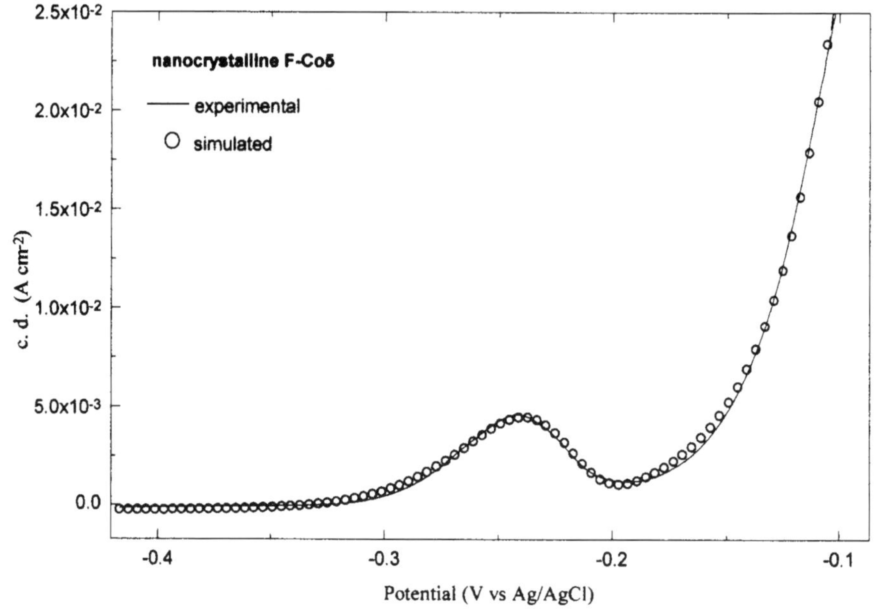

Fig. 12 *Experimental polarisation curve and simulated curve, obtained on basis of eqn (1), compared for F–Co5 nanocrystalline alloy. Data used for simulation are:* k_k: *1.4 × 10*$^{-4}$ *Am*$^{-2}$*;* c_k: *2 mol L*$^{-1}$*;* b_k: *0.482 × 10*$^{-2}$ *Am*$^{-2}$*;* k_{am}: *769.23 × 10*$^{-3}$ *Am*$^{-2}$*; Θ: 0.03 × 10*$^{-2}$*;* k_n: *0.838 × 10*^{6} *Am*$^{-2}$*; ν: 3.33 × 10*$^{-4}$ *Vs*$^{-1}$*; δ: 2914.497;* b_n: *41.436 V*$^{-1}$*;* E_o: *-0.417 V;* b_{am}: *33.155 V*$^{-1}$*.*

amorphous alloys were annealed at 600°C for 30 min, above the first crystallisation peak determined by DSC, we assume that with this annealing treatment the main grain size is stabilised. F–CoX(1–5) alloys contain the same at.% of Nb and Cu. Thus, the differences obtained in simulation for the surface coverage before the anodic polarisation, Θ_0, may be explained by the influence that the at.% of cobalt has on the grain growth during the annealing treatment. It has been also reported that the addition of chromium to FINEMET alloys changes the mean grain size of nanocrystals after the primary crystallisation [20]. This study did not attempt the quantification of the volume fraction of the individual phases, and supplementary DSC and X-ray diffractograms of the alloys corresponding to different positions on the DSC scan are necessary to extend the investigation.

4. Conclusions

From the experimental results obtained in the study of the electrochemical behaviour of the metallic glasses with a general composition $Fe_{73.5-X}Cu_1Nb_3B_9Si_{13.5}Co_{X(1-5)}$, the following conclusions have been obtained:

- The results obtained for the anodic voltammetric charges, in the generation of oxide layers in 0.5N KOH solution, led to the conclusion that the electrochemical behaviour of the amorphous alloys is not sensibly modified by the addition of small quantities of cobalt. It seems, however, that the different electrochemical behaviour of F–Co4 amorphous alloy is due to the non-homogeneous structure of the alloy since after annealing at 400°C, a temperature below the onset of nanocrystallisation, F–CoX(1–3,4) showed similar voltammetric charges. When the samples were annealed at 600°C for 30 min, the increased at.% of cobalt resulted in a gradual decrease in voltammetric charges.

- It was found that the addition of Co 1–5 at.% to $Fe_{73.5-X}Cu_1Nb_3B_9Si_{13.5}Co_{X(1-5)}$ alloys does not modify sensibly the polarisation curves in 2N HCl solutions. Although Co has not been found to change significantly the thermal behaviour of F–CoX(1–5) alloys, it is evident from the experimental results that even small differences in composition were enough to cause significant differences in the polarisation curves of nanocrystalline alloys. The results obtained by computer simultion are in good agreement with experimental data and enable us to describe qualitatively the preferential anodic dissolution of a nanocrystalline phase within the amorphous residual matrix.

5. Acknowledgements

The authors gratefully acknowledge financial support of this work by U.P.V. (112263–EC234/97) and M.E.C. (PBT97–1119–CO2–02).

References

1. Y. Yoshizawa, S. Oguna and K. Yamauchi, *J. Appl. Phys.*, 1988, **64**, 6044.
2. A. Pierna, F. F. Marzo and A. Lorenzo, *J. Braz. Chem. Soc.*, 1995, **6**, 55.
3. I. Chattoraj, K. Ram Mohan Rao, S. Das, A. Mitra, *Corros. Sci.*, 1999, **41**, 1.
4. Anna Slawska-Waniewska, *Proc. 4th International Workshop on Non-Crystalline Solids*, Spain, 3, 1994, World Scientific Publishing (1995), p.485.
5. J. M. Greneche and A. Slawska-Waniewska, *Proc. 5th International Workshop on Non-Crystalline Solids*, Spain, World Scientific Publishing (1997), p.233.
6. F. F. Marzo, A. R. Pierna, A. Altube, M. Sistiaga, A. Lorenzo, A. Salaverria and A. Fernández, *Mat. Sci. Forum*, 1998, **289–292**, 1047.
7. J. Filipecki, E. Jakubczyk and Z. Mandecki, *Proc. 4th Int. Workshop on Non-Crystalline Solids*, 1994, World Scientific Publishing (1995), p. 367.
8. A. Altube, A. Pierna, M. Sistiaga, F. F. Marzo and A. Lorenzo, *Proc. 5th Int.Workshop on Non-Crystalline Solids*, World Scientific Publishing (1997), p. 569.
9. F. F. Marzo, A. R. Pierna , A. Lorenzo, A. Altube and M. Sistiaga, *Mater. Sci. Eng. A*, 1998, **228**, 205.
10. B. P. Zhang, H. Habazaki, A. Kawashima, K. Asami, K. Higara and K. Hashimoto, *Corros. Sci.*, 1991, **32**, 443.
11. B. -M. Im, M. Komori, B. -T. Lee, E. Aoyagi, E. Akiyama, H. Habazaki, A. Kawashima, K. Asami, K. Hjiraga and K. Hashimoto, *Corros. Sci.*, 1995, **37**, 1411.
12. Z. Szklarska-Smialowska, *Pitting Corrosion of Metals*, NACE (1986), p.14.
13. L. D. Burke, *J. Electroanal. Chem.*, 1986, **198**, 346.

14. H. Habazaki, Y.-P. Lu, A. Kawashima and K. Hashimoto, *Corros. Sci.*, 1991, **32**, 1227.
15. H. Habazaki, S. Q. Ding, A. Kawashima, K. Asami, K. Hashimoto, A. Inoue and T. Masumoto, *Corros. Sci.*, 1989, **291**, 319.
16. J. Filipecki, K. Manzecki, C. F. Conde and A. Conde, *J. Mater. Sci.*, 1998, **33**, 2171.
17. D. D. McDonald and D. Owen, *J. Electrochem. Soc.*, 1973, **120**, 317.
18. D. Gaena, A. A. El Miligy and W. J. Lorenz, *J. Appl. Electrochem.*, 1974, **4**, 337.
19. H. Gomez Meier, J. R. Vilche and A. J. Arvia, *J. Electroanal. Chem.*, 1982, **134**, 251.
20. L. D. Burke and O. J. Murphy, *J. Electroanal. Chem.*, 1980, **109**, 379.
21. M. Millan, C. F. Conde and A. Conde, *J. Mater. Sci.*, 1995, **30**, 3591.
22. I. Chattoraj and A. Mitra, *Scripta Mater.*, 1998, **39** (6), 755.
23. A. R. Yavari, G. Fish, S. K. Das and L. Davis, *Mater. Sci. Eng.*, 1994, **A181/A182**, 1415.
24. R. Yavari, *Proc. 4th Int. Workshop on Non-Crystalline Solids,* 1994, World Scientific Publishing (1995), p. 35.
25. C. F. Conde, M. Millan and A. Conde. *J. Magn. Mat.*, 1994, **21**, 414.
26. L. Péter, G. Láng, L. Kiss, and J. Szalma. *J. Appl. Electrochem.*, 1996, **26**, 403.
27. Y. Yoshizawa, S. Oguna and K. Yamauchi, *J. Appl. Phys.*, 1988, **64**, 6044.
28. Y. Yoshizawa and K. Yamauchi, *Mater. Trans. Jpn. Inst. Met.*, 1990, **31**, 307.

Part 5

Corrosion Testing and Evaluation in Special Environments

22
The Anodic Behaviour of Iron and Steel in Very Alkaline Aqueous Solutions and in Cement Mortar

K. VIDEM
Centre for Materials Research, University of Oslo, Gaustadalleen 21, N-0349 Oslo, Norway

ABSTRACT
The kinetics of the anodic reaction of pure iron and steel were investigated by cyclic voltammetry, potentiostatic pulses and EIS. Steel was exposed to KOH solutions at pH 13 for more than two years, to cement mortar for one year and to the concrete of a large bridge for 17 years. Iron and steel are passive in oxygenated, strongly alkaline environments without aggressive ions. The passive current was independent of potential in tests of 36 h and was from 0.4×10^{-7} to 10^{-7} A cm^{-2} for pure iron. Capacitance measurements showed that the passive film became unstable when the potential was lowered below the formation potential. The capacitance was too high to originate from a physical capacitor with the film material as dielectric. Delayed electrochemical reactions created a phase shift between current and potential resembling the effect of a capacitance with a magnitude varying with frequency and with very high values for steel suffering chloride-induced corrosion. Due to these sluggish reactions, the polarisation resistance varied substantially with the method of measurements. These reactions, rather than diffusion in the environment, are also responsible for constant phase angle lines with EIS in alkaline solutions, cement mortar and concrete.

1. Introduction

This study is a part of a programme with the long range objective to contribute to a better understanding of the behaviour of steel in very alkaline environments and in concrete. Earlier reports have treated techniques for assessing the corrosion rate of steel in concrete, based on electrochemical experiments in the laboratory and in the field [1–3]. The present laboratory investigation is focused on properties of the corrosion film and the memory of earlier polarisation history. Powerful techniques for this purpose are electrochemical impedance spectroscopy (EIS) and cyclic voltammetry. As one cannot easily retrieve samples embedded in concrete for inspection and mass loss determination, reliable electrochemical measurements and interpretations are important. However, the electrochemistry of iron in alkaline environments is intricate and is difficult to explore as the electrodes are so easily disturbed by polarisation. The heterogeneous nature of concrete with a complex electric conductance [4] introduces additional problems. To distinguish better between problems originating from the electrochemistry of steel in alkaline environments and those caused by the cement mortar, studies were also carried out in aqueous solutions. The advantage of this is in having an environment with known composition and ohmic conductance and the possibility to vary mass transfer by agitation.

After a brief description of the experimental techniques, studies of iron and steel exposed to aqueous solutions, cement mortar and concrete are presented. The discussion takes up the properties of the corrosion films, the unusual type of passivity and the interpretations of a.c. measurements.

2. Experimental

Specimens of super pure iron had a single flat exposed surface of 1.2 cm^2. The other surfaces were embedded in evacuated epoxy. The surface was given a metallographic polish with 3 μm diamond particles. The final step in the surface treatment was pickling for two minutes in 10 mass % HCl, then rinsing in tartaric acid, in dilute ammonia solution and drying with alcohol. This was done immediately before the specimens were introduced to the test environments. The pure iron electrodes were reused after renewed polishing and etching. Steel specimens (C0.07, Si0.02, Mn0.029 and remainder < 0.003%) with a working area of about 10 cm^2, were cut from 1 mm mild steel sheet. This steel was chosen for its low content of alloying elements and impurities. Specimens were etched as described above just before exposure in solutions or casting in cement mortar The etching procedure with many steps and with tartaric acid as complexing agent removes rust and oxides without introducing new rust, as would be the case when HCl is rinsed off with water. This report treats exposures carried out under the following conditions:

- short exposures in KOH solutions deoxygenated by continuous bubbling with 99.998% N_2;
- 0.3M KOH in contact with air, 34 months exposure;
- 0.3M KOH + 0.1M NaCl in contact with air, 51 months exposure;
- cement mortar without added chloride, 11 months exposure;
- cement mortar with chloride added as 3% of the cement mass, 11 months exposure;
- cut reinforcement bars in the original concrete of a large coastal bridge, 17 years exposure.

The programme also included experiments in synthetic concrete pore water, consisting of 0.2M KOH, 0.1M NaOH saturated with $Ca(OH)_2$ with various levels of Cl^- although these experiments are not included as the results appeared to be similar to those with KOH and Cl^-. Nitrogen flushing was conducted for 18 h to remove oxygen before the specimens entered the cell. The cells for the exposures in aerated solutions had a permanent steel reference electrode at a distance of 2 mm from the specimens and an AISI 316 counter electrode. The solutions were renewed at intervals. Furthermore, the electrodes in cement mortar had reference electrodes embedded at about 2 mm distance to avoid problems with high resistance drop and the potential drift that would occur with reference electrodes at the mortar surface. The resistance between the reference electrode and the test electrode was about 3 Ω in the cells for solutions and about 300 Ω for the mortar slabs. Details of the test cells in the Gimsøystraumen bridge, consisting of cut reinforcement bars, reference electrodes and counter electrodes are provided in a separate report [5]. The potentials of the various reference electrodes were checked with a calomel electrode and values transferred to the saturated calomel scale (SCE). Conventional analogue equipment as well as a Gamry CMS 100 digital potentiostat and a Gamry 900 EIS instrument were applied for cyclic voltammetry, potentiostatic pulses, galvanostatic pulses and electrochemical impedance spectroscopy (EIS).

3. Results

3.1. Deoxygenated 0.3M KOH

Figure 1 shows the current density of steel as a function of potential for cyclic voltammetry in deoxygenated 0.3M KOH, starting from –0.8 V. Anodic peaks at about –1.05 and –0.85 V in the second scan in the positive direction were attributed to oxidation of hydrogen, as they were found only when the steel had been polarised at

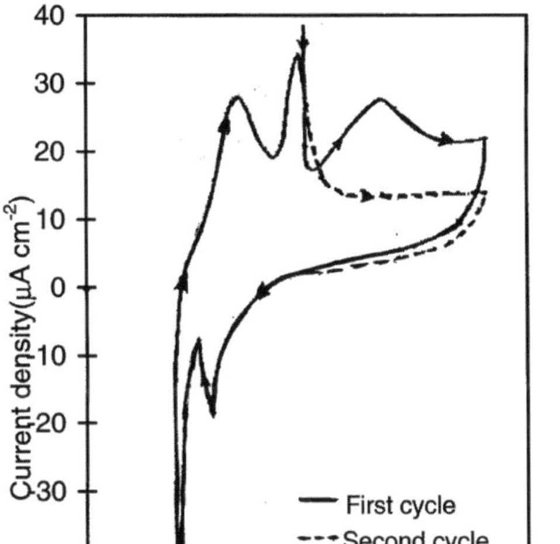

Fig. 1 The current density of steel as a function of potential for cyclic voltammetry in deoxygenated 0.3M KOH, starting from –0.8 V (SCE).

potentials sufficiently low to give hydrogen evolution. Figure 2 shows the electrode capacitance in 0.3M deoxygenated KOH, measured at 41 Hz, as a function of potential. The same data are plotted in Fig. 3 as reciprocal capacitance. In this experiment the potential was altered in steps of 10 mV every 10 s, giving an average scan rate of 1 mVs^{-1}. It is seen that the capacitance passed a maximum at –0.8 V and then decreased continuously to the apex potential 0.2 V. The capacitance indicates that the anodic current at potentials above –0.8 V was primarily used for formation of an anodic oxide film. The plot from the back-scan shows that the corrosion film formed at 0.2 V became unstable when the potential was lowered. The current was cathodic below –0.8 V and this led to a large increase of the capacitance indicating substantial changes of the electrode. The capacitance decreased slightly by polarisation below –1 V, probably because of some coverage by hydrogen. It is seen from Fig. 1 that the current above –0.8 V was much lower in the second cycle, as this film was present from the previous cycle. However, the capacitance is far too high to originate from a capacitor with the film material as dielectric. With a dielectric constant of 20, film thickness less than one atomic diameter is required to match the capacitance values shown in Fig. 2.

Figure 4 shows the current density of pure iron at 0 V as a function of time in a 120 h test. More data of this type were obtained from potentiostatic exposures of 36 h in the region –0.6 to 0.3 V. The current density was from 0.38×10^{-7} to 1.2 10^{-7} A cm^{-2} when the tests were concluded. The current density was independent of potential with about 0.4×10^{-7} A cm^{-2} as the most frequent value. The passive current was still diminishing with time after 36 h. No effect of KOH concentration in the range 0.1 to 3M was detectable from such experiments. To explore further the effect of pH, the KOH concentration was changed after 18 h, keeping other variables constant.

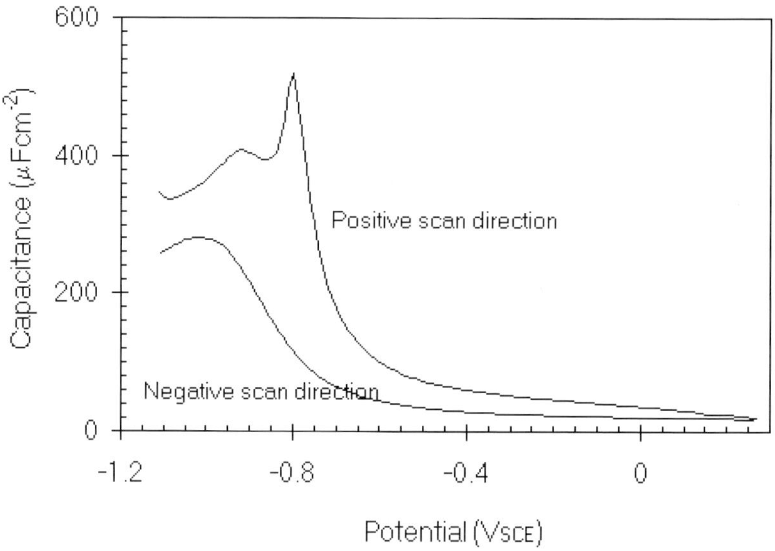

Fig. 2 *Electrode capacitance in deoxygenated 0.3M KOH, measured at 41 Hz, as a function of potential.*

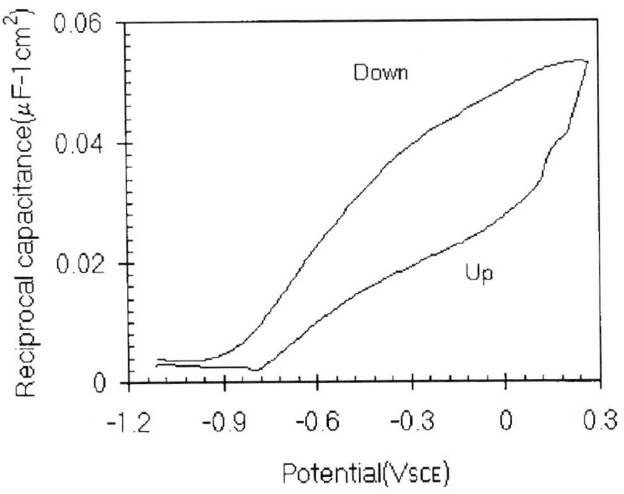

***Fig.* 3** *The data of Fig. 2 plotted as reciprocal capacitance.*

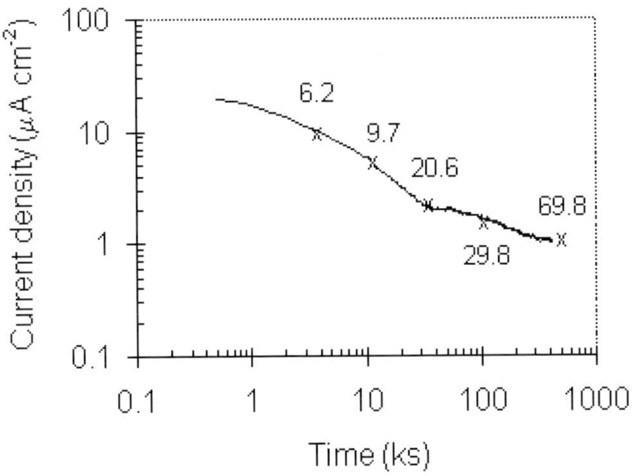

***Fig.* 4** *The current density of pure iron deaerated 0.3M KOH at 0 V(SCE) as a function of time in a test of 120 h. The numbers are the integrated charge in (mC cm^{-2}) at different times.*

Increasing the concentration from 0.03M to 0.3M. did not affect the current. Raising the concentration above 0.6M during the potentiostatic exposure led to a small current increase that faded away in less than an hour. The numbers at Fig. 4 are the integrated charge in mC cm^{-2} at different times. The charge indicated thicker films than usual for passivity. For a $Fe(OH)_2$ film with a density of 3.1 g cm^{-3} a charge of 1 mC corresponds to a film thickness of 1.6 nm. If magnetite is the film material, the thickness will be the half of this. It is assumed that all the oxide produced adheres — a reasonable assumption due to the low solubility of iron ions in the solution. Thus, the charge indicates passive films with thicknesses of 1 to 15 nm.

Figure 5 shows the current density as a function of time for a pure iron electrode that first had been polarised anodically for 36 h at –0.3 V and then subjected to a sudden 20 mV potential increase. The potential pulses led to an immediate current rise followed by a decrease with time. When switched back to –0.3 V, the current initially fell significantly below the steady state value at this potential. Results with potential steps of different magnitude are given in Fig. 6. As the current density decreased with time, different polarisation periods led to plots with different slopes. It is also seen that these plots differed from Tafel curves by being straight lines in a diagram with a linear current density scale.

3.2. Aerated Solutions, Cement Mortar and Concrete

Similar pulse experiments were carried out for steel exposed for a long time under aerated conditions. The shape of the curves resembled those in Figs 5 and 6, apart from the fact that the current density for a given potential step differed for the various

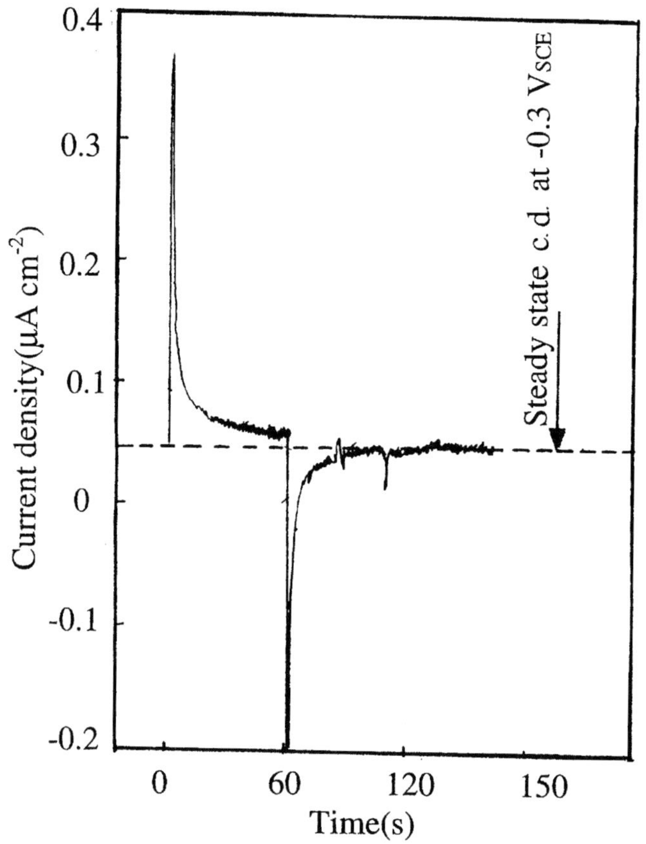

Fig. 5 The current density as a function of time for an electrode that had first been polarised anodically 36 h at–0.3 V (SCE) and then subjected to a sudden 20 mV potential increase.

environments. Polarisation resistances obtained from a 20 mV anodic step in various environments are shown in Fig. 7 with values calculated from the current at different times after the initiation of the potential step.

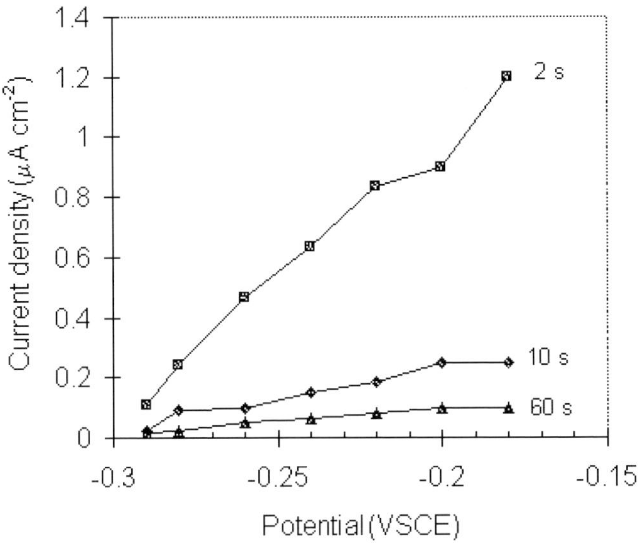

***Fig. 6** The current density in 0.3M KOH after 2, 10 or 60 s as a function of potential for an electrode that had first been polarised anodically 36 h at –0.3 V(SCE) and then subjected to potential increases.*

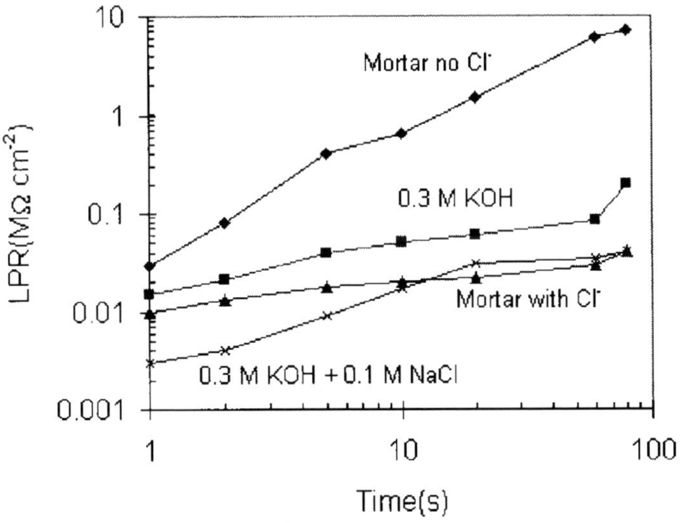

***Fig. 7** Polarisation resistance obtained from 20 mV anodic steps with values calculated from the current at different times after the initiation of the potential step.*

Figures 8–11 show Nyquist diagrams for these electrodes at the open circuit corrosion potential. This potential exhibited large variations from week to week. Two typical examples of EIS data for cut reinforcement bars exposed in a coastal concrete bridge are shown in Figs 12 and 13. The Nyquist diagrams apply to the total electrode without correction for the surface area. The numbers on the Figures are the frequency in Hz. None of the results in solutions were affected by agitation. Results from other measurements in these solutions are found in previous reports [3,6,7]. Only the plot for the chloride-free cement mortar resembled the beginning of a large semi-circle. To clarify the continuation of the plot, measurements at lower frequency are desirable,

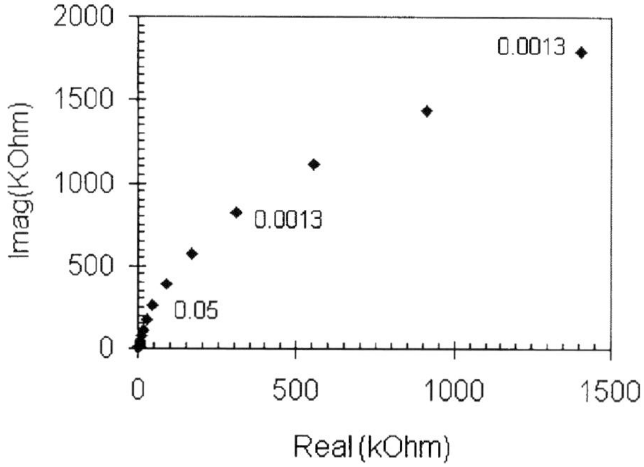

Fig. 8 Nyquist diagram at the corrosion potential (–0.058 V(SCE)) for a 7 cm² steel electrode exposed to cement mortar without chloride for 11 months.

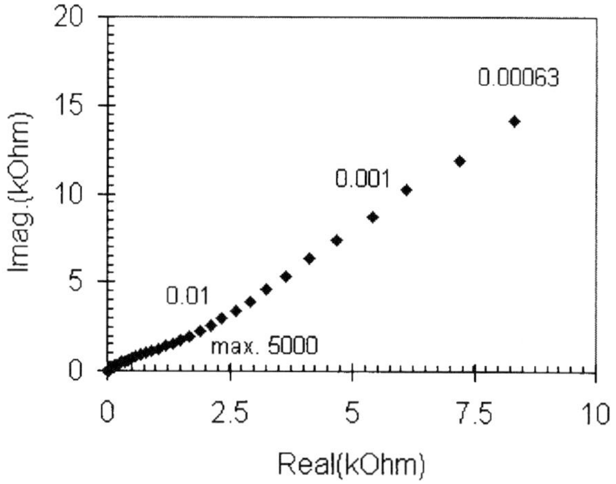

Fig. 9 Nyquist diagram at the corrosion potential (–0.15 V(SCE)) for a 14 cm² steel electrode exposed to aerated 0.3M KOH +0.1M NaCl for 51 months. The electrode had spots with thick rust layers.

but hardly possible as the measurements already required two days. The two plots for cement mortar with chloride appears to start as a semi-circle at the high frequency, changing to a constant phase angle line at low frequencies. Chloride reduced impedance values for the same frequency by more than an order of magnitude. The remaining diagrams resemble constant phase angle lines from the beginning.

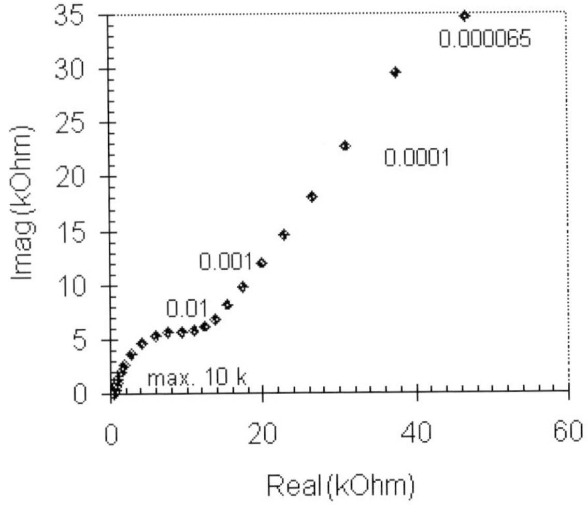

***Fig. 11** Nyquist diagram at the corrosion potential (–0.368 V(SCE))for a 7 cm² steel electrode exposed in cement mortar with 3% chloride of the cement weight for 11 months.*

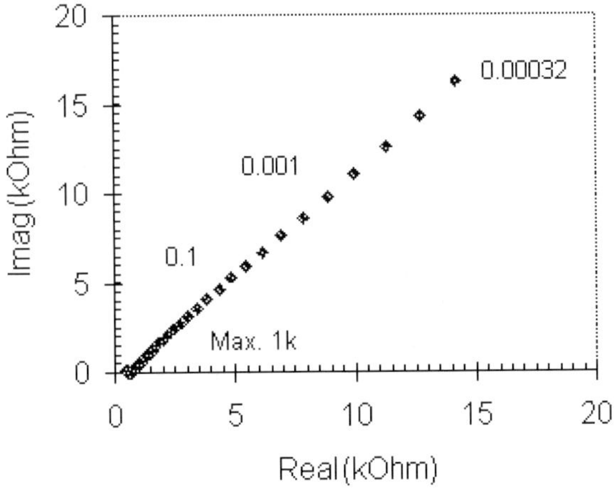

***Fig. 12** Nyquist diagram at the corrosion potential for a 40 cm² cut reinforcement bar exposed for 16 years in original concrete of a coastal bridge at a location with a low chloride content.*

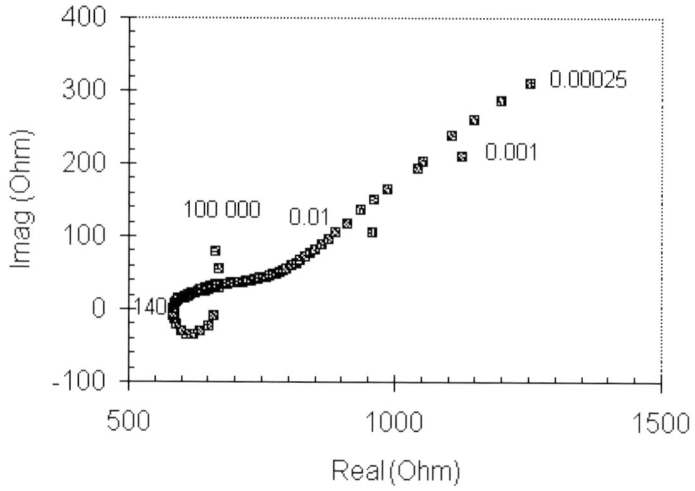

Fig. 13 *Nyquist diagram for a 25 cm² cut reinforcement bar exposed for 16 years in the original concrete of a coastal bridge. Site with chloride-induced passivity breakdown and localised corrosion.*

4. Discussion

4.1. The Peaks in Cyclic Voltammetry

The voltammogram in Fig. 1 applies for the specific steel used in this study. The anodic peak at –1.05 V in Fig. 1 occurred only when the electrode had been exposed to conditions with hydrogen evolution. This peak is due to oxidation of hydrogen. The cathodic peak at about –1.1 V is related to the anodic peak at –0.85 V. The experiments indicate that the anodic reaction at about –0.85 V is the first event. The cathodic reaction at about –1.1 V is a following reduction step. This cathodic peak was small or absent when the anodic scan started above –0.7 V. while anodic scans starting below –0.85 V gave large cathodic peaks at about –1.1 V. The anodic peak at –0.85 V may be caused by oxidation of some iron from the two-valent to the three-valent state, with the reverse reaction at about –1.1 V. However, the potential values do not agree with the Pourbaix diagram [8].

4.2. Passivity and Stability of the Passive Film

As described, iron and steel passivate in alkaline environments in absence of chloride. In the classical model for passivity a substantial current reduction takes place when the passivation potential is exceeded. However, no passivation potential could be detected from the current in cyclic voltammetry. As shown in Fig. 3, the reciprocal capacitance increased with the potential. An increase of reciprocal capacitance with potential is in harmony with a growing anodic film. However, a quantitative relationship hardly exists as the magnitude of the capacitance shows that it does not originate from a physical capacitor with the film material as dielectric. Reciprocal capacitance for the positive scans increased above –0.8 V. According to the Pourbaix diagram FeO becomes stable above this potential [8]. This agreement may be incidental as the capacitance in Fig. 3 has an uncertain origin.

No reactivation potential was observed from the current in cyclic voltammetry. A defined reactivation potential hardly exists. Instead the film became gradually unstable when the potential was lowered from the formation potential, as shown by the capacitance in Figs 2 and 3. Despite this instability, the earlier period at high potential maintained an effect as the current density in the second anodic scan in cyclic voltammetry was much lower than in the first.

The reactions responsible for the instability at potential decreases are not known. According to Pourbaix [8] the solubility of $HFeO_3^-$ is $10^{-5.7}$M at pH 13. To reduce dissolution further, the nitrogen flushed solutions were deliberately saturated with the corrosion product by being exposed to the metal for a long time before the experiments started. Therefore, bulk dissolution appears unlikely. Formation of pores when the potential was lowered is regarded as more reasonable. Changes of valence are discussed later. The charge used for film build up suggests a porous film. It is seen from Fig. 4 that the charge increased from 6.2 to 20.6 mC cm^{-2} in the period from 5000 to 50 000 s. The reciprocal capacitance increased only about 5% in the same period [9]. Furthermore, the capacity remained nearly unaltered after days' exposure at constant potential even though the current density was in the 10^{-7} A cm^{-2} region.

4.4. Time Effects of Small Polarisation Currents

The response of the electrodes to potential steps was an immediate current rise followed by a decrease with time. As understood from Figs 5 and 6 this applied to alkaline solutions with and without oxygen and for electrodes in cement mortar. For the steel in the concrete bridge the current density after a 20 mV step from the corrosion potential was followed for 28 500 s and was still decreasing [5]. The current decrease with time in chloride-free environments was so pronounced that the polarisation resistance became high and undefined with a potentiostatic polarisation time of more than 20 min.

A potential step back to the original potential led to current in the reverse direction (Fig. 5). These curves have some resemblance to charging and discharging of capacitance through a resistor. The current response to the step back to the original potential was affected by the duration of the potential pulse. A longer pulse led to higher backward current and more backward charge before the original current density was restored. As these experiments were carried out with automatic compensation of the resistance between the working electrode and reference electrode, the charging current for a single interfacial capacitance should diminish very quickly. Therefore, capacitance in the physical sense cannot explain this behaviour. Even though the performance shown in Figs 5 and 6 cannot be modelled with a single capacitance, it can be taken care of by considering the electrode as equivalent to two or more RC-links in series, one with a short and others with longer time constants. This applies also for other types of electrochemical experiments for iron and steel in these environments [1–3,5,6]. Millard *et al.* [11] had to use equivalent circuits with up to five RC-links with different time constants in series to model galvanostatic pulses and EIS. However, it is questionable what is gained by such modeling if these capacitors do not exist in the physical sense. Slow electrochemical reactions, rather than real capacitance, are responsible for phenomena observed with potentiostatic pulses with EIS and with galvanostatic pulse in previous studies [6,7]. The nature of the reactions responsible for sluggish changes of the electrodes is not fully clarified. To explain the results of EIS, Andrade *et al.* [12] were the first to suggest that valence changes of the iron in the corrosion film occurred for steel in cement mortar. This has been discussed further in a new paper [13]. Recently Joiret *et al.* have [14] presented results from Raman spectroscopy confirming the existence of redox-reactions between Fe^{2+} and Fe^{3+} in the oxide films formed on iron in KOH solutions. In contrast to the typical redox-reaction, the transition between Fe^{2+} and Fe^{3+} takes place in a wide potential region. To take care of this one has to consider a corrosion film with a gradual change in stoichiometry, with mainly Fe^{2+} close to the metal interface and more Fe^{3+} in the outer parts. As the amount of iron ions that can change valence is small, it is understood that a small charge can have a large effect. Another effect may also be considered. Thus, it was concluded that the passive film was porous and so as anodic polarisation produces oxide, it is to be expected that it affects the pore structure.

4.5. Constant Phase Angle Lines

The slow electrochemical steps, treated above, complicate the interpretation of EIS. Constant phase angle lines occurred in the Nyquist diagrams except in Fig. 8. Transport of oxygen in concrete is slow and due to the small amount of pore water its composition is easily affected by the electrode reactions. Therefore, it is tempting to attribute constant phase angle regions in concrete to diffusion in the environment, as has been done by several authors [15–17]. The slopes of constant phase angle lines in the Nyquist diagrams vary from 0.8 to 1.2, but values outside this occurred in other experiments. A slope of unity is obtained when the potential changes reversibly with respect to the concentration of the diffusing species. Myrdal and Videm [18] have examined the potential of steel in KOH with varying oxygen concentrations and found that it does not follow the Nernst equation. Therefore, a deviation from a slope of unity is acceptable for diffusion control in these environments. Another test of Warburg impedance is to examine whether the calculated diffusion rates are independent of frequency. This fitted reasonably well, apart from the data of Fig. 9. However, calculations according to Warburg impedance theory gave unreasonable results. The calculated diffusion rate in 0.3M KOH was two orders of magnitude lower than for chloride-containing cement mortar. The starting point for the calculations is unrealistic despite the existence of a constant phase angle line in Fig. 10, as the corrosion of steel in aerated 0.3M KOH is far too slow to be limited by diffusion in the solution. A fully convincing argument against Warburg impedance in this specific case is the lack of any effect of agitation. From the various observations it is concluded that the redox reactions of the corrosion product and possibly also a porous film create a pseudo capacitance giving the constant phase angle lines.

4.6. The Polarisation Resistance

The Stern–Geary [19] mathematical treatment is based on rate control by electron transfer for a single anodic and a single cathodic reaction and that the only effect of polarisation is to change the rates of these reactions. All common methods for determination of LPR should then give the same result. However, when current is consumed by valence changes of the corrosion product, the relationship between polarisation resistance and corrosion rate cannot be treated by any established theory. Figure 7 shows values for the polarisation resistance varying with the duration of potentiostatic pulses. Polarisation resistance obtained by different methods for steel in cement mortar are found in Table 1 [1]. The values deviate by more than a decade depending on the method of determination. Thus, no general constant for relating the corrosion rate to the polarisation resistance exists for steel in concrete. Another question is whether the different methods distinguish correctly between a low and high corrosion rate. The ratio between LPR in mortar with and without chloride in Table 1 varies between 8.9 and 157, with the highest values obtained by slow measurements. In agreement with this, Gonzalez *et al.* have stated that cyclic voltammetry more easily identifies localised corrosion with low scanning rates [20]. This is somewhat surprising as slow methods allow more time for sluggish polarisation induced phenomena. In an attempt to explain this, it is suggested that the disturbing polarisation induced phenomena become exhausted with time, thereby making the data for corrosion rate more apparent. As seen from Table 1 the various

Table 1. *Linear polarisation resistance obtained by various method for steel exposed 2 months in cement mortar without chloride and with mixed-in 5% chloride of the cement mass [1]. The corrosion potentials were –275 and –484 mV respectively*

METHOD	Main variable	Polarisation resistance/KΩ cm² Mortar without Cl^-	Mortar 5% Cl^-	Ratio
Potentiostatic pulse 20 mV	1.5 s polarisation	133	14.8	8.9
	5 s "	967	21.0	46
	10 s "	1300	24.2	53
Cyclic voltammetry	1.8 m Vs^{-1} scan rate	3500		
	0.72 m "	3900	24.8	157
	0.36 m "	5300	28.2	187
	0.06 m "	5800	42.3	137
Galvanostatic pulse	0.29 μA cm^{-2}		14	
	0.0026 "		1360	100
	EIS	40?	2?	20?

? Very uncertain values

methods do distinguish between harmless and severe corrosion. However, a quantitative determination of the corrosion rate is hardly possible.

5. Conclusions

The passive current for diamond polished, super pure iron in 0.3M KOH was independent of potential in the range studied (–0.5 to 0.3 V) and after 36 h was from 0.4.10–7 to 10^{-7} A cm^{-2}. It had still not reached steady state. Defined passivation- and reactivation potentials hardly exist for iron and steel in these environments.

A phase shift between potential and current resembling the action of a very large capacitance originated from polarisation induced changes of the electrodes. This led to an a.c. capacitance far too high to originate from a physical capacitor with the film material as dielectric.

The thickness of the passive film increased with increasing formation potential. The charge consumed during build up of the passive film indicate film thickness up to 15 nm, suggesting the possibility of porous films.

The passive films became unstable when the potential was lowered from the formation potential. As the capacitance is not linked to the film thickness, the nature of the polarisation induced changes of the electrodes is not fully clarified. Valence changes of the iron ions in the corrosion film occur and possibly also changes of the pore structure.

The a.c. pseudo-capacitance increased with decreasing frequency. Chloride-induced localised corrosion could increase this pseudo-capacitance by one order of magnitude.

The constant phase angle line obtained by EIS did not originate from Warburg impedance, but from phase shifts caused by phenomena in the oxide film rather than from diffusion in the environment.

As potentiostatic polarisation led to currents that decreased with time, plots of current density vs potential as well as polarisation resistance differed depending on the measuring technique and test parameters.

With current consumed by valence changes of the corrosion product the Stern–Geary treatment relating corrosion rate to the linear polarisation resistance is not valid for iron and steel in these environments. As the polarisation resistance varied substantially with the measuring technique for a given specimen, a general relationship between polarisation resistance and corrosion rate does not exist. Chloride-induced corrosion led to a marked reduction of the linear polarisation resistance observed for all types of electrochemical measurements tried. This reduction varied significantly for the various methods. Slow measurements appeared to discriminate better between harmless and serious corrosion.

References

1. K. Videm, The anodic behaviour of iron and steel in synthetic concrete pore water and in cement mortar. *Corrosion '99*, Paper 573, NACE International, Houston, TX, 1999.

2. K. Videm, Studies of the electrochemistry of steel in concrete. *Proc. Int. Conf. on Corrosion and Rehabilitation of Reinforced Concrete Structures*. NRC, CNRC, NACE International, Orlando, Florida (December 1998).
3. K. Videm, The reliability of electrochemical techniques for assessing the severity of corrosion of steel in concrete. *Corrosion '98*, Paper 794, NACE International, Houston, TX, 1998.
4. M. Keddam, H. Takenouti, X. R. Novoa, C. Andrade and C. Alomnso, Study of the dielectric characteristics of cement paste. *Proc. EMCR97, 6th Int. Symp. on Electrochemical Methods of Corrosion Research*. University of Trento, Italy, August, 1997.
5. K. Videm and R. Myrdal, Instrumentation and condition assessment performed on Gimsøystraumen bridge. *Int. Conf. on Repair of Concrete Structures*. SND, NPRA, Rescon. Svolvær, Norway , May 1997, p. 375.
6. K. Videm, Experience with galvanostatic pulse technique and other methods to assess rebar corrosion. *Corrosion '99*, Paper 279, NACE International, Houston, TX, 1997.
7. K. Videm, Field and laboratory experience with electrochemical methods for assessing corrosion of steel in concrete, *Mater. Sci. Forum*, 1998, **289–292**, 3–15.
8. M. Pourbaix, *Atlas of Electrochemical Equilibria*. Pergamon Press, 1966, p.307–321.
9. K. Videm, Studies of the anodic reactions of iron and steel in KOH solutions and in cement mortar. *196th Meeting of The Electrochemical Society*, Honolulu, Hawaii, October 1999, p. 606.
10. K. Videm, Uniform Corrosion and Pitting of Aluminium. Thesis, The Norwegian University for Science and Technology (NTNU), (1974).
11. S. G. Millard, K. R. Gowers and J. H. Bungey, Galvanostatic pulse techniques: A rapid method of assessing corrosion rates of steel in concrete structures. *Corrosion '95*, Paper 525, NACE International, Houston, TX, 1995.
12. C. Andrade, L. Soler and X. R. Novia, Advances in electrochemical impedance, *Mater. Sci. Forum*, 1995, **192–194** , 843–856.
13. C. Andrade, F. Bolzoni, M. Cabeza, X. R. Nóvoa and M. C. Pérez, Measurement of steel corrosion in concrete by electrochemical techniques; Influence of the redox processes in oxide scales, in *Electrochemical Approach to Selected Corrosion and Corrosion Control Studies* (1st Joint Symposium, 1999, P. L. Bonora and F. Deflorian, eds). Publication No. 28 in the European Federation of Corrosion Series. Publ. by The Institute of Materials, London, UK, 2000 (this volume, p.332–343).
14. S. Joiret, M. Keddam, X. R. Nóvoa, A. Collazo and M. C. Pérez, Studies by Raman spectroscopy of the inhibitive action of nitride ions on iron corrosion in alkaline solution, *50th ISE Meeting*, Pavia, Italy, p. 433, 1999.
15. N. G, Thompson, K. H. Lawson and J. A. Beavers, Monitoring cathodic protected concrete structures with electrochemical impedance techniques, *Corrosion '87*, Paper 139, NACE International, Houston, TX, 1987.
16. D. D. MacDonald, Y. A. El-Tantawy, R. C. Rocha-Filho and M. Urguidi-MacDonald, Evaluation of Electrochemical Impedance Techniques for Detecting Corrosion of Rebar in Reinforced Concrete: Strategic Highway Research Program, National Research Council, Washington, DC. Report SHRP-ID/UFR-91-524 (1991).
17. A. A. Sagues, S. C. Kranc and E. I. Moreno, Evaluation of electrochemical impedance with constant phase angle component from the galvanostatic step response of steel in concrete, *Electrochim. Acta*, 1996, **41**, 1239.
18. R. Myrdal and K. Videm, Evaluation of corrosion of steel reinforcement in concrete from measurements of embedded reference electrodes. *Corrosion '95*, Paper 512, NACE International, Houston, TX, 1995.
19. M. Stern and A. L. Geary, *J. Electrochem. Soc.*, 1957, **104**, 56.
20. J. A. Gonzalez, A. Molina, M. L. Escudero and C. Andrade, Errors in the electrochemical evaluation of very small corrosion rates — I. Polarization resistance method applied to corrosion of steel in concrete, *Corros. Sci.*, 1985, **25**, (10), 917–930.

23
Electrochemical Study of Iron Corrosion in Various Concentrations of Polyethylene Glycol (PEG 400) Solutions

E. GUILMINOT, F. DALARD and C. DEGRIGNY*
LEPMI, 1130 rue de la Piscine Domaine universitaire, 38 402 Saint Martin d'Hères, France
* Laboratoire Arc'Antique, 26 rue de la Haute-Forêt, 44 300 Nantes, France

ABSTRACT
Conservation of archaeological composite artefacts (waterlogged wood / iron) requires wood stabilisation in various concentrations of polyethylene glycol (PEG 400) for several weeks or several months. When the artefact is immersed, iron corrosion occurs in PEG 400 solutions. Our electrochemical study compares the corrosion of iron in various concentrations of PEG 400 solutions. The conductivity of the electrolyte changes with water content: the most conductive solutions are between 10% and 30% (v/v) of PEG 400. The evolution of potential–time and the value of the open potential (E_{cor}) of iron vary with water content. The interpretation of electrochemical measurements is difficult because of the high impedance of these solutions (10^4–10^5 Ω cm^{-1}). It is necessary to compensate the ohmic drop to investigate the polarisation curves. The anodic current of iron is low in 70–100% PEG 400 solutions, but there is corrosion of iron in 10–50% PEG 400 solutions. Interactions between water and PEG increase the conductivity and favour the corrosion of iron. When the PEG content is high, the high viscosity and the adsorption of PEG on the metallic interface limit the corrosion of iron.

1. Introduction

Among archaeological objects, waterlogged iron–wood composites bring specific conservation problems because the metallic parts often cannot be separated from the wood. Conservation of archaeological composite artefacts requires a treatment that favours the most fragile material, i.e. the waterlogged wood. This generally involves impregnating the object with polyethylene glycol (PEG 400) for stabilising and consolidating the wood. These treatments can last for several weeks or several months [1]. When the artefact is immersed in various concentrations of PEG 400 solutions, corrosion of iron occurs [2].

The glycol ether–H_2O system has already been studied by using ethylene glycol monoethyl ether ($C_2H_5OC_2H_4OH$) [3–5]. The physicochemical properties of this system depend on the H_2O content which influences the corrosion behaviour of iron [3]. To investigate the physicochemical properties of the polyethylene glycol PEG 400–H_2O system, the conductivity, viscosity and the dissolved oxygen content of various mixtures were determined.

The corrosion of iron in various PEG–H_2O mixtures was studied by electrochemical methods such as potential–time measurements, anodic or cathodic polarisation curves and polarisation resistance, R_p, measurements. This electrochemical study is compared with the mass loss method.

2. Experimental

2.1. Materials and Solutions

The iron was subjected to a 40 h heat treatment at 1050°C with Fe_2O_3, so its structure is similar to archaeological iron. The main impurity is Mn in the form of manganese sulfide (by EDS microanalysis). For the electrochemical measurements, the working electrode of annealed iron or Pt had a diameter of 2 mm. The samples of Pt were polished with diamond paste (3 μm) and rinsed in ethanol with ultrasonic action. The samples of iron prior to each set of electrochemical experiments were polished on silicon carbide paper of 1000, 2400, 4000 grades and with diamond paste (1 μm); and then rinsed in ethanol with ultrasonic action. For the mass loss experiments, 1.5 × 1 × 0.1 cm specimens of iron were cut; these were either samples of polished iron or samples with corrosion products. Iron was exposed to the atmosphere for ten years; the corrosion products were analysed by XRD and found to be Fe_3O_4 and FeO(OH).

The polyethylene glycol used was of the 400 grade — supplied by Acros. Its molecular formula is $HOCH_2(CH_2OCH_2)nCH_2O$ ($n \approx 7$–8). It had a maximum H_2O content of 0.2%. Distilled water was used to make up the solutions. The PEG–H_2O mixtures were prepared in volume to total volume (v/v) ratios.

2.2. Conductivity Measurements

The conductivity of various PEG–H_2O mixtures was determined using electrochemical impedance measurements, and a double-walled, three-electrode Metrohm cell placed in a Faraday cage at 25°C. The reference electrode was a calomel electrode with saturated KCl solution (SCE) (E_{SCE} = 0.245V vs standard hydrogen electrode SHE). In order to reduce any chloride contamination of the PEG solutions, the connection to the cell was made via an extension filled with the solution under study. The counter electrode was a platinum grid. The working electrode was a rotating disc electrode with rotation speed fixed at 500 rev/min. The experimental set-up included the following: a frequency response analyser Solartron 1260 and an electrochemical interface Solartron 1287 controlled by Zplot–Zview software applications. Following stabilisation of the potential, impedance measurements were carried out at the corrosion potential in the frequency range 50 kHz to 1 Hz with sinusoidal disturbance amplitude of 10 mV.

Electrochemical impedance spectroscopy makes it possible to determine in a precise way the electrolyte resistance R_e and therefore the conductivity. This method can separate the electrolyte resistance from adsorption resistances. Precise R_e values (Fig. 1) corresponding to the geometry of our electrochemical cell were obtained and also used to correct the polarisation curves for the ohmic drop.

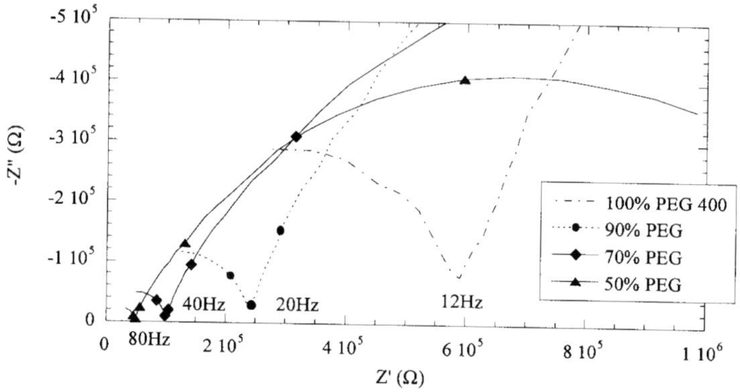

Fig. 1 *The* R_e *measurement by electrochemical impedance spectroscopy in various PEG–water mixtures at 25 °C.*

2.3. Dissolved Oxygen Content Measurements

The dissolved oxygen content of various PEG–H_2O mixtures was determined using a YSI 5000 dissolved oxygen meter at 25°C.

2.4. Mass Loss Method

Tests were conducted in eight solutions (distilled water and 10, 20, 30, 40, 50, 70, 90, 100 v/v% PEG solutions) which were held at 50°C for four months. For each experiment, there were three beakers containing 75 mL of the solution: a blank solution, a sample of polished iron and a sample of iron with corrosion products. After these experiments, all samples were cleaned in distilled water and weighed without dissolving corrosion products.

2.5. Electrochemical Methods

An EG&G potentiostat Model 273A and an electrochemistry software Model 352 were used as well as the cell described above. The following set of experiments was undertaken for each PEG–H_2O mixture with the polished iron (or Pt):

- Monitoring of corrosion potential (E_{corr}) for 5 h (for 1 h on Pt).
- Cathodic polarisation from +50 mV vs E_{corr} to –3 V vs SCE at scan rate of 1 mV s^{-1}.
- Anodic polarisation from –50 mV vs E_{corr} to +3 V vs SCE at scan rate of 1 mV s^{-1}.
- the polarisation resistance (R_p) values were calculated from the anodic polarisation curves taken at a scan rate of 1 mV s^{-1}. The potential range studied varied from –50 mV vs open circuit to + 50 mV vs open circuit.

To deal with the ohmic drop (*iR*) compensation on the anodic and cathodic curves, the *post factum* method from the values of the impedance measurements was used.

3. Results and Discussion

3.1. Conductivity Measurements

The conductivity of the distilled water was 6.25×10^{-7}S cm^{-1} (Fig. 2). It increased quickly to a maximum of about 4×10^{-5}S cm^{-1} at 20% PEG and thereafter decreased, dropping to 10^{-6}S cm^{-1} in the pure PEG solutions. A similar pattern had been observed in the glycol ether–H_2O mixtures [3] and also in acetic acid–H_2O or formic acid–H_2O mixtures [8]. The maximum in the conductivity is probably due to an increase in the concentrations of ionic species produced by the following reactions:

$$2HOCH_2(CH_2OCH_2)_nCHOH \leftrightarrow HOCH_2(CH_2OCH_2)_nCHO^- + HOCH_2(CH_2OCH_2)_nCHOH_2^+ \quad (1)$$

$$HOCH_2(CH_2OCH_2)_nCHOH + H_2O \leftrightarrow HOCH_2(CH_2OCH_2)_nCHO^- + H_3O^+ \quad (2)$$

$$HOCH_2(CH_2OCH_2)_nCHOH + H_2O \leftrightarrow HOCH_2(CH_2OCH_2)_nCHOH_2^+ + OH^- \quad (3)$$

When one adds H_2O to PEG, the quantity of ionic species produced by the H_2O–PEG interactions increases and so the conductivity increases. When the solution contains more than 90% of H_2O, the conductivity decreases. The dissociation of the PEG continues, but the number of ionic species produced by the reactions (1), (2) and (3) becomes small compared to the quantity of H_2O molecules.

3.2. Dissolved Oxygen Content Measurements

The dissolved oxygen depends on the H_2O content (Fig. 3). The dissolved oxygen content is 8 mg L^{-1} in distilled water. It decreases regularly with the quantity of H_2O to reach a value of 2 mg L^{-1} in pure PEG. Oxygen is less soluble in PEG than in distilled water.

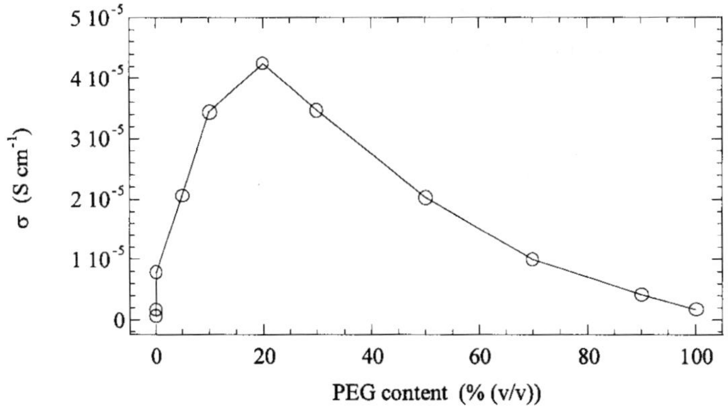

Fig. 2 Conductivity σ of various PEG–water mixtures at 25°C.

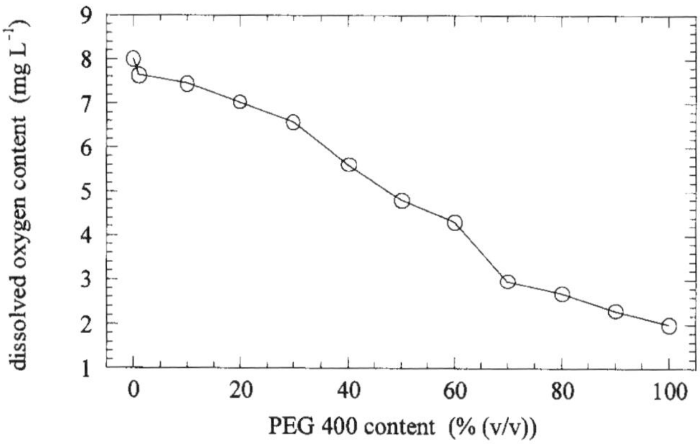

Fig. 3 *Dissolved oxygen content of various PEG–water mixtures at 25 °C.*

3.3. Mass Loss Method

The mass loss measurements also change with the PEG content (Fig. 4). In distilled H_2O, the mass of corroded iron increases because the corrosion products (Fe_3O_4 and FeO(OH) by XRD) continue to increase. The higher mass increase for the polished sample tends to show that the corrosion products naturally formed in the atmosphere limit the corrosion of iron. In the PEG–H_2O mixtures, the mass loss of polished iron increases to reach a maximum in 30% PEG. The XRD analysis shows the existence of the iron oxides: Fe_3O_4, FeO(OH) and Fe_2O_3. In the case of the corroded iron samples, the mass loss remain significant from 10 to 50% PEG. The iron oxides observed by XRD are Fe_3O_4 and Fe_2O_3. In the solutions of 50 and 70% PEG, the mass loss of corroded iron is more important than for polished iron mainly because of an alteration of corrosion products by complexation or by solvation [3,9,10]. In highly concentrated

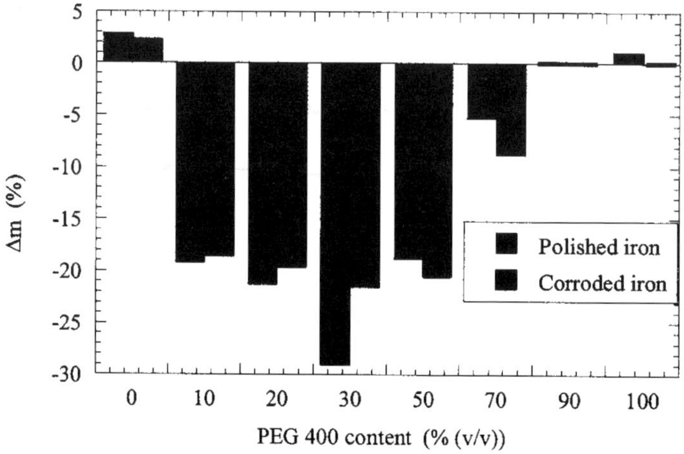

Fig. 4 *Mass loss Δ_m of polished iron or corroded iron in the various PEG–water mixtures at 50 °C during 4 months.*

solutions (90 and 100% PEG), the mass loss is low for the polished and corroded iron. No corrosion seems to appear on polished iron samples. Corrosion products are retained on the corroded iron sample; so complexation or solvation may not occur.

3.4. Electrochemical Methods

The corrosion potential E_{corr} of polished iron varied with the PEG content (Fig. 5). From 0 to 40% PEG solutions, it increased by approximately 200 mV before stabilising. This evolution of potential generally characterises the formation of a passive layer. At 50% of PEG, the corrosion potential decreased before its stabilisation, this behaviour could be due to the alteration of the passive layer. In this case, the area of anodic sites would increase [5]. The decrease of corrosion potential can also be the result of a decrease in the concentration of iron ions in the solution. The iron ions could be fixed by solvation or by a complex reaction [3,9,10]. For the 60% PEG solution, the corrosion potential initially decreased, then increased before stabilising. Above 70% of PEG, the corrosion potential increased and tended to stabilise, again corresponding to the behaviour of a passive layer. Although the small quantity of dissolved oxygen makes it difficult to form iron oxides, the adsorption of PEG molecules on the metal could constitute a protective coating [11]. The variation of the corrosion potential according to the PEG content (Fig. 6) indicates the different equilibria between iron and PEG–H_2O mixtures. Thus, from 0 to 20% PEG, the corrosion potential characterises mainly the equilibrium between iron and H_2O, the slight decrease is due to the dissolved oxygen decrease; from 20 to 50% PEG, the potential changes as a result of reactions of solvation or complexation; above 60% PEG, a new equilibrium is established between iron and the PEG solvent.

The cathodic polarisation of polished iron or Pt shows similar reactions, i.e. the reduction of dissolved oxygen and of water. The dissolved oxygen limiting current depends on the dissolved oxygen concentration, the viscosity, and the diffusion coefficient. Viscosity increases with the PEG content and the dissolved oxygen content decreases (Fig. 3). These evolutions accord with the results of polished iron

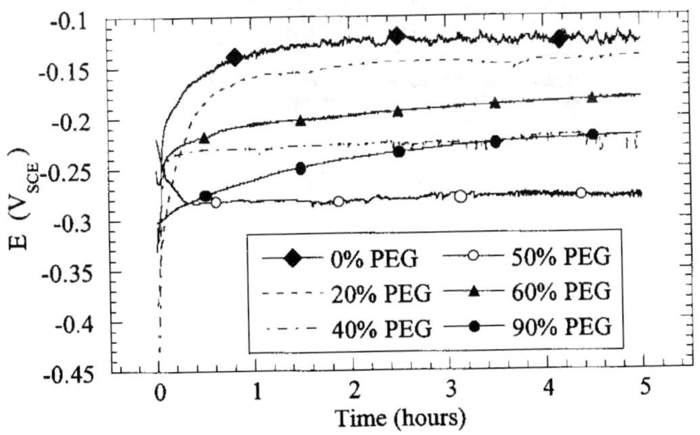

Fig. 5 Potential–time of polished iron for various PEG–water mixtures at 25°C.

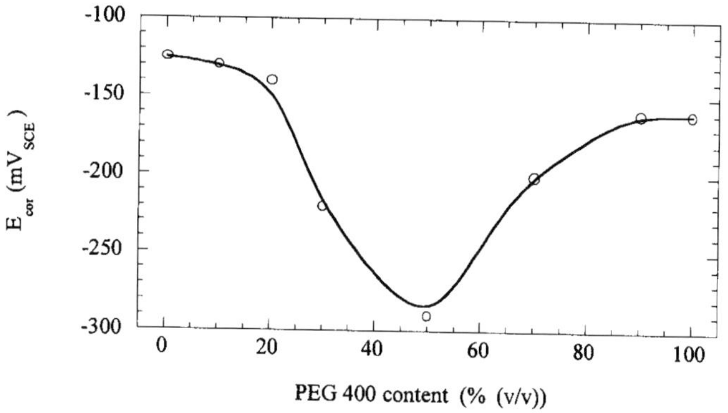

Fig. 6 *Corrosion potential of polished iron in the various PEG–water mixtures at 25 °C.*

polarisation (Fig. 7): that is to say an increase in PEG content is followed by a decrease of the dissolved oxygen reduction current. The anodic polarisation of Pt shows that PEG oxidation does not occur at 25°C, and H_2O oxidises only towards 1 V (SCE). On the iron polarisation curves (Fig. 8), the iron oxidation current is significant ($i \approx 5 \times 10^{-5}$ A cm^{-2}) until 50% PEG. For a concentration of 60% or more, the oxidation current remains low (about 10^{-6} A cm^{-2}).

The polarisation resistance R_p is inversely proportional to the corrosion current [12]. Its evolution in PEG–H_2O mixtures shows a minimum at 10–30% PEG (Fig. 9) which is why the iron corrosion reaches a maximum in these solutions. The R_p resistance increases with the increase of PEG and reaches values of $10^6 \Omega$ for highly concentrated solutions, the iron corrosion is then insignificant.

As already shown, the amount of iron corrosion varies with the PEG content. In distilled water or pure PEG, the iron corrosion is low. In distilled water, the corrosion products are stable [13] and limit the corrosion. In pure PEG, iron seems to be protected by a layer of PEG molecules. The maximum corrosion of iron occurs in the solutions from 10 to 30% PEG content. According to the conductivity measurements, the H_2O–PEG contains different concentrations of ionic species depending on the

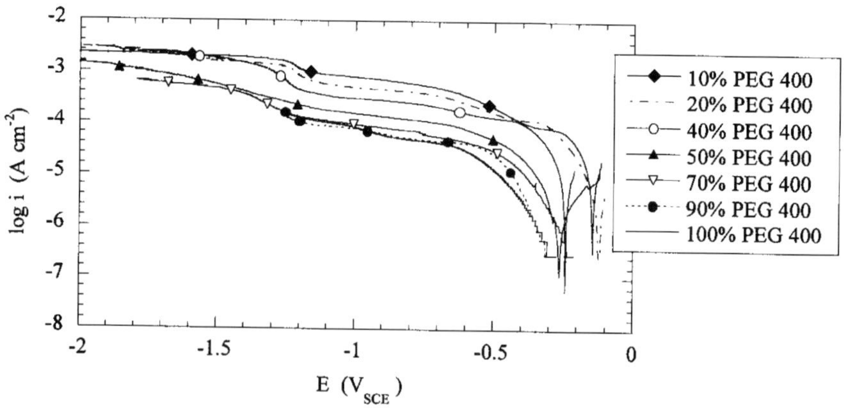

Fig. 7 *Cathodic polarisation of iron in the various PEG–water mixtures at 25 °C.*

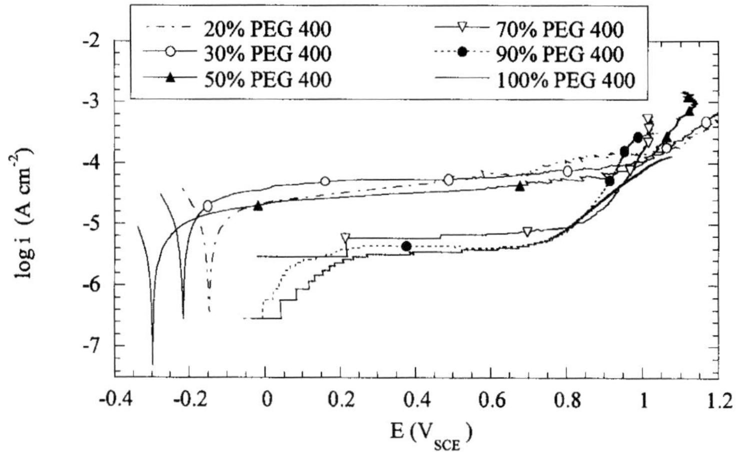

***Fig.** 8 Anodic polarisation of polished iron in the various PEG–water mixtures at 25°C.*

PEG content which modifies the stability of corrosion products. For example, Fe_2O_3 is more stable in presence of PEG. The H_2O–PEG mixtures promote the iron corrosion, and, in fact, the polarisation resistance measurements show the maximum corrosion of polished iron to occur between 10 and 30% PEG. The anodic polarisation curves and the mass loss measurements both indicate that the iron corrosion remains significant from 10 to 50% PEG, even if the quantity of dissolved oxygen — the "corrosion motor" — decreases with H_2O content. The maximum corrosion of iron corresponds to the peak conductivity.

As the iron corrosion is certainly caused by the PEG–H_2O interactions, the corrosion current cannot be proportional to the conductivity of the solution. The PEG modifies the diffusion of iron ions in solution by solvation or complexation [3,9,10], the solvation of a cation being facilitated by the free electron pairs of the oxygen atoms distributed along the PEG chain. This strong capacity for solvation of the PEG is

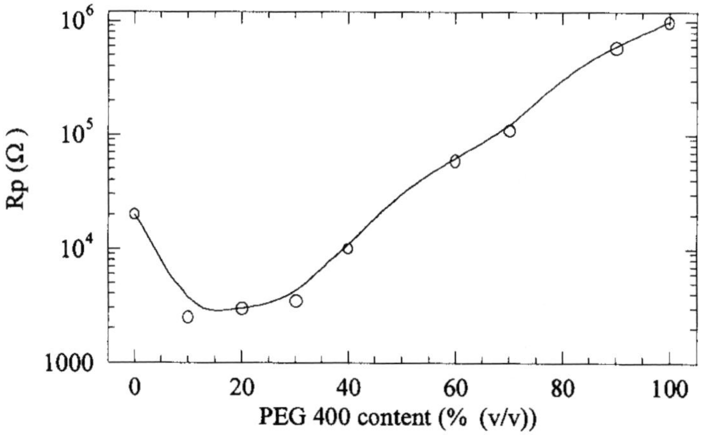

***Fig.** 9 Polarisation resistance of polished iron in the various PEG–water mixtures at 25°C.*

responsible for the formation of many complexes. The consumption of iron ions facilitates the corrosion of polished iron. As the corrosion products in pure PEG solutions had not disappeared during the mass loss measurements, the PEG molecule could not be responsible for the complexation and solvation of iron. So the ionic species produced by H_2O–PEG interactions should have accentuated the iron corrosion, and the more probable reaction is that of complexation with the anion $HOCH_2(CH_2OCH_2)_nCHO^-$. These ionic species could also influence the iron oxidation by Frumkin's effect. The PEG–H_2O interactions can also modify the iron corrosion by changing the quantity of free H_2O in PEG. The PEG–H_2O structure changes considerably with different H_2O contents. The interactions of PEG with H_2O have been investigated by NMR [14], infra red spectroscopy [15] and Raman spectroscopy [16] which have shown that for high H_2O contents, a H_2O molecule has only a hydrogen bond with an oxygen atom of a PEG chain; for low H_2O contents, the two hydrogens of H_2O molecule have a hydrogen bond bridge with two oxygen atoms of a same or different PEG molecule. The H_2O is freer in low PEG contents, and can be more reactive so accentuating the iron corrosion.

4. Conclusions

This electrochemical study of iron corrosion in aqueous PEG solutions shows that the iron behaviour depends on the PEG content as follows:

- In distilled H_2O, corrosion products are formed but the iron corrosion remains low.

- On the other hand, in PEG–H_2O mixtures, the iron corrosion is significant. It reaches a maximum in the 10–30% PEG solution. The PEG–H_2O interactions facilitate the iron corrosion.

- In pure PEG, iron is not corroded because the dissolved oxygen content is low and the PEG molecules are probably adsorbed on the metallic surface to form a passive coating.

This study of iron corrosion in PEG–H_2O mixtures is at present being continued by other electrochemical methods (i.e. electrochemical impedance spectroscopy) in order to understand the role of the PEG–H_2O interactions.

The highest corrosion of iron occurs for PEG contents commonly used in conservation of archaeological waterlogged wood. Thus, the treatment of archaeological composite artefacts may damage irreversibly the iron parts. To conserve entirely the composite artefact, the iron must be protected during the impregnation of waterlogged wood. Addition of a corrosion inhibitor to the PEG–H_2O mixtures can protect the iron from this phenomenon.

5. Acknowledgements

the authors would like to acknowledge ARC Nucleart laboratory (Grenoble, France) for their interest in this research, and Dr Maniguet for XRD analysis.

References

1. D. Grattan, *Studies in Conservation*, 1982, **27**, 124–136.
2. L. S. Selwyn, D. A. Rennie–Bisaillon and N. E. Binnie, *Studies in Conservation*, 1993, **36**, 180–197.
3. S. E. Faidi, G. Jones and J. D. Scantlebury, *Electrochim. Acta*, 1987, **32**, 947–953.
4. S. E. Faidi and J. D. Scantlebury, *J. Electrochem. Soc.*, 1989, **136**, 990–995.
5. S. E. Faidi and J. D. Scantlebury, *Proc. 6th SEIC*, Ann. Univ. Ferrara N.S., Sez, V, 1985, Suppl. No. 8, 1187–1204. Publ. Univ. Ferrara, Italy.
6. G. Barral, J. P. Diard, B. Le Gorrec, Lu Dac Tri and C. Montella, *J. Appl. Electrochem.*, 1985, **15**, 913–924.
7. G. Barral, J. P. Diard, B. Le Gorrec, Lu Dac Tri and C. Montella, *J. Appl. Electrochem.*, 1987, **17**, 695–701.
8. I. Sekine and K. Senoo, *Corros. Sci.*, 1984, **24**, 439–448.
9. M. Armand, *Adv. Mater.*, 1990, **2**, 278–286.
10. D. F. Shriver, B. L. Papke, M. A. Ratner, R. Dupon, T. Wong and M. Brodwin, *Solid State Ionics*, 1981, **5**, 83–88.
11. E. Guilminot, J. J. Rameau, F. Dalard, C. Degrigny and X. Hiron, *J. Appl. Electrochem.*, 2000, **30**, 21–28.
12. D. Landolt, Traité des matériaux 1993, 12, 131. Ed. Presses Polytechniques et Universitaires Romandes, Lausanne.
13. M. Pourbaix, *Atlas d'Equilibres Électrochimiques*, 1963, 312–313, Ed. Gauthier–Villars, Paris.
14. S. Lüsse and K. Arnold, *Macromolecules*, 1996, **29**, 4251–4257.
15. R. Begum and H. Matsuura, *J. Chem. Soc. Faraday Trans.*, 1997, **93**, 3839–3848.
16. V. Crupi, M. P. Jannelli, S. Magazu, G. Maisano, D. Majolino, P. Migliardo and R. Ponterio, *J. Mol. Struct.*, 1996, **381**, 207–212.

24
Micro-MHD Effect on Metal Corrosion in a High Magnetic Field

M. ASANUMA and R. AOGAKI*
Department of Information Technology, Gunma Polytechnic College, 918,Yamanachyo, Takasaki, Gunma, 370-1213, Japan
*National Research Laboratory for Magnetic Science (NRLM), JST, 1-1-56, Shibashimo, Kawaguchi, Saitama, 333-0848, Japan, and Department of Product Design, Polytechnic University, 4-1-1, Hashimotodai, Sagamihara, Kanagawa, 229-1196, Japan

ABSTRACT
The micro-MHD effect on metal dissolution in acid solution induced by a vertical magnetic field was theoretically examined.

Metal dissolution is sometimes affected by a high magnetic field; the dissolution rate often decreases with the magnetic field strength. From the magnetohydrodynamic formulation of the non-equilibrium fluctuations, two kinds of the fluctuations of the fluid motion with different scales of length were distinguished, i.e. macroscopic and microscopic fluctuations that are essential for the phenomena. In the vicinity of the metal surface, the microscopic fluctuations form numerous vortexes to disturb the development of the non-equilibrium fluctuations in the reaction processes. The macroscopic fluctuations, which generate a pair of large-scale rotations, play an important role in mediating these fluctuations in the fluid motion and the reaction processes.

Accordingly, the average dissolution rate was predicted to decrease in proportion to the reciprocal of the 2/3rd order of the magnetic flux density, which was experimentally confirmed by measurements of copper dissolution in nitric acid under high magnetic field.

1. Introduction

With the recent development of the superconducting magnet, many new findings concerning high magnetic fields have been reported. The magnetic field effect on corrosion has been one of the most interesting phenomena to receive attraction. As a result, a number of mechanisms for the corrosions involving metal dissolution in magnetic fields have been proposed; many of them were sometimes contradictory, so that it seemed impossible to understand such effects from a strictly scientific viewpoint [1–5]. For example, people sometimes exaggerated the energies activated by the magnetic fields. However, the quantities acquired by each chemical species are only 1/1000th of the thermal energies even high magnetic fields of the order of 10 T (tesla), and do not affect the reaction. In heterogeneous magnetic fields, however, non-magnetic materials receive magnetic forces, which are not so strong, but which can sometimes induce levitation or convection (magnetic levitation or magnetic convection) [6,7]. Furthermore, even in a homogeneous magnetic field, the Lorentz force generated by the interaction between an electrolytic current and the magnetic field can drastically enhance the mass transfer process, leading to the well known magnetohydrodynamic (MHD) effect [8].

Shinohara and Aogaki have recently presented a new mechanism called the micro-MHD effect for the case of copper dissolution. They pointed out that circulating currents in the electrochemical local cells formed at the interface between the metal and the electrolyte solution are of great importance; the currents interact with the magnetic field to produce localised Lorentz forces, which generate a lot of microscopic convection flows [8, 9]. Such microscopic flows suppress the development of the metal dissolution process. This mechanism is based on the assumption that some kind of fluid flow does hinder the metal dissolution, and this is supported by experiments with a copper rotating disc electrode in nitric acid; when significant reduction in the reaction rate has been observed with increasing rotation speed [10].

As shown in Fig. 1, these authors also found that with copper dissolution in nitric acid solution in a homogeneous high magnetic field, a pair of large-scale rotating motions emerged over the copper surface [10]. These large-scale motions were thought to result from the microscopic MHD flows.

In this paper, by using non-equilibrium fluctuation theories in electrochemistry [11, 12] and magnetohydrodymanics [13], we shall attempt to construct a theoretical basis to describe the mechanism of the micro-MHD effect in metal dissolution in a homogeneous magnetic field vertical to the metal surface.

Fig. 1 *Photo of large scale rotations during copper dissolution in a magnetic field of 6 T (tesla) (after 50 s from the beginning) [10]. The copper surface area was* 5×5 *mm*^{2}*, and the solution contained 3 mol m*$^{-3}$ *nitric acid. Solution depth was 1 mm. The fluid motion was visualised by a carbon colloid.*

2. Theoretical

2.1. Basic Equations

In the chemical dissolution of metal such as copper in nitric acid, the whole process proceeds via electrochemical and chemical reactions on the metal surface, i.e. the metal dissolution takes place at the anodic sites with acid reduction at the cathodic sites. Each pair of sites forms an electrochemical local cell. The currents flowing in the local cell also circulate through the solution phase. Therefore as shown in Fig. 2, such currents produce a Lorentz force in a magnetic field, which in turn induces a localised convection cell.

These microscopic convection flows occur through the mechanism of the non-equilibrium fluctuations, and the macroscopic rotations are thought to arise from microscopic convection cells. In this paper, we shall show that the microscopic convection flows are composed of numerous vortexes and that the macroscopic rotations are necessary for the production of the vortexes that hinder the dissolution process.

The electromagnetic nature of an electrolyte solution can be described by the Maxwell equations, whereas the fluid motion follows the Navier-Stokes equation. With the displacement currents ignored, the Maxwell equations are

$$\nabla \cdot \vec{B} = 0 \tag{1}$$

$$\nabla \times \vec{B} = \mu_0 \vec{j} \tag{2}$$

$$\nabla \times \vec{E} = -\frac{\partial \vec{B}}{\partial t} \tag{3}$$

where $\nabla \equiv (d/dx, d/dy, d/dz)$ in a Cartesian coordinate system, μ_0 the magnetic permeability, and $\vec{E}$ and $\vec{B}$ are the electric field intensity and magnetic flux density,

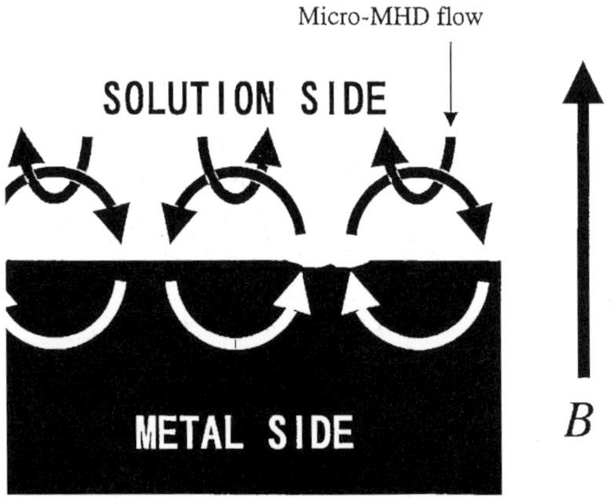

Fig. 2 Micro-MHD flows occurring at the metal surface in an acid solution under high magnetic field.

respectively. $\vec{j}$ is the electrolytic current density, which is expressed by the following equation depending on the ionic concentrations.

$$\vec{j} = \sigma_0(\vec{E} + \vec{u} \times \vec{B}) - F\sum_k z_k D_k \nabla C_k \tag{4}$$

where σ_0 is the electric conductivity, $\vec{u}$ is the fluid velocity, and is $\vec{u} \times \vec{B}$ the induced electric field, F is the Faraday constant, z_k the valency of the kth ion, D_k its diffusion coefficient and C_k is the concentration. The Lorentz force induced by this current density under the magnetic field is

$$\vec{F} = \vec{j} \times \vec{B} \tag{5}$$

Considering eqns (1)–(5), we can derive the Navier-Stokes equation for the motion of an electrolyte solution as follows [13],

$$\frac{\partial \vec{u}}{\partial t} + (\vec{u} \cdot \nabla)\vec{u} - \frac{1}{\rho\mu_0}(\vec{B} \cdot \nabla)\vec{B} = -\nabla\left(\frac{P}{\rho} + \frac{|\vec{B}|^2}{2\mu_0\rho}\right) + \nu\nabla^2\vec{u} \tag{6}$$

where P is the pressure, ρ is the density, and ν is the kinematic viscosity. Moreover, for an incompressible fluid, the continuity equation

$$\nabla \cdot \vec{u} = 0 \tag{7}$$

holds. Then, using eqns (1)–(5) and eqn (7), we can derive the following equation for the transfer of the magnetic flux density [13].

$$\frac{\partial \vec{B}}{\partial t} + (\vec{u} \cdot \nabla)\vec{B} - (\vec{B} \cdot \nabla)\vec{u} = \eta\nabla^2\vec{B} \tag{8}$$

where $\eta \equiv 1/(\sigma_0\mu_0)$ is called the resistivity. The mass transport process of the kth ion in the presence of a large amount of supporting electrolyte is expressed by the convective diffusion equation

$$\frac{\partial C_k}{\partial t} + (\vec{u} \cdot \nabla)C_k = D_k\nabla^2 C_k \tag{9}$$

where C_k is the concentration of the kth ion, and D_k is the diffusion coefficient.

2.2. Macroscopic Rotations

The two kinds of motions can be expressed by the microscopic and macroscopic non-equilibrium fluctuations in hydrodynamics, respectively. In the present paper, for the first case, the fluctuations in the magnetic field vertical to the metal surface are examined. Consequently, from eqns (1)–(9) the following five equations are obtained [13].

$$\frac{\partial b_z^*}{\partial t} = \eta\nabla^2 b_z^* + B_0\frac{\partial w^*}{\partial z} \tag{10}$$

$$\frac{\partial j_z^*}{\partial t} = \eta \nabla^2 j_z^* + \frac{B_0}{\mu_0}\frac{\partial \zeta^*}{\partial z} \tag{11}$$

$$\frac{\partial \zeta^*}{\partial t} = \nu \nabla^2 \zeta^* + \frac{B_0}{\rho}\frac{\partial j_z^*}{\partial z} \tag{12}$$

$$\frac{\partial}{\partial t}\nabla^2 w^* = \nu \nabla^4 w^* + \frac{B_0}{\rho\,\mu_0}\frac{\partial}{\partial z}\nabla^2 b_z^* \tag{13}$$

$$\frac{\partial c_k^*}{\partial t} = D_k \nabla^2 c_k^* - L_k w^* \tag{14}$$

where '*' means the macroscopic fluctuation and B_0 is the vertical homogeneous magnetic flux density, and b_z^*, j_z^* and w^* are the z-components of the magnetic flux density fluctuation, the current density fluctuation and the velocity fluctuation, respectively. The z-component of the vorticity fluctuation that newly appears implies the horizontal rotation induced by the vertical magnetic field. Then, c_k^* is the concentration fluctuation of the kth ion, and L_k is the average concentration gradient in the z-direction, i.e.[11],

$$L_k \equiv \left\{ \frac{d\langle C_k(z)\rangle}{dz} \right\}_{z=0} \tag{15}$$

where '$\langle\ \rangle$' means the area average over the metal surface, and $z = 0$ indicates the metal surface.

These macroscopic fluctuations are solved on the assumption that they can be expressed as two-dimensional waves. Accordingly, we finally obtain the equation for the macroscopic rotating motion as follows [14](see Appendix 1),

$$\Omega^* = -\tfrac{1}{2}\left(\frac{B_0\delta}{\rho\nu}\right) J^*(0) z e^{-a^* z} \tag{16}$$

where Ω^* is the amplitude of the z-component of the macroscopic vorticity fluctuation, $J^*(0)$ is the amplitude of the macroscopic current density fluctuation at the surface, which is, as shown later, equalised to the average current density over the dissolved surface, δ is the thickness of the convective diffusion layer, and z and a^* are the non-dimensional distance and the macroscopic non-dimensional wave number measured by δ as a unit length, respectively. From the physical meaning of Ω^*, it is concluded that, at least, a pair of macroscopic rotations in the opposite directions emerges with angular velocity of $1/2\Omega^*$.

2.3. Microscopic Vortexes

The microscopic fluctuation of the fluid motion can grow by linking up with the macroscopic rotations, the process being schematically represented in Fig. 3.

As to additional effects of the rotations, the Coriolis acceleration is the most important; the centrifugal component disappears together with the pressure term. The following equations are thus derived.

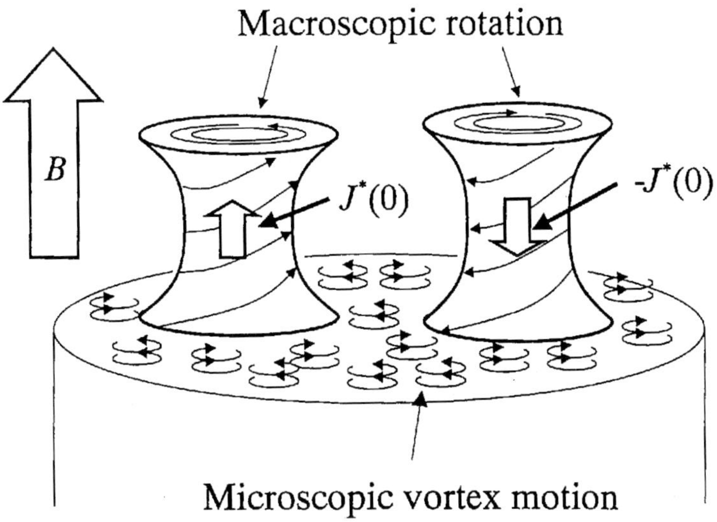

Fig. 3 *Macroscopic rotations and the microscopic vortexes induced by a vertical homogeneous magnetic field.*

$$\frac{\partial b_z}{\partial t} = \eta \nabla^2 b_z + B_0 \frac{\partial w}{\partial z} \tag{17}$$

$$\frac{\partial j_z}{\partial t} = \eta \nabla^2 j_z + \frac{B_0}{\mu_0}\frac{\partial \zeta}{\partial z} \tag{18}$$

$$\frac{\partial \zeta}{\partial t} = \nu \nabla^2 \zeta + \frac{B_0}{\rho}\frac{\partial j_z}{\partial z} + \Omega^* \frac{\partial w}{\partial z} \tag{19}$$

$$\frac{\partial}{\partial t}\nabla^2 w = \nu \nabla^4 w + \frac{B_0}{\rho \mu_0}\frac{\partial}{\partial z}\nabla^2 b_z - \Omega^* \frac{\partial \zeta}{\partial z} \tag{20}$$

$$\frac{\partial c_k^i}{\partial t} = D_k^i \nabla^2 c_k^i - L_k^i w \tag{21}$$

where b_z, j_z, w and ζ are the microscopic fluctuations with the same meanings as the macroscopic fluctuations; c_k^i, D_k^i and L_k^i mean c_k, D_k and L_k in the partial reaction i (anodic (a) or cathodic (c) reaction) , respectively. These equations are solved with the following boundary conditions.

2.3.1. Conditions for magnetohydrodynamic fluctuations

All the fluctuations vanish at a distance far away from the surface.

$$b_z = j_z = \zeta = w = c_k^i = 0 \quad \text{for} \quad z = \infty \tag{22}$$

The motion of the solution is stopped at the surface.

$$\zeta = w = 0 \ \text{ for } \ z = 0 \tag{23}$$

2.3.2. Conditions for electrochemical fluctuations

In an electrolyte solution, the electric conductivity is not so high that the electromagnetically induced potential can be ignored. The current density fluctuation in the solution is thus described by the fluctuations of the concentration overpotential and the activation overpotential,

$$\vec{j}_i = -\sigma_0 \nabla(\eta_i + \phi_i) \tag{24}$$

where the subscript i implies the cathodic(c) or anodic(a) partial reaction, and η_i and ϕ_i mean the fluctuations of the concentration overpotential and the activation overpotential, respectively.

Furthermore, the current density fluctuation is conserved in the solution, i.e.

$$\nabla \cdot \vec{j}_i = 0 \tag{25}$$

The z-component of the current density fluctuation is related to the mass flux at the surface by the following,

$$\left(\vec{j}_i\right)_z = -z_i F D_c^i \left(\frac{\partial c_c^i}{\partial z}\right)_{z=0} \tag{26}$$

where z_i is the valency, c_c^i is the concentration fluctuation of the active cation in the partial reaction i, and D_c^i is the diffusion coefficient. To discriminate between the components in the anodic and cathodic partial reactions, the superscript i is introduced for c_c and D_c.

Then, the fluctuation of the concentration overpotential at the surface is expressed as [11–16, 17]

$$\eta_i(0) = \frac{RT}{z_i F} \frac{c_c^i(0)}{\left\langle C_c^i(0)\right\rangle} \tag{27}$$

The fluctuation of the activation overpotential ϕ_i is described by

$$\phi_i = \frac{RT}{z_i F A_i} \left\{ j_i^R - \frac{\left\langle J_i^R \right\rangle}{\left\langle J_i^{ex} \right\rangle} j_i^{ex} \right\} \tag{28}$$

where $\left\langle J_i^{ex} \right\rangle$ and j_i^{ex} are the average exchange current densities of the partial reaction i and its fluctuation, respectively, and $\left\langle J_i^R \right\rangle$ and j_i^R are the average reaction current density of the partial reaction i and its fluctuation. $\left\langle J_i^R \right\rangle$ and A_i are written as follows,

$$\left\langle J_i^R \right\rangle = \left\langle J_i^R \right\rangle_a - \left\langle J_i^R \right\rangle_c \tag{29}$$

and

$$A_i \equiv \alpha_i \left\langle J_i^R \right\rangle_a + \beta_i \left\langle J_i^R \right\rangle_c \tag{30}$$

where

$$\left\langle J_i^R \right\rangle_a \equiv \left\langle J_i^{ex} \right\rangle \exp\left(\frac{\alpha_i z_i F}{RT} \langle \Phi_i \rangle \right) \tag{31}$$

$$\left\langle J_i^R \right\rangle_c \equiv \left\langle J_i^{ex} \right\rangle \exp\left(-\frac{\beta_i z_i F}{RT} \langle \Phi_i \rangle \right) \tag{32}$$

where α_i and β_i are the transfer coefficients of the partial reaction i, and R and T are the gas constant and the absolute temperature, respectively.

The reaction current density is equalised to the z-component of the current density at the surface; therefore

$$j_i^R = \left(\vec{j}_i \right)_z \quad \text{for} \quad z = 0 \tag{33}$$

Finally, supposing that the metal dissolution is rate-determining, we introduce, as the simplest model, the following random instability process for the anodic partial reaction. This is based on the mechanism for the catalytic metal dissolution [18]; the total reaction proceeds through the specific adsorption of the anion such as Cl^- and NO_3^- on to a metal surface. The simplified partial anodic reactions in the present case are thus formally written as [16, 17]

$$M(metal) + mX^-(sol) \rightarrow (M^{z_a+}X_m^-)_{ad} + z_a e \tag{34}$$

$$(M^{z_a+}X_m^-)_{ad} \rightarrow M^{z_a+}(sol) + mX^-(sol) \tag{35}$$

where m is the coordination number of the anion, *sol* means the solution phase, and subscript *ad* indicates adsorption, M and M^{z_a+} refer to the metal atom and its cation, X^- is the anion, and $(M^{z_a+}X_m^-)_{ad}$ indicates a kind of complex formed by the specific adsorption of the anions on to the metal atom.

If the first reaction (eqn (34)) controls the total reaction rate and the metal dissolution is determined by the concentration of the anion (this was supported by the fact that for copper dissolution in nitric acid, the dissolution rate increases in proportion to the copper nitrate concentration [19]). On the other hand, we assume that the reaction in eqn (35) is much faster than the other rate processes including reactions and mass transfers. Therefore, according to the steady state method in chemical kinetics, the reaction is approximately regarded to be in a dissociation equilibrium; between the dissociating metal ions and anions, with the concentration fluctuations of the anion $c_a^a(0)$ and the metal ion $c_c^a(0)$ related by the following relation,

$$c_c^a(0) = mc_a^a(0) \tag{36}$$

in which the superscript a means the anodic partial reaction, and the subscripts a and c indicate the anion and cation, respectively.

The electrode reaction is assumed to follow the following apparent rate reaction,

$$j_a^{ex} = z_a F k_a^0 \left\{ mc_a^a(0) \right\} \tag{37}$$

where j_a^{ex} is the exchange current density fluctuation in the dissolution reaction, and k_a^0 is the rate constant at the corrosion potential. Substituting eqn (36) into eqn (37), we can obtain the equation

$$j_a^{ex} = z_a F k_a^0 c_c^a(0) \tag{38}$$

The above discussions can be applied to the average value, i.e.

$$\left\langle J_a^{ex} \right\rangle = z_a F k_a^0 \left\langle C_c^a(0) \right\rangle \tag{39}$$

From the preceding paper [16], these processes make the reaction system unstable, so that the fluctuations develop with time. The micro-MHD flows composed of numerous vortexes suppress such enhanced states of the fluctuations. Here, we limit the discussion to only the anodic partial reaction because in this case, the metal dissolution, i.e. the anodic reaction, is rate-determining, so that the discussion on the cathodic partial reaction can be neglected.

Using the boundary conditions mentioned above, all the equations eqns (17)–(21) are solved in the form of two-dimensional waves in steady state, e.g. the z-component of the velocity fluctuation is given by

$$w = W(z)\exp\left\lfloor i(a_x x + a_y y)\right\rfloor \tag{40}$$

The z-component of the vorticity fluctuation is written as

$$\zeta = Z(z)\exp\left\lfloor i(a_x x + a_y y)\right\rfloor \tag{41}$$

The concentration fluctuation of the cation (metal ion) in the partial anodic reaction (metal dissolution) is

$$c_c^a = \Theta_c^a(z)\exp\left\lfloor i(a_x x + a_y y)\right\rfloor \tag{42}$$

where a_x and a_y are the x- and y-components of the non-dimensional wave number a measured by the thickness of the convective-diffusion layer, respectively, W, Z and Θ_c^a are the amplitudes of the fluctuations. These amplitudes are expressed as the series of z as follows,

$$W(z) = \left(\sum_{i=0}^{\infty} \alpha_i z^i\right) z^2 e^{-az} \tag{43}$$

$$Z(z) = -\frac{B_1^* \delta}{\nu}\left(\sum_{i=0}^{\infty} \gamma_i z^i\right) z e^{-az} \tag{44}$$

$$\Theta_c^a = C_{1c} e^{-az} - \left(\frac{\Delta C_c^a \delta}{D_c^a}\right)\left(\sum_{i=0}^{\infty} \delta_i z^i\right) e^{-az} \tag{45}$$

where α_i, γ_i and δ_i are the coefficients of the expansion terms, and B_1^* and C_{1c} are arbitrary constants (see eqn (a1.15) in Appendix 1 and eqn (a2.26) in Appendix 2) and ΔC_c^a is the concentration difference of the cation in the anodic partial reaction between the bulk and the surface.

Finally, the amplitude of the microscopic anodic current density fluctuation at the surface is approximately obtained as follows (see Appendix 2),

$$J^a(0) = -\frac{1}{32a^8}\left(\frac{J^*(0)\delta}{\nu}\right)\left(35a^5 + \frac{363T_0}{32a^2}\right)A_0 \tag{46}$$

where A_0 is an arbitrary constant, and T_0 is a newly introduced non-dimensional number to describe the vortex motion of the microscopic fluctuation, which in this case is written as

$$T_0 \equiv \frac{\delta^6 J^*(0)^2 B_0^2}{4\rho^2\nu^4} \tag{47}$$

Here, $J^*(0)$ is the amplitude of the macroscopic current density fluctuation, which is equalised to the average current density (see Appendix 3). Therefore, we obtain

$$J^*(0) = \left(\frac{\pi}{2}\right)^2 z_a F D_c^a \frac{\langle C_c^a(0)\rangle}{\delta} \tag{48}$$

where in the case when the anodic dissolution is rate-determining, $\langle C_c^a(0)\rangle$ is considered to be approximately equal to the saturated concentration of the metal salt.

2.4. Suppression of Dissolution Rate

Substituting eqns (43)–(45) together with eqns (40)–(42) into eqns (17)–(21), we can determine the coefficients of eqns (43)–(45) within a first approximation to derive the following characteristic equation for the microscopic fluctuations [14] (see Appendix 4).

$$T_0 = \frac{6\alpha_0\delta - 35\beta_0 a}{\frac{241}{480}\alpha_0\delta + \frac{363}{32}\beta_0 a}a^7 \tag{49}$$

where α_0 and β_0 are defined as

$$\alpha_0 \equiv (2/\pi)^2 \nu(1-\alpha_a)\langle J_{corr}\rangle \sin\theta \tag{50}$$

$$\beta_0 \equiv z_a F D_c^{a^2} \langle C_c^a(0)\rangle \tag{51}$$

where $\langle J_{corr}\rangle$ is the average corrosion current density determined by the Tafel lines, and θ is the phase difference of the two kinds of the concentration fluctuations in the anodic and cathodic partial reactions.

Then, after introducing the dimensional wave number k ($\equiv a/\delta$), eqn (49) is rewritten as

$$\frac{\tilde{T}_0}{\delta^3} = \frac{6\alpha_0 - 35\beta_0 k}{\frac{241}{480}\alpha_0 + \frac{363}{32}\beta_0 k} k^7 \tag{52}$$

where

$$\tilde{T}_0 \equiv \frac{1}{4}\left(\frac{\pi}{2}\right)^4 \frac{(z_a F D_c^a)^2 \left\langle C_c^a(0)\right\rangle^2 B_0^2}{\rho^2 \nu^4} \tag{53}$$

The right hand side of eqn (52), as shown in Fig. 4, changes with k, having a maximum value, which corresponds to the minimum thickness δ of the convective diffusion layer. The derivative of the right hand side of eqn (52) is equalised to zero, so that the minimum value of δ is obtained.

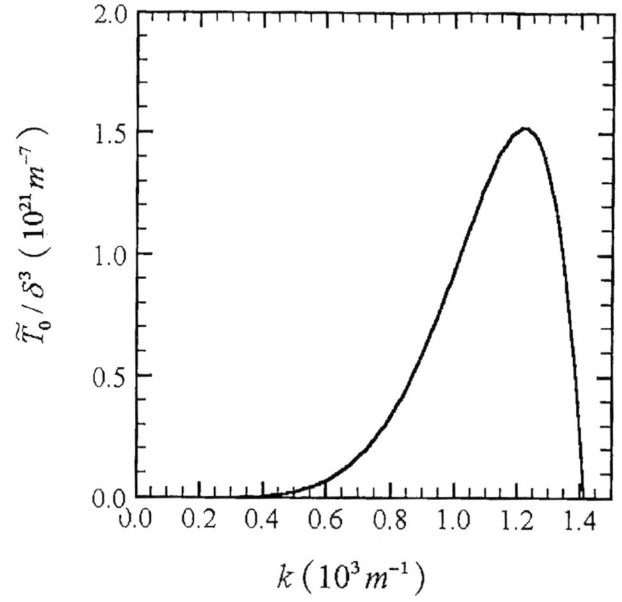

Fig. 4 $\tilde{T}_0/\delta^3$ *vs* k *plot when copper dissolution is rate-determining*.*

* $\tilde{T}_0$ *is the parameter given by eqn (53),* δ *is the thickness of the convective diffusion layer, and* k *is the dimensional wave number of the microscopic fluctuation.*

The data used are $B_0 = 1$ T, $\langle j_z^*(0)\rangle = 2.0 \times 10^2\ Am^{-2}$ *(estimated from the experimental result),* $\langle C_c^a(0)\rangle = 8.3 \times 10^3\ mol\ m^{-3}$ *(saturated concentration of copper nitrate at 25°C),* $\langle J_{corr}\rangle = 1.32\ Am^{-2}$ *(estimated value by the Tafel plots),* $D_c^a = 5 \times 10^{-10} m^2 s^{-1}$, $\rho = 10^3\ kgm^{-3}$, $\nu = 10^{-5} m^2 s^{-1}$, $\alpha_a = 0.75$ *(referring to the copper dissolution in a copper sulfate solution [20],* $\theta = \pi/2$, $z_a = 2$, *and* $F = 96\ 500\ coul\ mol^{-1}$.

$$\delta_{min} = 1131.8 \frac{(z_a F D_c^a)^3 \left\langle C_c^a(0) \right\rangle^3 B_0^{2/3}}{\rho^{2/3} \nu^{4/3} (\nu / D_c^a)^{7/3} (1-\alpha_a)^{7/3} \left\langle J_{corr} \right\rangle^{7/3} \sin\theta^{7/3}} \tag{54}$$

Substituting δ_{min} for eqn (54) in eqn (a3.3) in Appendix 3, we can finally obtain the dissolution current density under the condition when the metal dissolution is the rate-determining step as follows,

$$\left\langle j_z^*(0) \right\rangle = 8.835 \times 10^{-4} \frac{\rho^{2/3} (\nu / D_c^a)^{11/3} (1-\alpha_a)^{7/3} \left\langle J_{corr} \right\rangle^{7/3} \sin\theta^{7/3}}{(z_a F)^2 D_c^{a2/3} \left\langle C_c^a(0) \right\rangle^2 B_0^{2/3}} \tag{55}$$

where θ is, as mentioned above, the phase difference between the concentration fluctuations of the active cations in the anodic and cathodic partial reactions; in the case of $\theta = 0$, the active sites of both reactions are very near, so that any circulating current and thus the micro-MHD effect do not appear. For $\theta \neq 0$, the circulating currents leak into the solution phase, and eqn (55) indicates the decrease of the macroscopic current density in proportion to the reciprocal of the 2/3rd order of the magnetic flux density (this equation is correct only in the range of high magnetic field). This means that many vortexes occurring in the neighborhood of the metal surface disturb the unstable growth of the concentration fluctuations.

3. Discussion

The prediction of eqn (55) was, as shown in Fig. 5, confirmed by plotting the copper dissolution rate in a nitric acid solution against the –2/3rd order of the magnetic flux density.

All the data were rewritten using the preceding experimental results [9]. The dissolution rate, as expected from eqn (55), decreases with the magnetic flux density, by a straight line with a slope of 2.14×10^{-2} $gcm^{-2}h^{-1}T^{2/3}$.

Considering this result, we can estimate the value of the convective diffusion layer thickness δ_{min} with the following data: $B_0 = 1$ T, $\left\langle j_z^*(0) \right\rangle = 2.0 \times 10^2$ Am^{-2} (estimated from the experimental result), $\left\langle C_c^a(0) \right\rangle = 8.3 \times 10^3$ mol m^{-3} (saturated concentration of copper nitrate at 25°C), $\left\langle J_{corr} \right\rangle = 1.32$ Am^{-2} (estimated value from the Tafel plots), $D_c^a = 10^{-10} m^2 s^{-1}$, $\rho = 10^3$ kgm^{-3}, $\nu = 10^{-5} m^2 s^{-1}$, $\alpha_a = 0.75$ (referring to the copper dissolution in a copper sulfate solution [20], $\theta = \pi/2$ (completely different phase), $z_a = 2$, and $F =$ 96 500 coul mol^{-1}.

Then, δ_{min} at 1 T is calculated as 8.0×10^{-4}m, and the corresponding non-dimensional wave number a is 439.2. From these results, it can be said that the thickness of the convective diffusion layer is the order of 1 mm, and the size of a single vortex is the order of 0.01 mm. The derived value of the convective diffusion layer thickness is reasonable considering the slow velocity of the rotation in Fig. 1.

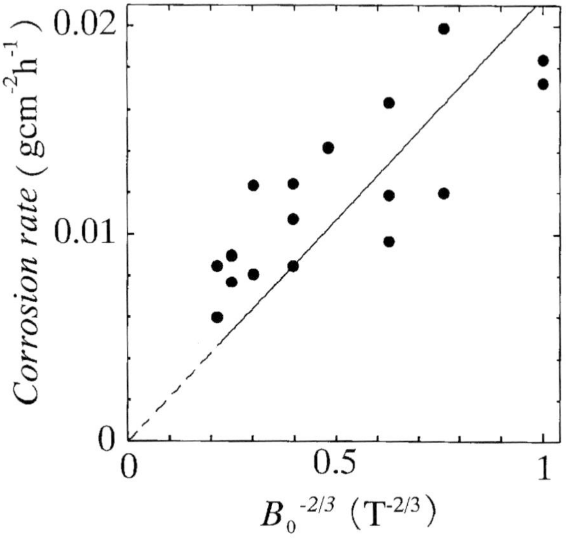

Fig. 5 *Copper dissolution rate against magnetic flux density [9].* B_0 *indicates the homogeneous magnetic flux density, and T is tesla. Solution contained 3 mol* m^{-3} *nitric acid. Temperature was kept at 11±1 °C.*

4. Conclusions

The micro-MHD effect on metal dissolution has for the first time been theoretically analysed by using the non-equilibrium fluctuations and magnetohydrodynamic equations. In this paper, we have treated the case when the homogeneous magnetic field is applied vertically to the metal surface and the metal dissolution process is rate-determining. The result was that it became obvious that the macroscopic and microscopic fluctuations play mutually important roles, i.e. the micro-MHD flows induced by the interaction between the currents circulating in the local cells and the numerous vortexes generated by the magnetic field. These vortexes influence the mass transfer process of the dissolved metal ion by modifying the macroscopic rotation.

As a result, it was predicted as a first order approximation that the total dissolution rate decreases with the reciprocal of the 2/3rd order of the magnetic flux density. This prediction was verified by the data for copper dissolution in a nitric acid solution. However, in the present case, the equation is applicable only to the case of a high magnetic field. We are now attempting to derive the equation for the low magnetic field together with the numerical approach using computer techniques.

5. Acknowledgement

The authors thank Dr Kenji Shinohara at NRLM-JST for providing the good photograph and the experimental data and for stimulating and fruitful discussions.

Appendix 1

Equations (10)–(14) can be solved by the following two-dimensional plane waves,

$$w^* = W^*(z)\exp\left[i\left(k_x^* x + k_y^* y\right) + p^* t\right] \tag{a1.1}$$

$$\zeta^* = \Omega^*(z)\exp\left[i\left(k_x^* x + k_y^* y\right) + p^* t\right] \tag{a1.2}$$

$$b_z^* = K^*(z)\exp\left[i\left(k_x^* x + k_y^* y\right) + p^* t\right] \tag{a1.3}$$

$$j_z^* = J^*(z)\exp\left[i\left(k_x^* x + k_y^* y\right) + p^* t\right] \tag{a1.4}$$

$$c_k^* = \Theta_k^*(z)\exp\left[i\left(k_x^* x + k_y^* y\right) + p^* t\right] \tag{a1.5}$$

where $W^*(z)$, $\Omega^*(z)$, $K^*(z)$, $J^*(z)$ and $\Theta_k^*(z)$ are the amplitudes of the macroscopic fluctuations, i is $\sqrt{-1}$, k_x and k_y are the x- and y-components of the wave number k^*, and p^* is the amplitude factor.

Assuming δ and δ^2/ν as the representative length and the representative time (δ is the convective-diffusion layer thickness), we can introduce the non-dimensional wave number $a^* \equiv k^*\delta$, the non-dimensional amplitude factor $\sigma^* \equiv p^*\delta^2/\nu$ and other non-dimensional numbers $p_{1k} \equiv \nu/D_k$ and $p_2 \equiv \nu/\eta$. At the same time, the non-dimensional differential operator $D \equiv \delta^{-1}d/dz$ is defined, where z means the non-dimensional z-coordinate measured with δ.

Substituting eqns (a1.1)–(a1.5) into eqns (10)–(14), and using the non-dimensional parameters, it follows that

$$\left(D^2 - a^{*2} - p_2\sigma^*\right)K^* = -\left(\frac{B_0\delta}{\eta}\right)DW^* \tag{a1.6}$$

$$\left(D^2 - a^{*2} - p_2\sigma^*\right)J^* = -\left(\frac{B_0\delta}{\mu_0\eta}\right)D\Omega^* \tag{a1.7}$$

$$\left(D^2 - a^{*2} - \sigma^*\right)\Omega^* = -\left(\frac{B_0\delta}{\rho\nu}\right)DJ^* \tag{a1.8}$$

$$\left(D^2 - a^{*2}\right)\left(D^2 - a^{*2} - \sigma^*\right)W^* = -\left(\frac{B_0\delta}{\mu_0\rho\nu}\right)D\left(D^2 - a^{*2}\right)K^* \tag{a1.9}$$

$$\left(D^2 - a^{*2} - p_{1k}\sigma^*\right)\Theta_k^* = \left(\frac{\Delta C_k\delta}{D_k}\right)W^* \tag{a1.10}$$

where ΔC_k is the concentration difference of the kth ion at the bulk and the surface, i.e.

$$\Delta C_k \equiv C_k(z=\infty) - \langle C_k = (0)\rangle \tag{a1.11}$$

where $C_k(z = \infty)$ and $\langle C_k(0) \rangle$ mean the bulk concentration and the average surface concentration of the kth ion, respectively.

In eqns (a1.6)–(a1.10), K^* and W^* have a closed relation with eqns (a1.6) and (a1.9), whereas J^* and Ω^* are completely determined by eqns (a1.7) and (a1.8). This means that the current density (J^*), i.e. the Lorentz force does not generate the velocity (W^*), but generates the vorticity (Ω^*). Since Θ^* has a relation only with W^* in eqn (a1.10), the current density does not directly affect the concentration, which seems reasonable for the macroscopic motions outside the convective-diffusion layer. From eqns (a1.7) and (a1.8), we can obtain the vorticity equation as follows,

$$\left\{\left(D^2 - a^{*2} - \sigma^*\right)\left(D^2 - a^{*2} - p_2\sigma^*\right) - QD^2\right\}\Omega^* = 0 \qquad \text{(a1.12)}$$

where Q is a non-dimensional number to express the electromagnetic induction.

$$Q \equiv \frac{\sigma_0 B_0^2 \delta^2}{\rho\nu} \qquad \text{(a1.13)}$$

In electrochemical systems, the electric conductivity σ_0 is not so high as that of metals, so that Q can be neglected. Then, assuming the steady state, we can obtain the following equation

$$\left(D^2 - a^{*2}\right)^2 \Omega^* = 0 \qquad \text{(a1.14)}$$

The equation is solved with the boundary conditions; (1) $\Omega^* \to 0$ and $D\Omega^* \to 0$ as $z \to 0$. (2) $\Omega^* = 0$ for $z = 0$. In particular, it follows that

$$\Omega^* = B_1^* z e^{-a^* z} \qquad \text{(a1.15)}$$

where B_1^* is an arbitrary constant. Furthermore, eqn (a1.8) is rewritten for the steady state condition as follows,

$$\left(D^2 - a^{*2}\right)\Omega^* = -\left(\frac{B_0\delta}{\rho\nu}\right) DJ^* \qquad \text{(a1.16)}$$

This equation can be integrated with the condition that J^* and $\Omega^* \to 0$ as $z \to \infty$ in the following,

$$D\Omega^*(0) - a^{*2}\int_{\infty}^{0} \Omega^* \, dz = -\left(\frac{B_0\delta}{\rho\nu}\right) J^*(0) \qquad \text{(a1.17)}$$

After substituting eqn (a1.15) into eqn (a1.17), we can finally obtain the equation for Ω^*,

$$\Omega^* = -\tfrac{1}{2}\left(\frac{B_0\delta}{\rho\nu}\right) J^*(0) z e^{-a^* z} \qquad \text{(a1.18)}$$

Appendix 2

Equations (17)–(21) also can be solved in the forms of the following plane waves.

$$w = W(z)\exp\left[i\left(k_x x + k_y y\right) + pt\right] \quad \text{(a2.1)}$$

$$\zeta = Z(z)\exp\left[i\left(k_x x + k_y y\right) + pt\right] \quad \text{(a2.2)}$$

$$b_z = K(z)\exp\left[i\left(k_x x + k_y y\right) + pt\right] \quad \text{(a2.3)}$$

$$(\vec{j}_a)_z = J^a(z)\exp\left[i\left(k_x x + k_y y\right) + pt\right] \quad \text{(a2.4)}$$

$$c_c^a = \Theta_c^a(z)\exp\left[i\left(k_x x + k_y y\right) + pt\right] \quad \text{(a2.5)}$$

where $(\vec{j}_a)_z$ and $J^a(z)$ refer to the z-component of the anodic current density and its amplitude, respectively. Except that the amplitude of ζ changes from Ω to Z, these equations are essentially consistent with eqns. (a1.1) to (a1.5) eliminating '*'. With the same procedures as Appendix 1, we can make the equations non-dimensional as follows,

$$\left(D^2 - a^2 - p_2\sigma\right)K = -\left(\frac{B_0\delta}{\eta}\right)DW \quad \text{(a2.6)}$$

$$\left(D^2 - a^2 - p_2\sigma\right)J^a = -\left(\frac{B_0\delta}{\mu_0\eta}\right)DZ \quad \text{(a2.7)}$$

$$\left(D^2 - a^2 - \sigma\right)Z = -\left(\frac{B_0\delta}{\rho\nu}\right)DJ^a - \left(\frac{\Omega * \delta}{\nu}\right)DW \quad \text{(a2.8)}$$

$$\left(D^2 - a^2\right)\left(D^2 - a^2 - \sigma\right)W = -\left(\frac{B_0\delta}{\mu_0\rho\nu}\right)D\left(D^2 - a^2\right)K + \left(\frac{\Omega * \delta^3}{\nu}\right)DZ \quad \text{(a2.9)}$$

$$\left(D^2 - a^2 - p_{1k}\sigma\right)\Theta_c^a = \left(\frac{\Delta C_c^a\delta}{D_c^a}\right)W \quad \text{(a2.10)}$$

where ΔC_c^a means the concentration difference of the cation in the anodic partial reaction between the bulk and the surface. From the fact that steady rotations are experimentally observed, assuming that there is no oscillatory mode, we can confine the solutions to the stationary mode with $\sigma = 0$. Equations (a2.6)–(a2.10) are rewritten as follows,

$$\left(D^2 - a^2\right)K = -\left(\frac{B_0\delta}{\eta}\right)DW \quad \text{(a2.11)}$$

$$\left(D^2 - a^2\right)J^a = -\left(\frac{B_0\delta}{\mu_0\eta}\right)DZ \quad \text{(a1.12)}$$

$$\left(D^2 - a^2\right)Z = -\left(\frac{B_0\delta}{\rho\nu}\right)DJ^a - \left(\frac{\Omega * \delta}{\nu}\right)DW \tag{a2.13}$$

$$\left(D^2 - a^2\right)^2 W = -\left(\frac{B_0\delta}{\mu_0\rho\nu}\right)D\left(D^2 - a^2\right)K + \left(\frac{\Omega * \delta^3}{\nu}\right)DZ \tag{a2.14}$$

$$\left(D^2 - a^2\right)\Theta_c^a = \left(\frac{\Delta C_c^a \delta}{D_c^a}\right)W \tag{a2.15}$$

Then, from eqns (a2.11) and (a2.14),

$$\left\{\left(D^2 - a^2\right)^2 - QD^2\right\}W = \left(\frac{\Omega * \delta^3}{\nu}\right)DZ \tag{a2.16}$$

from eqns (a2.12) and (a2.13),

$$\left\{\left(D^2 - a^2\right)^2 - QD^2\right\}Z = -\left(\frac{\Omega * \delta}{\nu}\right)\left(D^2 - a^2\right)DW \tag{a2.17}$$

are obtained.

Disregarding Q owing to the low electric conductivity, we can rewrite eqns (a2.16) and (a2.17) as follows:

$$\left(D^2 - a^2\right)^2 W = \left(\frac{\Omega * \delta^3}{\nu}\right)DZ \tag{a2.18}$$

$$\left(D^2 - a^2\right)^2 Z = -\left(\frac{\Omega * \delta}{\nu}\right)\left(D^2 - a^2\right)DW \tag{a2.19}$$

Moreover, with the function F,

$$F \equiv \left(D^2 - a^2\right)W \tag{a2.20}$$

equations (a2.18) and (a2.19) are reduced to

$$\left\{\left(D^2 - a^2\right) + TD^2\right\}F = 0 \tag{a2.21}$$

where T is defined as

$$T \equiv \left(\frac{\Omega * \delta^2}{\nu}\right)^2 \tag{a2.22}$$

It should be noticed that T is not a constant, but a function of z because Ω^* is a function of z as shown in eqn (a1.15) in Appendix 1. Therefore, it is too complicated to solve analytically the equations mentioned above, so that we have to use the method of series expansion in the following procedure.

Firstly, we obtain F in eqn (a2.21). Then, from eqn (a2.20), W is derived. Furthermore, using eqn (a2.19), we decide the function form of Z, and from eqn (a2.15), Θ_c^a is determined. At the same time, J^a is obtained by substituting Z into eqn (a2.13). Since in electrochemical systems, the fluctuation of the magnetic flux density does not play important roles because of low electric conductivity, the solution of K is not pursued.

The boundary conditions to obtain the solutions are as follows:

(1) F, W, Z and $\Theta_c^a \rightarrow 0$ as $z \rightarrow 0$.

(2) $W = DW = Z\ (D^2 - a^2)F = 0$ for $z = \infty$.

The next step is to expand the solutions to the series of z; at first, F is written in the following form,

$$F \equiv \sum_{i=0}^{\infty} A_i z^i \tag{a2.23}$$

Equation (a2.23) is substituted into eqn (a2.21) to determine the coefficient A_i with the boundary conditions. Then, the resultant expression for F is substituted into eqn (a2.20) to derive the series for W:

$$W = \left(\sum_{i=0}^{\infty} \alpha_i z^i \right) z^2 e^{-az} \tag{a2.24}$$

Then, substituting W for eqn (a2.24) in eqn (a2.19), we obtain

$$Z = -\frac{B_1^* \delta}{\nu} \left(\sum_{i=0}^{\infty} \gamma_i z^i \right) z e^{-az} \tag{a2.25}$$

where A_i, α_i and γ_i are the coefficients of the series. Θ_c^a is derived from the substitution of eqn (a2.24) into eqn (a2.15), i.e.

$$\Theta_c^a = C_{1c} e^{-az} - \left(\frac{\Delta C_c^a \delta}{D_c^a} \right) \left(\sum_{i=0}^{\infty} \delta_i z^i \right) e^{-az} \tag{a2.26}$$

where C_{1c} is an arbitrary constant, and δ_i is the coefficient of the series.

Finally, using eqn (a2.13), with eqns (a2.24) and (a2.25), the approximated equation of J^a at the surface is obtained.

$$J^a(0) = -\frac{1}{32a^8}\left(\frac{J^*(0)\delta}{\nu}\right)\left(35a^5 + \frac{363T_0}{32a^2}\right)A_0 \qquad \text{(a2.27)}$$

where A_0 is the coefficient of the zero order term of z in the function F in eqn (a2.23).

Appendix 3

The macroscopic current density fluctuation at the surface is expressed as

$$j_z^*(0) = J^*(0)\exp\left[i\left(k_x^* x + k_y^* y\right)\right] \qquad \text{(a3.1)}$$

The dissolution current density can be obtained for the range, $0 \le x \le \pi / k_x^*$ and $0 \le y \le \pi / k_y^*$, so that the average value is given by

$$\left\langle j_z^*(0)\right\rangle = \frac{\left|J^*(0)\int_0^{\pi/k_x^*}\int_0^{\pi/k_y^*}\exp\left[i\left(k_x^* x + k_y^* y\right)\right]dxdy\right|}{\left(\pi/k_x^*\right)\left(\pi/k_y^*\right)} = \left(\frac{2}{\pi}\right)^2 J^*(0) \qquad \text{(a3.2)}$$

It is also written as

$$\left\langle j_z^*(0)\right\rangle = z_a F D_c^a \frac{\left\langle C_c^a(0)\right\rangle}{\delta} \qquad \text{(a3.3)}$$

Therefore, from eqns (a3.2) and (a3.3), we obtain the following expression

$$J^*(0) = \left(\frac{\pi}{2}\right)^2 z_a F D_c^a \frac{\left\langle C_c^a(0)\right\rangle}{\delta} \qquad \text{(a3.4)}$$

Appendix 4

To obtain the fluctuation of the anodic current density, eqn (24) is substituted into eqn (25). Consequently,

$$\nabla^2(\eta_a + \phi_a) = 0 \qquad \text{(a4.1)}$$

is obtained. From the boundary condition eqn (22), the expression for the current density fluctuation satisfying eqn (a4.1) is given as

$$(\vec{j}_a)_z = J^a(0)e^{-kz}\exp\left[i\left(k_x x + k_y y\right) + pt\right] \qquad \text{(a4.2)}$$

On the one hand, substituting eqns (26), (27), (33), (38) and (39) into eqn (28), we obtain

$$\eta_a + \phi_a = \frac{RT\left(A_a - \left\langle J_a^R \right\rangle\right)}{z_a F A_a \left\langle C_c^a(0) \right\rangle} c_c^a(0) - \frac{RTD_c^a}{A_a}\left(\frac{\partial c_c^a}{\partial z}\right)_{z=0} \tag{a4.3}$$

In considering sufficiently high anodic polarisation in eqn (29),

$$\left\langle J_a^R \right\rangle = \left\langle J_a^R \right\rangle_a = \left\langle J_{corr} \right\rangle \tag{a4.4}$$

holds. From eqn (30), A_a is also simplified in the following,

$$A_a = \alpha_a \left\langle J_a^R \right\rangle_a = \alpha_a \left\langle J_{corr} \right\rangle \tag{a4.5}$$

where $\langle J_{corr} \rangle$ is the average corrosion current density determined by the Tafel plots. Under the boundary condition of eqn (22), eqn (24) is integrated from $z = 0$ to ∞ as follows,

$$\int_{\infty}^{0} \frac{1}{\sigma_0} \left(\vec{j}_a\right)_z dz + \eta_a(0) + \phi_a(0) = 0 \tag{a4.6}$$

where $\eta_a(0)$ and $\phi_a(0)$ indicate η_a and ϕ_a at $z = 0$.

Substituting eqns (a4.2) and (a4.3) into eqn (a4.6), and making the resultant equation non-dimensional, we obtain

$$-\frac{\delta J^a(0)}{\sigma_0 a} + \frac{RT(\alpha_a - 1)}{\alpha_a z_a F \left\langle C_c^a(0) \right\rangle} \Theta_c^a(0) - \frac{RTD_c^a}{\alpha_a \langle J_{corr} \rangle \delta} D\Theta_c^a(0) = 0 \tag{a4.7}$$

On the other hand, making eqn (26) non-dimensional, we can derive the following amplitude equation.

$$J^a(0) = -z_a F D_c^a \delta^{-1} D\Theta_c^a(0) \tag{a4.8}$$

Therefore, substitution of eqn (a4.8) into eqn (a4.7) leads to

$$\Theta_c^a(0) = -\tilde{A}_a D\Theta_c^a(0) \tag{a4.9}$$

where

$$\tilde{A}_a = \frac{z_a F D_c^a \left\langle C_c^a(0) \right\rangle}{\delta(1-\alpha_a)\langle J_{corr} \rangle} \tag{a4.10}$$

For simplicity, $\alpha_a z_a F \langle J_{corr} \rangle \delta << \sigma_0 RTa$ is assumed.

From eqn (a2.26) in Appendix 2,

$$\Theta_c^a(0) = C_{1c} - \left(\frac{\Delta C_c^a \delta}{D_c^a}\right)\delta_0 \tag{a4.11}$$

$$D\Theta_c^a(0) = -aC_{1c} - \left(\frac{\Delta C_c^a \delta}{D_c^a}\right)a\delta_0 \tag{a4.12}$$

are obtained. δ_0 is the coefficient of the zero order term of z in eqn (a2.26) in Appendix 2, i.e.

$$\delta_0 = \frac{1}{16a^8}\left(\frac{3a^4}{2} - \frac{241T_0}{1920a^3}\right)A_0 \tag{a4.13}$$

where T_0 is a non-dimensional number defined by eqn (47), and A_0 is the coefficient of the zero order term of z in the function F in eqn (a2.23) in Appendix 2.

Finally, substituting eqns (a4.11)–(a4.13) into eqn (a4.9), we can derive eqn (49) for the case where , i.e.

$$T_0 = \frac{6\alpha_0\delta - 35\beta_0 a}{\frac{241}{480}\alpha_0\delta + \frac{363}{32}\beta_0 a}a^7 \tag{a4.14}$$

References

1. M. Sagawa, *Trans. Jpn Inst. Met.*, 1982, **23**, 28.
2. A. Chiba, M. Okada and T. Ogawa, *Fushoku Bshoku Toronkai Koenshu*, 1986, **33**, 113 (in Japanese).
3. S. S. Dushikin and V. N. Evstratov, *Magnetic Treatment of Water in Chemical Enterprise*, Japan-Soviet News Agency, Wakayama, 1986.
4. M. A. Grhabashy, *Corros. Prev. Control*, 1988, **35**, 73.
5. A. Terlain and T. Dufrenoy, *J. Nucl. Mater.* , 1994, **212**, 1504.
6. Y. Ikezoe, N. Hirota, J. Nakagawa and K. Kitazawa, *Nature*, 1998, **393**, 749.
7. S. Kishioka and R. Aogaki, *Chem. Lett.*, 1999, 91–92.
8. R. Aogaki, A. Tadano and K. Shinohara, *Transfer Phenomena in Magnetohydrodynamic and Electroconducting Flows* (A. Almany, Ph. Marty and J. P. Thibault, eds). Kluwer, Netherlands, 1999, p169.
9. K. Shinohara and R. Aogaki, *Electrochemistry*, 1999, **67**, 126.
10. K. Shinohara and R. Aogaki, submitted to *Electrochim. Acta.*
11. R. Aogaki, *J. Electrochem. Soc.*, 1995, **142**, 2954.
12. M. Asanuma and R. Aogaki, *J. Chem. Phys.*, 1997, **106**, 9930.
13. S. Chandrasekhar, *Hydrodynamic and Hydromagnetic Stability*. Oxford University Press, London, 1961, Ch. 1.
14. The complete mathematical treatment will appear in the paper by M. Asanuma and R. Aogaki, submitted to *J. Chem. Phys.*
15. M. Asanuma and R. Aogaki, *J. Electroanal. Chem.*, 1995, **396**, 241.

16. R. Aogaki and M. Asanuma, *Mater. Sci. Forum*, 1998, **289–292**, 8335.
17. R. Aogaki, *Modern Aspects of Electrochemistry*, 33 (R. E. White, J. O'M. Bockris and B. E. Conway, Eds). Plenum, New York, 1999, p. 217.
18. N. Sato, *Proc. 2nd Japan–USSR Corrosion Seminar*, JSCE, Tokyo, 1979, p.35.
19. M. Sato and R. Aogaki, *Mater. Sci. Forum*, 1998, **289–292**, 459.
20. E. Mattson and J. O'M. Bockris, *Trans. Faraday Soc.*, 1959, **55**, 1586.

25
Measurement of Steel Corrosion in Concrete by Electrochemical Techniques: Influence of the Redox Processes in Oxide Scales

C. ANDRADE, F. BOLZONI*, M. CABEZA†, X. R. NÓVOA† and M. C. PÉREZ†
Institute of Construction Sciences "Eduardo Torroja", CSIC, Apdo. 19002, Madrid, Spain
*Politecnico di Milano, Dipartamento di Chimica Fisica Applicata, Via Mancinelli, 7, 20131 Milano, Italy
†Universidade de Vigo. E.T.S.E.I.M., Lagoas-Marcosende, 9, 36280 Vigo, Spain

ABSTRACT
Alternating and direct current electrochemical measurements were performed on various types of reinforced concrete probes: atmosphere-aged and chloride-contaminated. The samples were selected so that the corrosion potential was equal to or more positive than –0.3 V vs SCE. From the a.c. results an interpretation is given of the commonly observed difficulty in reaching the steady state when performing d.c. measurements. The interpretation is based on the redox activity of the oxide layer in this potential range. A Linear Polarisation method is proposed that allows the measurement of the charge transfer resistance to be associated correctly to the corrosion process. The undesired contributions that need to be eliminated are the bulk concrete resistance and the resistance associated with Fe_3O_4/ Fe_2O_3 transformations.

1. Introduction

Corrosion of the reinforcing steel in concrete is an electrochemical process in which metallic iron is converted to the voluminous ferric oxide corrosion product [1]. Chloride attack and/or carbonation are the main causes of damage and early failure of reinforced concrete structures [2]. Because of the enormous economic costs related to the maintenance, restoration and replacement of concrete structures, rapid non-destructive inspection techniques have been developed. Good descriptions of the various methods currently in use for in situ evaluation of concrete structures are given in the literature [3]. The most widely used methods for controlling steel reinforcements are [4]: visual observation, gravimetric tests, potential mapping (PM), linear polarisation (so called polarisation resistance method, R_p), transient techniques (coulostatic, galvanostatic, potentiostatic), electrochemical noise analysis (ENA) and electrochemical impedance spectroscopy (EIS).

Potential mapping according to ASTM C876 standard is the most commonly used electrochemical technique to assess rebar corrosion [5] because, perhaps, it is easy to operate and it is the only standard method available. Nevertheless, some doubts arise about the real meaning of the measured electrode potentials due to the many different phenomena present in the concrete bulk that can influence the measured potential. Diffusion potentials, membrane potentials and streaming potentials can be cited among such factors [6].

Potential mapping gives only qualitative information. If quantitative data have to be obtained then it is necessary to polarise the steel rebar. The only exception is ENA which is a non-perturbing technique that, according to some authors [7] permits kinetic information to be obtained, although considerable uncertainties remain when the technique is applied to steel in concrete [8]. The electrochemical parameter needed to obtain the rebar corrosion rate is the corrosion current density, i_{corr}. The objective of the electrochemical methods mentioned above is the determination of the polarisation resistance, R_p, because this is related to corrosion rate through the classical Stern formula $i_{corr} = B/R_p$. This relationship is valid [9–11] if, among other conditions, R_p corresponds to the slope of the steady state polarisation curve (corrected for ohmic drop) at the corrosion potential, E_{corr}, provided this is far enough from the reversible potentials of the redox processes involved.

Extensive experimental work carried out since the 1970s by one of the authors (Andrade) [12] has permitted the right B value and sweep rate to fit electrochemical data to mass loss results to be determined, for each corrosion situation. These investigations made R_p measurements applicable in fieldwork, overcoming the problems of linearity and stationarity inherent in the technique. Nevertheless, the fundamental scientific reasons (in terms of process understanding) for those useful choices remain so far unknown.

Transient techniques (a.c. impedance, galvanostatic pulse, potentiostatic pulse) have also been used to evaluate corrosion of steel in concrete. These techniques require a model for the transfer function that describes the steel/concrete interface in order to extract the kinetic parameter related to the corrosion rate of the tested rebar. Several models have been proposed to account for the various relaxation phenomena observed at this interface, although the most commonly employed is the Randles circuit in which the electrolyte resistance is in series with the double-layer capacitance that discharges through the R_p [13]. In this paper results will be presented to illustrate the importance of oxide scale transformations in the electrochemical response of the steel/concrete interface and thus on the R_p significance.

2. Experimental

In order to obtain suitable electric field distribution, cylindrical probes have been used for measurements. The importance of the electrode arrangement for impedance measurements has been discussed elsewhere [14,15]. Two types of probe have been employed, according to the position of the reference and counter electrodes, which were either external, as depicted in Fig. 1(a), or embedded, as indicated in Fig. 1(b). Although the embedded reference electrode was activated titanium, all potentials in the text are referred to the saturated calomel electrode (SCE). The use of the Ti/TiO_2 electrode as reference electrode is described in the literature [16–19].

All the tested probes were prepared in a similar manner: a bare carbon steel bar was embedded in concrete made of ordinary Portland Cement. The concrete was prepared using 330 kg of cement/m^3 and a water/cement ratio of 0.6. However, for

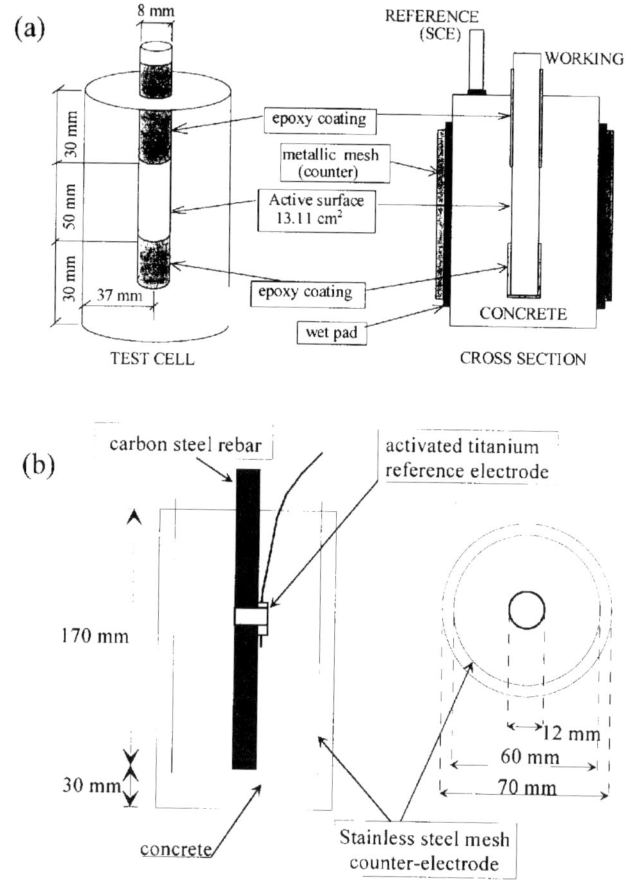

***Fig. 1** Schematic representation of the test probes employed for measurements. (a) external reference and counter electrodes; (b) embedded reference and counter electrodes.*

the set of probes having embedded reference and counter electrodes $CaCl_2$ was added to the mixing water in amounts representing up to 3% by weight of cement.

Ageing was also different. Chloride-containing probes were aged for six months in the laboratory atmosphere, while chloride-free probes were exposed to a rural atmosphere for eight years. In neither case did carbonation reach the steel rebars.

The a.c. and d.c. measurements were performed using standard electrochemistry instrumentation from Solatron. A 1286 potentiostat and a 1255 FRA, both controlled by a personal computer via IEEE were employed.

3. Results and Discussion

3.1. Atmosphere Aged Chloride Free Probes

Figure 2 shows the impedance spectrum obtained on one of the chloride-free probes aged for eight years in the atmosphere. The measurement was carried out on the as-received probe, i.e. immediately after being withdrawn from the exposure site on a sunny and dry day. The measured potential was, in these conditions, quite high, +150 mV. In the same Figure a detail of the high frequency region of the impedance spectrum is also presented. The meaning of the capacitive loops observed in this region has been discussed elsewhere [20,21] and the only point of interest here is that the impedance at about 10 Hz corresponds to the electrolyte resistance, i.e. the resistance to be compensated when performing linear polarisation measurements.

In the low frequency range (below 10 Hz), two time constants can be differentiated. We assume here that the constant at higher frequencies corresponds to the charge transfer resistance in parallel with the double layer capacitance, while the lowest part of the spectrum corresponds to redox processes in the oxide layer. It has been shown [22] that a process such as that given in eqn (1) can result in a RC behaviour in this frequency range, showing capacitance values of some mF cm^{-2}.

$$3\,Fe_3O_4 \rightleftarrows 4\,\gamma\text{-}Fe_2O_3 + Fe^{2+} + 2e^- \tag{1}$$

Moreover, the reversible transformation of magnetite to Fe_2O_3/FeOOH has been observed in alkaline solution [23,24] by *in situ* RAMAN spectroscopy, between about –0.3 and 0.5 V.

Assuming those two contributions to the impedance spectrum, eqn (2) can be employed to model the experimental data

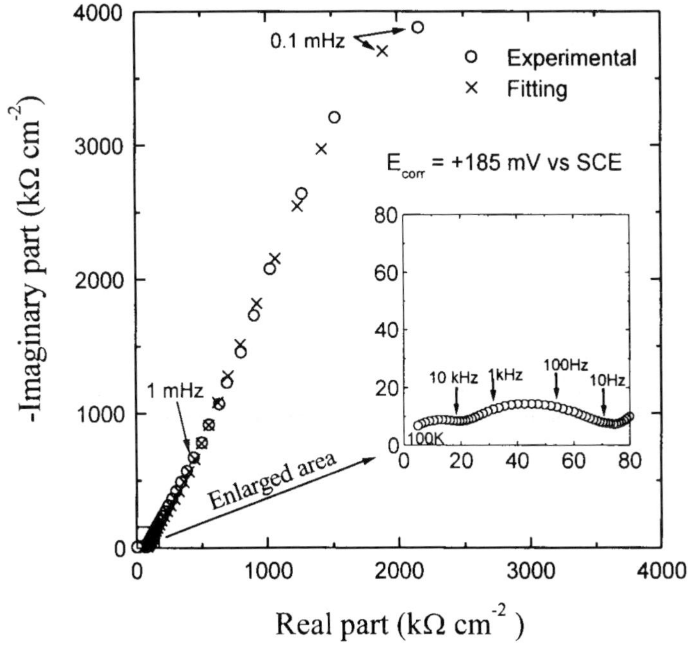

***Fig.* 2** *Complex plane impedance plot obtained on an eight year atmosphere-aged probe in dry conditions at E_{corr} = +185 mV. Dots correspond to the experimental data and crosses to the fitting to eqn (2). The best fitting parameters are given in Table 1. Only data below 10 Hz have been considered for fitting (see text for details).*

$$Z(\omega) = R_e + \frac{R_{ct}}{(R_{ct}/(Z_2(\omega)+R_{ct})) + (j\omega R_{ct}C_{dl})^{\alpha_1}} \quad Z_2(\omega) = \frac{R_2}{1+(j\omega R_2 C_2)^{\alpha_2}} \tag{2}$$

In eqn (2) R_e corresponds to the series resistance (or electrolyte resistance), $\omega = 2\pi f$, R_{ct} is the change transfer resistance for the corrosion process, C_{dl} the double layer capacitance and α_1, represents the degree of dispersion of this $R_{ct}C_{dl}$ time constant ($0 < \alpha_1 \leq 1$). The term $Z_2(\omega)$ accounts for the redox impedance of the process given in eqn (1). Here C_2 is not a geometric or double layer capacitance, but a redox capacitance, in the sense of battery capacitance, R_2 is related to the kinetics of the redox process, and α_2 measures the degree of dispersion of the R_2C_2 time constant.

In Table 1 the values of parameters in eqn (2) best fitting to the data in Fig. 2 (below 10 Hz) are given. The fitting procedure is a simplex method already described [25]. It should be noted that for this system the characteristic frequency of the $R_{ct}C_{dl}$ time constant is in the mHz range. The C_{dl} values depend on the state of the electrode surface but they remain in the μF cm^{-2} range. The charge transfer resistance, R_{ct}, is related to the corrosion rate, i_{corr}, through eqn (3).

$$i_{corr} = \frac{B}{R_{ct}} \tag{3}$$

If B is taken as 26 mV, the R_{ct} value in Table 1 gives i_{corr} = 0.06 μA cm^{-2}, a corrosion current density well below the threshold usually considered for a passive rebar (0.1 μA cm^{-2}).

The very low characteristic frequency corresponding to the R_2C_2 values (f_2 in Table 1) indicates that the process described in eqn (1) is very slow, which is typical for solid-phase transformations. This can explain the usual finding in the literature [5] concerning the practical impossibility of reaching the steady state when the interface is polarised by an electrical signal.

The data given in Fig. 3 support the argument that the R_2C_2 time constant is related to the eqn (1) process. The polarisation curve, starting in the cathodic region, gives a zero current potential (corrosion potential) close to –50 mV and proceeds anodically to +400 mV. At this potential the direction of the potential sweep is reversed and a new zero current potential is found close to +325 mV. From this point the current

Table 1. *Best fitting parameters of data in Figs 2, 4 and 5 to the impedance model given in eqn (2). The fitting procedure was a SIMPLEX-based home made method [25]*

Data	R_e (kΩ cm^2)	R_{ct} (kΩ cm^2)	C_{dl} (μF cm^{-2})	α_1	f_1 (mHz)	R_2 (kΩ cm^2)	C_2 (mF cm^{-2})	α_2	F_2 (μHz)
Fig. 2	83.0	464	101	0.788	3.4	2×10^9	2.9	0.673	2.6
Fig. 4	57.6	230	112	0.730	6.2	31.5×10^3	15.5	1.0	326
Fig. 5	2.7	21.7	406	0.800	18.1	24.3×10^3	21.5	0.80	303

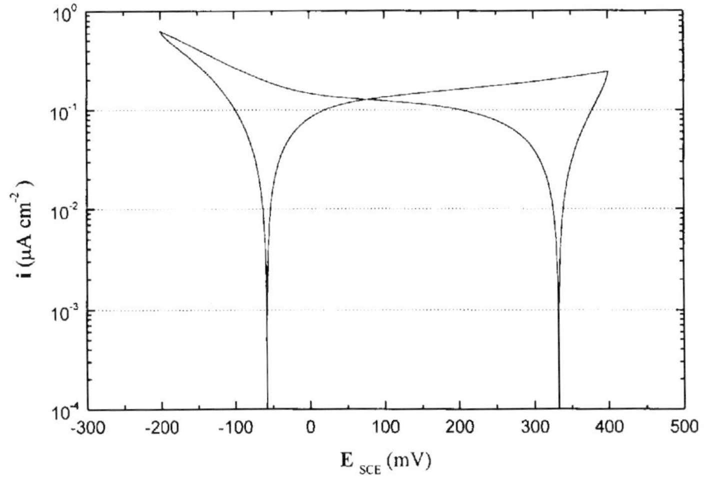

***Fig.** 3 Polarisation curve obtained on an eight-year atmosphere-aged probe at 1 mV s^{-1} sweep rate.*

remains cathodic but shows two different slopes, respectively above and below the corrosion potential. The Tafel line in the cathodic range which shows a slope of about 270 mV/decade, close to that measured for aerated pore solutions [26], can be attributed to oxygen reduction. However, in the potential range of +50 to +250 mV, the Tafel line has a seven times higher slope, which is evidence for the existence of a different and slower cathodic process. Those observations suggest that in the forward scan the oxidation of some species occurs in the range of about +50 to +400 mV.

These oxidised species are partially reduced during the reverse scan given rise to the observed cathodic slope processes that can easily be interpreted if eqn (1) is taken into account. During the forward scan magnetite oxidises to γ–Fe_2O_3, while in the reverse scan Fe^{2+} ions are re-incorporated into the oxygen lattice resulting in the reduction of γ–Fe_2O_3 to form magnetite. Both compounds have the same oxygen lattice and thus magnetite can reversibly transform to γ–Fe_2O_3 [27].

Figure 4 shows the impedance spectrum for the same aged probe as in Fig. 2, but after one day wetting of its external surface by using a damp sponge. With respect to Fig. 2, a small modification of the higher part of the spectrum is observed which can be explained in terms of changes in the concrete resistivity [14], which is higher for the dry concrete (higher R_e value in Table 1). However, the most significant changes are in the lowest frequency range. The Fe^{2+}/Fe^{3+} process corresponding to eqn (1) disappears as a redox process giving rise to an almost undetectable R_2C_2 time constant. Nevertheless, the corrosion process kinetics remain roughly unchanged. The obtained R_{ct} value (see Table 1) slightly decreases but it remains in the typical range for a passive surface.

The disappearance of the low frequency arc with hydration can be interpreted in terms of the electrical conductivity of the passive layer, which decreases as hydration increases. For the process in eqn (1) to take place a good electrical conductivity in the passive film is necessary and this is only achieved in dry conditions. This aspect has been considered elsewhere [28,29] to interpret the effect of the oxide layer on macrocell

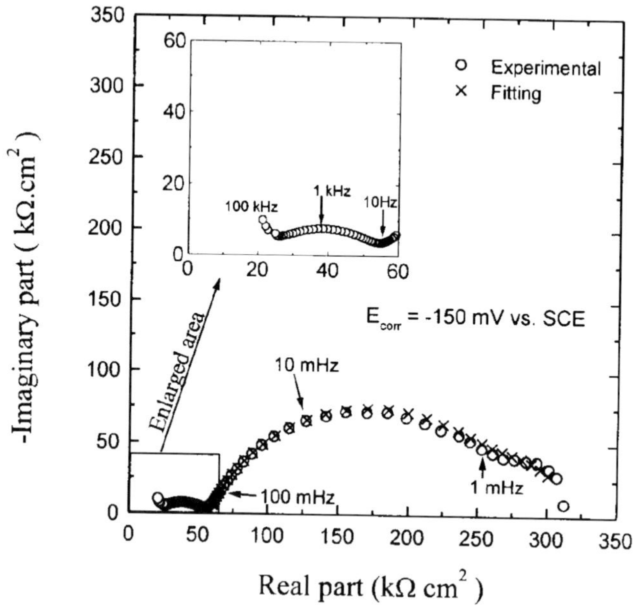

Fig. 4 Complex plane impedance plot obtained on a eight year aged probe after one day wetting at E_{corr} *= –150 mV. Dots correspond to the experimental data and crosses to the fitting to eqn (2). The best fitting parameters are given in Table 1. Only data below 10 Hz have been considered for fitting (see text for details).*

activity. When the oxide layer shows conductivity character, the activity of the Fe^{2+}/ Fe^{3+} couple shifts the corrosion potential to the noble direction (E_{corr} = +185 mV for Fig. 2, dry condition). However, if this conductivity decreases then the redox process does not occur electrically connected to the substrate and the corrosion potential shifts to the cathodic direction (E_{corr} = –150 mV for Fig. 4, wet condition).

3.2. Chloride-containing Probes

In order to analyse an active corrosion situation measurements on chloride-containing probes have also been performed. The behaviour for probes having 0.4% Cl^- is similar to that previously described for chloride-free probes. Figure 5 presents the impedance spectrum for one probe containing 3% Cl^- at its corrosion potential (–304 mV). As for chloride-free probes, the experimental results have been modelled using eqn (2). The parameters corresponding to the best fit are summarised in Table 1. As for chloride-free probes, R_e roughly (because it depends on the relative position of the reference electrode) corresponds to the mortar resistivity. It decreases as the mortar gets wet (Figs 2 and 4), and reaches the lower value when hygroscopic salt is present (Fig. 5). From the R_{ct} value the use of eqn (3) results in i_{corr} = 1.2 μA cm^{-2}, a value that clearly corresponds to an active corrosion situation.

The double layer capacitance is about four times higher than that for Cl^- free probes. Different factors can contribute to that high C_{dl} value. From a geometrical point of view, it can be considered that the corrosion process increases the electrode

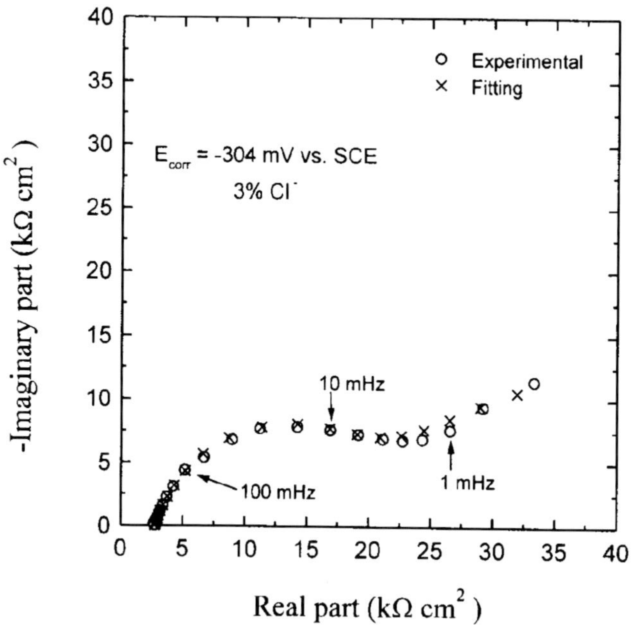

Fig. 5 *Complex plane impedance plot obtained on a six months ageing, 3% $CaCl_2$-containing probe in wet condition at $E_{corr} = -304$ mV. Dots correspond to the experimental data and crosses to the fitting to eqn (2). The best fitting parameters are given in Table 1. Only data below 10 Hz have been considered for fitting (see text for details).*

roughness and thus its effective surface. A chemical aspect is related to the change in pore solution chemistry, i.e. chlorides present in this solution tend to adsorb on the electrode surface, thus increasing the double layer charge. Finally, an electrochemical aspect must be considered, because redox processes occurring at more cathodic potentials [24,30] start to develop at about –0.3 V and the associated redox capacitance can contribute to the measured capacitance.

One interesting point to stress here is that, for the studied situations, the corrosion process is observed in the medium frequency range, between the bulk concrete response and the magnetite to ferric oxide transformations. According to Table 1 the characteristic frequency for this process (f_1 in Table 1) increases with corrosion rate but it remains around 10 mHz. Furthermore, from data in Table 1 and Figs 2, 4 and 5, it follows that R_{ct} corresponds closely to the real part of the impedance at the minimum (Figs 4 and 5) or inflexion point (Fig. 2) of the complex plane representations in the frequency range 100 to 1 mHz.

3.3. Polarisation Resistance Measurements

The above results support the idea that a correct estimation of the corrosion current density, i_{corr}, can only be performed in terms of R_{ct} through eqn (3). Any attempt of using eqn (3) by using d.c. R_p measurements will result in unrealistic values if R_p is defined as the zero frequency limit of the impedance spectrum, i.e. the slope of the stationary polarisation curve at E_{corr}. It is clear from Figs 2 and 4 that in some cases R_p

is non-measurable because the time to reach the steady state tends to infinity. It can only be obtained from fitting of the EIS spectra to circuit models: $R_p = R_{ct} + R_2$ (once R_e is corrected). Even in the case when it is measurable by d.c. polarisation, as Fig. 4 suggests, the measurement should be performed at a very low polarisation rate (corresponding to about 0.1 mHz). The R_p value thus obtained underestimates the corrosion rate because it includes R_2, a parameter not correlated to the corrosion rate.

One successful attempt to solve the problem of corrosion rate measurements was made years ago by defining the sweep rate able to give R_p (i_{corr}) values according to mass loss measurements [12]. Figure 6 reproduces this experimental choice. Two main aspects must be highlighted on this figure. First it assumes a non-stationary measurement, i.e. the condition for R_p measurements are not fulfilled. Thus, the resistance obtained is something with no special physical meaning but does fit into the experimental mass loss data. Secondly, the range of sweep rates suggested, from 2 to 10 mV min^{-1} (0.03 to 0.17 mV s^{-1}) correspond, over a total range of 40 mV (± 20 mV polarisation) to a frequency range of 0.8 to 4.2 mHz.

The real part of the impedance at about 1 mHz closely corresponds to the R_{ct} value in the three given examples (Fig. 2, 4 and 5), thus the resistance measured at the slowest sweep rate suggested by Andrade *et al.* [12] corresponds to R_{ct}; that is, this measurement correctly estimates the corrosion rate.

According to Figs 2, 4 and 5 the impedance (and thus the resistance) decreases as the frequency increases. The magnitude of the variation depends on the shape of the complex impedance plot and it is clear that the 0.8 to 4.2 mHz frequency range can

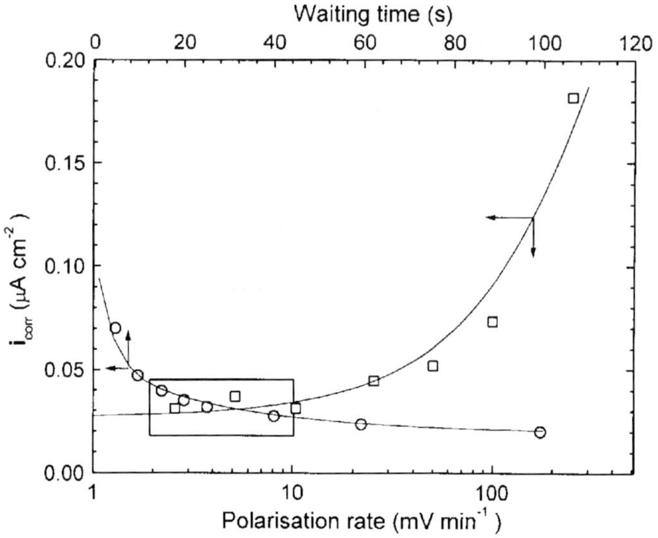

Fig. 6 *Variation of the corrosion rate as a function of the sweep rate (potentiodynamic test, ΔE = ± 10 mV) and the waiting time in potentiostatic step (ΔE = ± 10 mV). The highlighted zone corresponds to the sweep rate/waiting time that fits electrochemical data into weight loss measurements [12]. Sweep rate (squares) applies to potentiodynamic polarisation measurements and waiting time (circles) to potentiostatic step method.*

result in considerable uncertainty in the measured corrosion rate. Figure 7 represents an attempt to solve the problem of choosing the right sweep rate to correctly estimate R_{ct} by d.c. measurements.

If the electrode is cyclically polarised around the desired potential (E_{corr}) using a sinusoidal signal of the type $E = |\Delta E| \sin(\omega t)$ then the response of the electrode will be $I = |\Delta I| \sin(\omega t + \varphi)$, φ being the phase difference between potential and current. Thus, the impedance, Z, can be obtained from the classical Lissajous figure, i.e. $|Z| = |\Delta E| / |\Delta I|$, and $\sin(\varphi) = I_{E=0}/I_{E\,max}$. In this way, a frequency sweep between 0.5 and 5 mHz will allow the determination of the minimum phase angle (situations such as in Figs 4 and 5) or the inflexion angle (situation such as in Fig. 2) to which the real part of the impedance ($|Z| \cos(\varphi)$) corresponds to $R_e + R_{ct}$. Thus, it is necessary to know R_e in order to obtain R_{ct}. An easy way to do this is, according to Figs 2 and 5, to perform the same measurement but at about 10 Hz. The real part of the impedance at this frequency corresponds to R_e.

The equipment usually employed for d.c. measurements uses triangular waves to generate the potential sweep thus, it was considered useful to analyse whether cyclic voltammetry using the classical $|\Delta E| = \pm 10$ mV (or ± 20 mV as in Fig. 7) can be treated as Lissajous figure or not. Figure 7 shows this kind of plot for the same probe as in Fig. 4 and the corresponding significant parameters are given in Table 2. As can be seen, the phase angle increases and the resistance decreases when frequency (sweep rate) increases from 1 to 20.8 mHz (5 to 100 mV min^{-1}), just as expected from the shape of the impedance plot in Fig. 4. Moreover, according to the data in Table 1, the resistance measured at 1 mHz (minimum of the phase angle) closely corresponds to $R_e + R_{ct}$.

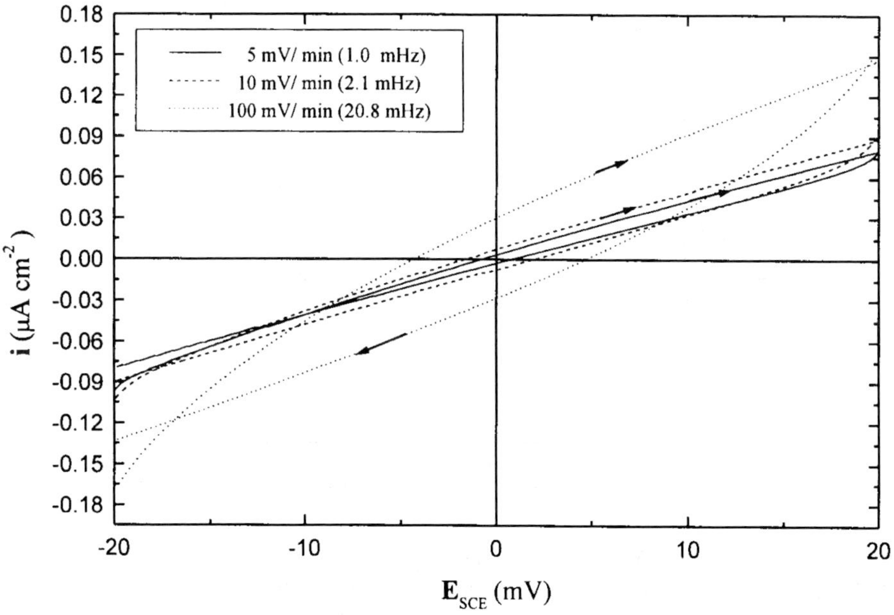

Fig. 7 *Linear polarisation plots obtained at different sweep rate on a probe under similar conditions as for Fig. 4. The relevant parameters obtained from this representation are given in Table 2.*

Table 2. Characteristic parameters obtained from linear polarisation on an atmosphere-aged specimen in the same experimental conditions as for Fig. 4.

Sweep rate mV min^{-1}	Sweep rate (frequency, mHz)	Phase angle (Degree)	R (kΩ cm^2)
5	(1.0)	6	260
10	(2.1)	8	220
100	(20.8)	20	124

4. Conclusions

The presented results, as well as the data in the cited literature, suggest that for electrode potentials more noble than –0.3 V (SCE) the electrochemical response of steel rebars embedded in concrete is largely influenced by a redox process occurring in the oxide layer. This redox reaction masks the corrosion reaction so that steady state (R_p) measurements lead to an underestimation of the corrosion rate.

A linear polarisation (LP) method is proposed that allows the correct estimation of the charge transfer resistance corresponding to the corrosion process, i.e. the corrosion rate, for electrode potentials more noble than –0.3V (SCE).

References

1. N. P. Mailvaganam, *Repair and Protection of Concrete Structures*. CRC Press, Boca Raton, FL, USA, 1992, 146.
2. B. Elsener, *Mater. Sci. Forum*, 1995, **192–194**, 857–866.
3. P. K. Mukherjee and N.P. Mailvaganam, *In situ evaluation of concrete structures, in Repair and Protection of Concrete Structures*. CRC Press, Boca Raton, FL, USA, 1992, 191–225.
4. P. Rodríguez, E. Ramírez and J. A. González, *Mag. Concrete Res.*, 1994, **46** (167), 81–90.
5. K. Videm, Field and laboratory experiences with potentiostatic polarisation and potentiokinetic scans to assess the severity of corrosion of steel in concrete, in *Corrosion of Reinforcement in Concrete, Monitoring, Prevention and Rehabilitation* (J. Mietz, B. Elsener and R. Polder eds). Papers from EUROCORR '97. Publication No. 25 in European Federation of Corrosion Series published by The Institute of Materials, London, 1998, 104–121.
6. H. Arup and O. Klinghoffer, Junction potentials at a concrete/electrolyte interface, in *Corrosion of Reinforcement in Concrete, Monitoring, Prevention and Rehabilitation*, ibid., 31–39.
7. F. Mansfeld and H. Xiao, *J. Electrochem. Soc.*, 1993, **140** (8), 2205–2210.
8. U. Bertocci, A comparison of electrochemical noise and impedance spectroscopy for the detection of corrosion in reinforced concrete, in *Electrochemical Noise Measurement for Corrosion Applications*, ASTM STP1277 (J. R. Kearns, J. R. Scully, P. R. Roberge, D. L. Reichert and J. L. Dawson eds), ASTM, West Conshohocken, Pa, 1996, 39–58.
9. M. Stern and R.M. Roth, *J. Electrochem. Soc.*, 1957, **104** (6), 390–392.
10. M. Stern, *Corrosion*, 1957, **13**, 775t–781t.
11. M. Stern, *Corrosion*, 1958, **14**, 329t–332t.

12. C. Andrade, V. Castelo, C. Alonso and J. A. González, The determination of the corrosion rate of steel embedded in concrete by the polarization resistance and AC impedance methods, in *Corrosion Effect of Stray Currents and the Techniques for Evaluating Corrosion of Rebars in Concrete*, ASTM STP906, (V. Chaker ed), ASTM, Philadelphia. 1986, 46–63.
13. G. K. Glass, C. L. Page, N. R. Short and J.–Z. Zhang, *Corros. Sci.*, 1997, **39** (9), 1657–1663.
14. C. Andrade, L. Soler, C. Alonso, X. R. Nóvoa and M. Keddam, *Corros. Sci.*, 1995, **37** (12), 2013–2023.
15. C. Andrade, L. Soler and X.R. Nóvoa, *Mater. Sci. Forum*, 1995, **192–194**, 843–856.
16. K. Kinoshita and M. J. Madou, *J. Electrochem. Soc.*, 1984, **131** (5), 1089–1094.
17. T. Pastore, P. Pedeferri, G. Mussinelli and M. Tettamanti, *Proc. 11th Int. Corrosion Congr.*, Publ. NACE, Houston, TX, USA, 1990, 467–472.
18. L. Bertolini, F. Bolzoni, T. Pastore and P. Pedeferri, Atti del congresso Omaggio Scientifico a Renato Turriziani, Roma, 1992, Vol. 1, 11–22.
19. A. Sagüés, P. Castro, L. Maldonado, E.I. Moreno and J. Genescá, *Proc. 1st Mexican Symp. on Metallic Corrosion*, Merida, Mexico, 1994, Paper No. 64, 43–52.
20. M. Keddam, H. Takenouti, X. R. Nóvoa, C. Andrade and C. Alonso, *Cem. Concr. Res.*, 1997, **27** (8), 1191–1201.
21. C. Alonso, C. Andrade, M. Keddam, X. R. Nóvoa and H. Takenouti, *Mater. Sci. Forum*, 1998, **289–292**, 15–28.
22. C. Andrade, P. Merino, X. R. Nóvoa, M. C. Pérez and L. Soler, *Mater. Sci. Forum*, 1995, **192–194**, 891–898.
23. S. Joiret, M. Keddam, X.R. Nóvoa, M.C. Pérez and H. Takenouti, Study by Raman spectroscopy of the inhibitive action of nitrite ions on iron corrosion in alkaline solution, Paper No. 43, Symposium 7. *50th ISE Meeting*, Abstract in CD ROM, Pavia, 5–10 October, 1999.
24. S. Joiret, M. Keddam, H. Takenouti, X. R. Nóvoa and M. C. Pérez, *Proc. 8th Int. Symp. on Passivity of Metals and Semiconductors*, Jasper, Alberta, Canada, 1999, Paper No. 144.
25. C. M. Abreu, M. Izquierdo, M. Keddam, X. R. Nóvoa and H. Takenouti, *Electrochim. Acta*, 1996, 41 (15) 2405–2415.
26. J. L. Dawson and P. E. Langford, The electrochemistry of steel corrosion in concrete compared to its response in pore solution, in *The Use of Synthetic Environments for Corrosion Testing*, ASTM STP970 (P. E. Francis and T. S. Lee, eds). ASTM, Philadelphia. 1988, 264–273.
27. N. N. Greenwood and A. Earnshaw, *Chemistry of the Elements*, Pergamon Press, NY, 1990, p. 1255.
28. C. Alonso, C. Andrade, M. Izquierdo, X. R. Nóvoa and M.C. Pérez, *Corros. Sci.*, 1998, **40** (8) 1379–1389.
29. C. Alonso, C. Andrade, A. Barreiro, L. Espada and X.R. Nóvoa, *Mater. Sci. Forum*, 1998, **289–292**, 45–54.
30. X. R. Nóvoa, M. C. Pérez, C. Rangel and H. Takenouti, *4th Int. Symp. on Electrochemical Impedance Spectroscopy*, 1998, Rio de Janeiro, Brazil. Extended Abstracts, 300–303.

26
Influence of Instrumentation on the Error Structure of Impedance Measurements

S. L. CARSON, M. E. ORAZEM, O. D. CRISALLE
and L. H. GARCIA-RUBIO*
Department of Chemical Engineering, University of Florida, Gainesville, FL, 32611, USA
*Department of Chemical Engineering, University of South Florida, Tampa, FL, 33620, USA

ABSTRACT
This paper reviews the work conducted to-date on the identification, and quantification, of the error structure for electrochemical impedance spectroscopy measurements. It formulates key questions leading to the identification of error structures, and presents a novel approach based on the modelling of the instruments actually used for the measurement of complex quantities. The propagation of errors from the original measurements in the time domain to the desired complex variables in the frequency domain is analysed. It is concluded that, as long as the complex quantities are measured through similar physico–chemical principles, a general model for the propagation of error can be established. The use of the proposed model is demonstrated through applications to electrochemical impedance spectroscopy and to other complex quantities.

1. Introduction

The Kramers–Kronig relations (KKR) are frequency-dependent integral transformations that relate the real and imaginary components of material properties that are measured as complex numbers. Among them, the complex dielectric function, the complex refractive index, the complex viscosity, and the electrochemical and acoustic impedances are commonly used for the detailed characterisation of materials and to establish their performance. Reliable estimation of these complex functions requires measurements over a broad range of frequencies, often with several instruments. Typically, these complex numbers are measured in time domain and then converted into the frequency domain. The variety of instruments and techniques used for the measurements makes it necessary to identify the corresponding error structure on an almost individual basis. As a result it has been difficult to generalise the treatment of the propagation of errors through the Kramers–Kronig relations, and therefore, to establish the reliability, and consistency of the estimates of the real and imaginary components of the measured variables.

2. Background

Complex nonlinear least-squares (CNLS) regression techniques were developed in the late 1960s [1–3]. The CNLS approach provides a significant improvement over nonlinear least-squares (NLS) techniques because a common set of parameters can be estimated by simultaneous regression of the model to both real and imaginary data [4]. Since the Kramers–Kronig Relations (KKR) constrain the real and imaginary parts of the complex quantity, the appropriate regression strategy should allow for the correlation of the real and imaginary parts of the complex model through the KKR.

The quality of information obtained by regression of models to impedance spectra is enhanced by choice of an appropriate weighting strategy. While an accepted approach is to weight the regression by the inverse of the variance of the stochastic errors in the data, there has been uncertainty in the literature over the functionality that the variance should take [5–11]. The discussion of error structures in impedance spectroscopy had, therefore, been limited to *a priori* predictions of measurement noise based on instrument noise [12] and to standard assumptions concerning the error structure, e.g. that errors are constant or proportional. Zoltowski [5,8,9] and Boukamp [6] have advocated use of modulus weighting; whereas, Macdonald *et al.* [7,10,11] have advocated use of a modified proportional weighting strategy.

Agarwal *et al.* developed an experimental approach to assess the error structure of impedance measurements [13–17]. The measurement model approach, widely used for identification of error structures of spectroscopic measurements [18–20], was used to filter experimental bias errors from estimates of the standard deviation of electrochemical impedance measurements. Their work showed that it was possible to identify a general structure for the errors in impedance measurements. For example, the standard deviation of the real part of the impedance response was found to be equal to that of the imaginary part. This result could be compared to the standard weighting strategies discussed in the literature. Under modulus weighting, the variances of real and imaginary parts of the impedance are equal; whereas, under proportional weighting, they are not. The finding that the variances of the real and imaginary parts of the impedance were equal for systems that satisfied the Kramers–Kronig relations has been verified both experimentally [21–23] and analytically [24–26]. Nevertheless, it has been suggested that the observed equality of variances in the stochastic errors in the real and imaginary impedance is an artifact of the instrumentation used to measure the impedance spectra, caused by undesired correlation between the errors in the real and imaginary impedance [26].

The *a priori* use of modulus or proportional weighting entails an assumed model for the variance of impedance data. The analysis of errors described by Agarwal *et al.* allowed identification of the model for the variance from experimental data under the implicit assumption that stochastic errors in the impedance are normally distributed. Yet, in impedance spectroscopy, the fundamental measured quantities are time–domain signals, and the frequency–domain result is a derived quantity. Thus, while the work to date has made assumptions about the structure of frequency domain errors, the manner in which these frequency–domain errors develop from time–domain measurements has been largely uninvestigated. The recent application of systems identification techniques to impedance measurements has focused on the influence of time–domain noise on the closure error of Fourier integrals for individual spectra but does not address the stochastic variance for repeated measurements [27,28].

The objective of the present work was to explore the manner in which stochastic errors in time–domain signals propagate through the instrumentation to the frequency domain.* The technologies considered included frequency-response analysis, phase-sensitive detection, and Lissajous analysis. While the complex quantity simulated in this work is electrical impedance, the approach is general and applies to measurement of other material properties such as the complex dielectric function, the complex refractive index, the complex viscosity, and acoustic impedance.

3. Methods for Impedance Calculation

The objective of this work was to explore the manner in which noise in time–domain signals is transformed to the frequency domain. Frequency Response Analysis (FRA), Lissajous Parameterisation (LP), and Phase-sensitive detection (PSD) are commonly used to convert time–domain signals into the frequency domain.

3.1 Frequency Response Analysis

Frequency Response Analysis represents a specific application of Fourier signal analysis, used, for example, by the Solartron Instruments family of gain/phase analysers [29]. While Fourier analysis provides a spectral representation of a time–domain signal, frequency response analysis determines the coefficient of a specific term in the Fourier series representation of the signal. With the generation of the perturbation signal, an in-phase sine wave and a 90° out-of-phase sine wave are generated. The input signal (e.g. current for galvanostatically-modulated impedance spectroscopy) and the output signal (voltage) are multiplied against the in-phase and the out-of-phase sine waves producing four signals for analysis. Each product-of-signals is integrated over each cycle of the generated input signal. The current and voltage signals can be described by a complex representation where the real part of the current signal is

$$I_r(\omega) = \frac{1}{T}\int_0^T I_0 \sin(\omega t + \phi)\sin(\omega t)dt \tag{1}$$

the imaginary part of the current signal is

---

* A distinction is drawn in the present work between stochastic errors that are randomly distributed about a mean value of zero, errors caused by the lack of fit of a model, and experimental bias errors that are propagated through the model. The problem of interpretation of impedance data is therefore defined to be one of identification of experimental errors, which includes assessment of consistency with the Kramers–Kronig relations, and fitting, which entails model identification, selection of weighting strategies, and examination of residuals. The error analysis provides information that can be incorporated into regression of process models. The experimental bias errors, as referred to here, may be caused by a changing base line or by instrumental artifacts. The emphasis of the present work is on identification of stochastic experimental errors.

$$I_j(\omega) = \frac{1}{T}\int_0^T I_0 \sin(\omega t + \phi)\cos(\omega t)dt \tag{2}$$

the real part of the voltage signal is

$$V_r(\omega) = \frac{1}{T}\int_0^T V_0 \sin(\omega t + \phi)\sin(\omega t)dt \tag{3}$$

and the imaginary part of the voltage signal is

$$V_j(\omega) = \frac{1}{T}\int_0^T V_0 \sin(\omega t + \phi)\cos(\omega t)dt \tag{4}$$

The constant coefficients of the integrands are the amplitudes of the respective signals. The current is the variable I and the voltage is the variable V. Integration is carried out over an integer number of cycles. The impedance is calculated as the complex ratio of the complex representations of the output signal to the input signal, as shown in Fig. 1 (overleaf).

3.2. Lissajous Parameterisation

Prior to the development of fast microprocessors, estimation of the peaks and axis-intercepts of the Lissajous representation of the input and output signals on an oscilloscope was used commonly to determine the impedance. To simulate use of Lissajous analysis in a manner that accounted for uncertainty associated with noise in the time–domain signals, a sine wave

$$f(t) = A\sin(\omega t + \phi) \tag{5}$$

was fitted to each signal, where A, the amplitude, and ϕ, the phase lag, are estimated parameters [30]. The ratio of the amplitude for the voltage signal to the amplitude for the current signal provided the magnitude of the impedance transfer function at a given frequency. The difference between the phase lag for the output signal and the phase lag for the input signal provided the phase angle of the transfer function at the same frequency.

3.3 Phase-Sensitive detection

A lock-in amplifier can also be used with a potentiostat to measure the complex impedance. The method of its use is fundamentally different from the method of a frequency response analyser. Whereas the FRA performs an assessment of the Fourier coefficients of the input and output signals, the lock-in amplifier measures the amplitudes of the two signals and the phase angle of each signal with respect to some reference signal. Thus, the impedance is measured in polar, rather than Cartesian, form.

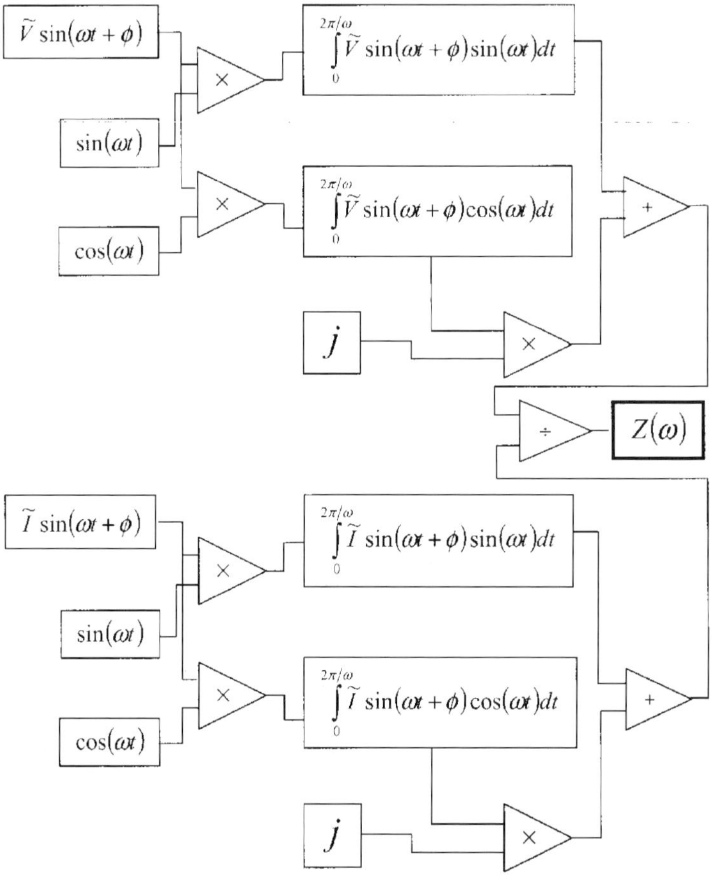

Fig. 1 *Algorithm for Assesment of Fourier coefficients. The impedence is calculated as the ratio of the complex representation of the voltage to the current.*

A general signal can be written in the form of eqn (5) as

$$A = A_0 \sin(\omega t + \phi_A). \tag{6}$$

A reference square wave of unity amplitude, generated at the same frequency as the sinusoidal perturbation, can be approximated by the Fourier series

$$S = \frac{4}{\pi}\sum_{n=0}^{\infty}\frac{1}{2n+1}\sin[(2n+1)\omega t + \phi_S]. \tag{7}$$

As was done in the Fourier assessment technique, the measured signal and reference signal are multiplied, resulting in a product signal

$$AS = \frac{4A_0}{\pi}\sum_{n=0}^{\infty}\frac{1}{2n+1}\sin(\omega t + \phi_A)\sin[(2n+1)\omega t + \phi_S]. \tag{8}$$

This can be rewritten using trigonometric identities to be

$$AS = \sum_{n=0}^{\infty} \frac{2A_0}{(2n+1)\pi} \left\{ \begin{matrix} \cos[-2n\omega + \phi_A - (2n+1)\phi_S] \\ -\cos[(2n+2)\omega t + \phi_A + (2n+1)\phi_S] \end{matrix} \right\}. \tag{9}$$

The product of the signals can be expanded via the trigonometric identity for the cosine of the sum of two angles to be

$$AS = \frac{2A_0}{\pi} \left[\begin{matrix} \cos(\phi_A - \phi_S) - \cos(2\omega t + \phi_A + \phi_S) \\ +\frac{1}{3}\cos(-2\omega t + \phi_A - 3\phi_S) - \frac{1}{3}\cos(4\omega t + \phi_A + 3\phi_S) \\ +\frac{1}{5}\cos(-4\omega t + \phi_A - 5\phi_S) - \frac{1}{5}\cos(6\omega t + \phi_A + 5\phi_S) + \ldots \end{matrix} \right]. \tag{10}$$

If the product signal is integrated over each cycle, it is clear that only the leading term of the series will have a non-zero value, thus,

$$\frac{\omega}{2\pi} = \int_0^{2\pi/\omega} AS dt = \frac{2A_0}{\pi} \cos(\phi_A - \phi_S). \tag{11}$$

The integral (11) will have a maximum value when the phase angle of the square wave is equal to the phase angle of the measured signal. In practice, the phase angle of the generated square wave is adjusted such that the integral is maximised. The phase angle of the square wave at the maximum value of the integral yields the phase angle of the measured signal. The maximum value of the integral provides the amplitude of the measured signal. This procedure is used on both signals such that the amplitude and phase angle of each signal is measured as shown in Fig. 2. The ratio of the voltage amplitude to the current amplitude is the impedance magnitude and the difference between the phase angles of the output and input signals is the impedance phase angle.

4. Simulation of Impedance Systems

Code was written in the LabView® G Language [31]. The program was designed to simulate an actual impedance experiment. The parameters of the simulation mimicked the experimental parameters of the Solartron Instruments family of frequency response and gain/phase analysers, i.e. SI 1250, SI 1254, and SI 1260 instruments. The interactive panel of the program provided a choice of the number of replicate scans to be made, the frequency range of the scan, the number of points in the scan, the auto-integration channel, the auto-integration mode, the delay before enforcing the auto-integration criterion, and the maximum time for integration.

A choice of transfer functions, or systems, to study was also provided. In this work, a simple single Voigt element was chosen. The Voigt element consisted of a 1 Ω resistor, R_{sol}, in series with a 1 Ω resistor, R, and a 100 μF capacitor, C, in parallel.

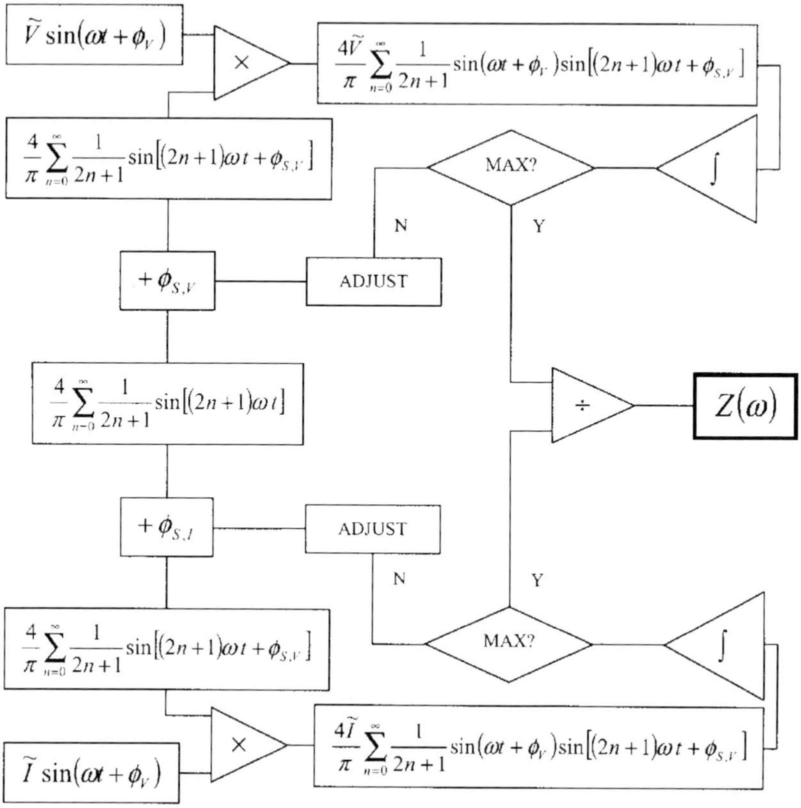

Fig. 2 Algorithm for phase-sensitive detection. The integral of the product of the measured signal and a square wave of the same frequency is maximised with respect to the phrase angle of the square wave.

The transfer function was written as

$$Z(\omega) = Z_r + jZ_j = R_{sol} + \frac{R}{1+\omega^2R^2C^2} + j\frac{-\omega RC}{1+\omega^2R^2C^2}. \tag{12}$$

Since this was a digital simulation, a sampling number, p, was an option also provided to the user. In the simulations discussed here, each cycle of each sine wave was sampled at 1024 samples/cycle. The effective digital sample rate, s, is related to the sine wave frequency, ω, by

$$s = p \cdot \omega \tag{13}$$

where s provides the upper frequency limit on the FFT of the sampled signal. Finally, the option of adding Gaussian-white noise to the input signal, the output signal, or both signals was provided. The noise level was set as a percentage of the input signal amplitude. In these simulations the Gaussian-noise level was 10% of the input signal amplitude unless otherwise noted.

Each technique was applied to each cycle of the input and output signal. The impedance estimation was completed when the auto integration criterion was met on the chosen channel. For short integration, convergence was achieved when the ratio of the standard deviation of the estimation of the magnitude to the magnitude over all cycles on the selected channel was less than 0.10. For long integration, convergence was achieved when the same ratio was less than 0.01. The simulation could be run without an auto integration criterion; the impedance at each frequency was measured until the auto integration time elapsed. The impedance was taken to be the mean of the impedances calculated at all cycles for the given frequency.

4.1 Time–domain Noise

Two forms of time–domain noise were considered. The first was a stochastic error signal added to the signals or the system. Thus, for galvanostatic impedance measurements, the signals could be described by

$$\tilde{V}(t) = V(\omega)\sin\left(\omega t + \phi_V\right) + \varepsilon_V \tag{14}$$

$$\tilde{I}(t) = I_{applied}\sin\left(\omega t + \phi_I\right) + \varepsilon_I \tag{15}$$

where ε_V and ε_I are stochastic additive errors to the true value of the potential and current signals, respectively.

The second type of noise was a stochastic error with a standard deviation that was proportional to the instantaneous value of the signal. This noise distribution was not intended to represent an experimentally observed noise distribution. Proportional noise was a mathematical construct in which stochastic error was added at each point in time to each signal amplitude and phase angle (or instantaneous value of the frequency). However, this type of error could be viewed as describing noise in the frequency control, because the error in the phase angle could be rewritten as the product of an error in frequency and time. For galvanostatically controlled impedance measurements, the signals could be described by

$$\tilde{V}(t) = \left(\hat{V}(\omega) + \varepsilon_V\right)\sin\left(\omega t + \hat{\phi}_V + \varepsilon_{\phi_V}\right) \tag{16}$$

$$\tilde{I}(t) = \left(I_{applied} + \varepsilon_I\right)\sin\left(\omega t + \hat{\phi}_I + \varepsilon_{\phi_I}\right). \tag{17}$$

Of course, the two types of errors could be combined to provide a general form for a signal with error by

$$\tilde{A}(t) = \left(\hat{A} + \varepsilon_A\right)\sin\left(\omega t + \hat{\phi}_A + \varepsilon_{\phi_A}\right) + \varepsilon_1. \tag{18}$$

This was rewritten such that

$$\begin{aligned}\tilde{A}(t) = {} & \hat{A}\sin\left(\omega t + \hat{\phi}\right) + \hat{A}\cos\left(\omega t + \hat{\phi}\right) \\ & + \varepsilon_1 + \varepsilon_2\sin\left(\omega t + \hat{\phi}\right) + \varepsilon_3\cos\left(\omega t + \hat{\phi}\right).\end{aligned} \tag{19}$$

where

$$\varepsilon_2 = \varepsilon_A \cos\left(\varepsilon_{\phi_A}\right) + A\left(\cos\left(\varepsilon_{\phi_A}\right) - 1\right) \tag{20}$$

$$\varepsilon_3 = \varepsilon_A \sin\left(\varepsilon_{\phi_A}\right) + A\left(\sin\left(\varepsilon_{\phi_A}\right) - 1\right). \tag{21}$$

In general, a noisy sinusoidal signal could contain the true sinusoid plus an out-of-phase component plus error proportional to the true sinusoid plus error proportional to the out-of-phase component plus additive error. The noise introduced for this paper was Gaussian, or normal, white noise. The values of this noise waveform had mean of zero and were evenly distributed around the mean [32]. A histogram of this type of error can be found in Fig. 3. In other work, the influence of colored noise will be discussed.

4.2 Application of Measurement Techniques

One cycle of the (clean or noisy) perturbation signal was constructed. The Fast Fourier Transform of the signal was then calculated. The response signal was obtained from the real inverse FFT of the product of the FFT of the perturbation signal and the impedance transfer function. A second random noise signal could be added to the response signal. Thus, if noise were added only to the input, then the response signal would contain noise transformed from noise in the perturbation via the impedance

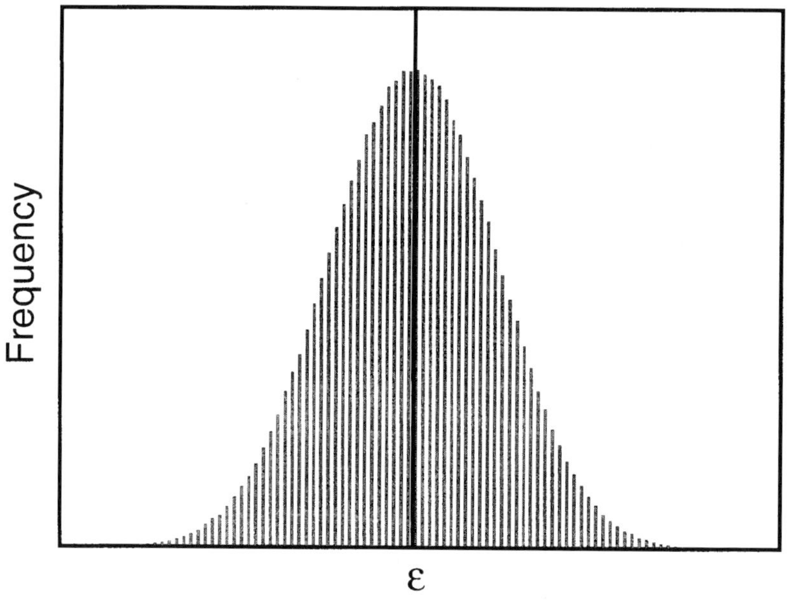

Fig. 3 Distribution of values for typical waveform of Gaussian white noise. Solid black line along middle of chart is equal to zero.

transfer function. If noise was added to the perturbation and response signals, then the response signal contained transformed noise and additive noise. It should be noted that this algorithm was tested on a perturbation impulse, and the response of the first order transfer function showed the expected transient behavior that would be associated with analytical Fourier transformation of the same impulse. Thus, in contrast to recent applications of systems identification techniques to propagation of noise through impedance measurements [28], the influence of the transfer function on the transformation of noise was taken into account.

The impedance spectra were calculated simultaneously for each of the techniques to calculate the impedance: assessment of the real and imaginary Fourier coefficients using trapezoidal quadrature, regression to the individual waveforms, and phase-sensitive detection. The pair of single cycles of the perturbation and response signals was then passed to the calculation techniques. The complex impedance was calculated for that cycle of the two signals. After three cycles of measurement, the auto-integration criterion was tested. The calculations were performed over a number of cycles sufficient to allow convergence of the auto-integration criterion to be achieved by each technique.

The FRA technique was applied twice. In one case, the trapezoidal quadrature rule was used to integrate the signals. In the other case, Bode quadrature was employed for signal integration [30]. The Bode quadrature rule is a higher order rule, and therefore generated a smaller error than the trapezoidal rule (order h^7 as compared to h^3, where h is the step-size of integration, $2\pi/1024$) [30].

5. Results

Four categories of simulations were conducted: no noise added (a test of precision for each technique), noise added only to the perturbation signal, noise added only to the response signal, and noise added to both the perturbation and response signals. The noise in each case was Gaussian with a standard deviation of 10% of the input perturbation. Thus the standard deviation of the additive noise to the response varied between 5 and 10% of the signal depending on the frequency of the measurement, since the impedance varied between one and two ohms. The impedance calculated from noise-free perturbation and response signals provided verification that each technique generated only deterministic errors, namely those associated with the numerical technique employed. The deterministic errors were on the order of the level of precision associated with the code compiler and microprocessor-set.

The calculated impedance response using each of the measurement algorithms was normally distributed. A representative distribution of frequency–domain errors of the real part impedance is presented in Fig. 4. For this simulation, the impedance response was calculated 500 times using the FRA technique at 1 kHz with the Gaussian noise of Fig. 3. added to the input signal and output signal in the time–domain. Similar results were seen at all frequencies for both real and imaginary components of the impedance. As presented elsewhere [31,33], the calculated impedance was normally distributed, even when coloured noise was added to the time domain signals. Thus the assumption implicit in the error analysis conducted by Agarwal *et al.* [13–17] is justified.

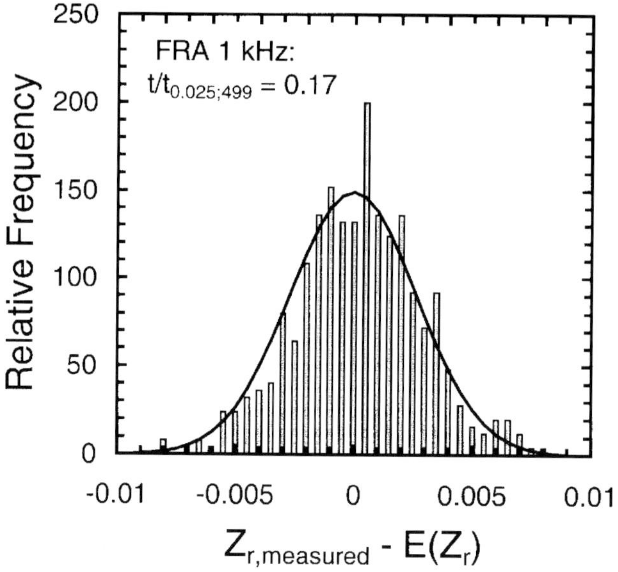

Fig. 4 Distribution of frequency–domain errors of the real impedence calculated using the FRA technique at 1kHz with Gaussian noise (see Fig. 3) added to the imput signal and output signal in the time–domain.

A typical distribution of $\log\left(\sigma_r^2/\sigma_j^2\right)$ is presented in Fig. 5 for a simulated measurement using the Lissajous technique. This term is normally distributed with a mean value equal to zero. The critical Student's *T*-test parameter, given by the ratio $t/t_{0.025/90}$, had a value less than unity. Thus the mean value of $\log\left(\sigma_r^2/\sigma_j^2\right)$ was equal

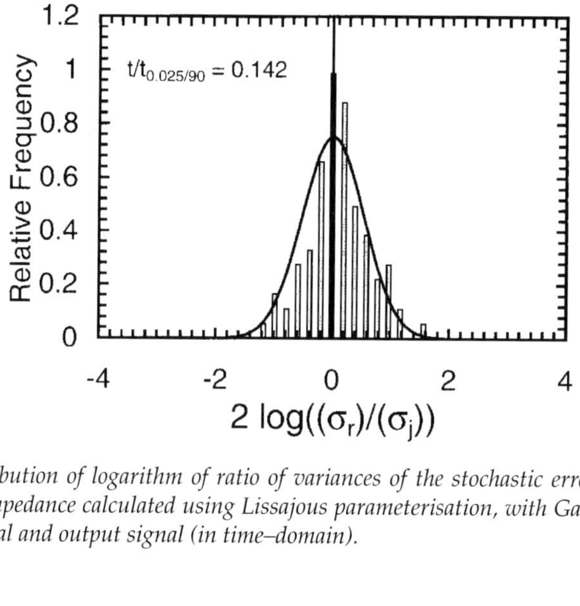

Fig. 5 Distribution of logarithm of ratio of variances of the stochastic errors in the real and imaginary impedance calculated using Lissajous parameterisation, with Gaussian noise added to input signal and output signal (in time–domain).

to zero, and $\sigma_r^2 = \sigma_j^2$. Similar results were seen in all cases where additive noise was added to input and/or output time–domain signals processed by Lissajous analysis or FRA.

The use of the Student's *T*-test represents a test, over the ensemble of measured frequencies, of whether the variance for real and imaginary parts of the impedance can be considered to be equal. The *F*-test statistic provides a measure of whether, frequency by frequency, the variance for the real part of the measurement can be considered to be different from the variance of the imaginary part. A typical application of the *F*-Test to data generated using the FRA algorithm is presented in Fig. 6. The ratio is generally distributed around one and most points fall within the 99% probability level. Therefore, the null hypothesis that the variances are equal can be accepted. For the FRA and the Lissajous parameterisation calculations, the variance in the stochastic errors in the real impedance were equal to the variance in the stochastic errors in the imaginary impedance.

The variances were not equal when the PSD calculation was applied to the same set of perturbation–response signals. The ratio of the variance in the real impedance to the variance in the imaginary impedance is presented in Fig. 7 as a function of frequency for typical PSD calculation. The ratio was not distributed around one and most points fall outside the 99% probability level. Therefore, the hypothesis that the variances are not equal can be accepted with 99% confidence. The equality was tested for each case and for each measurement technique at each frequency using the *F*-test, and over the spectrum using the Student's *T*-test. The results of the Student's *T*-test, shown in Fig. 8, were consistent with the results of the *F*-test for each case. As seen in Fig. 8, the variance of the real part of the impedance was less than the variance of the imaginary part, thus one component of the impedance was biased with respect to the other. The Student's *T*-test was consistent with this inequality of the variances. The results of these simulations verify that the instrumentation used to conduct impedance spectroscopy measurements influences the correlation between the

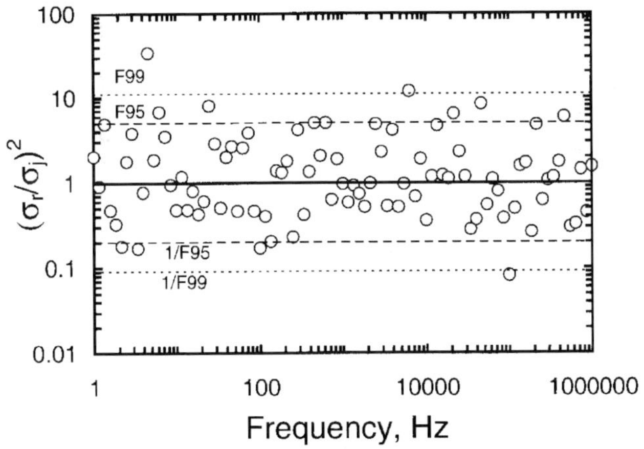

Fig. 6 *F-test for impedance calculated using the FRA technique (Bode quadrature) with Gaussian noise added to both the imput and output signal (in time–domain).*

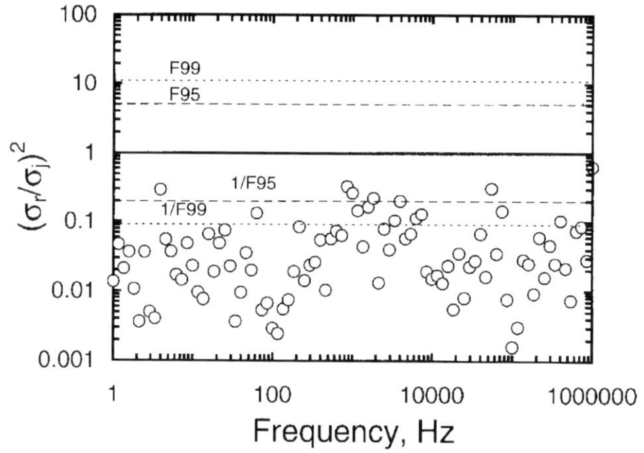

Fig. 7 F-*test for impedance calculated using the PSD technique with Gaussian noise added to both the imput and output signals.*

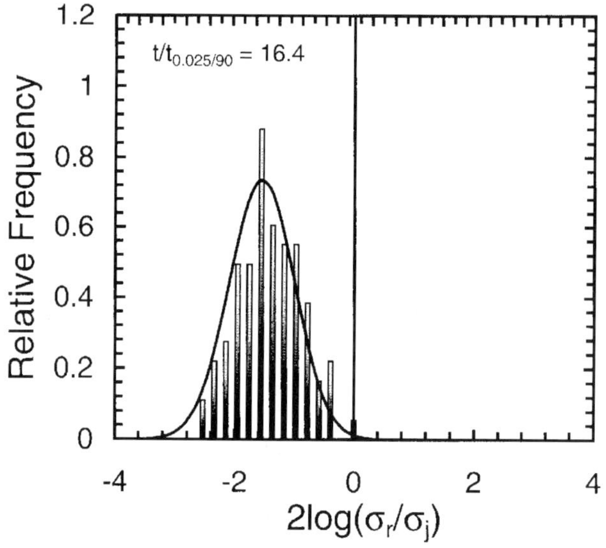

Fig. 8 Distribution of logarithm of ratio of variances of the stochastic errors in the real and imaginary impedance calculated using phase-sensitive detection with Gaussian noise added to input signal and output signal (in time–domain).

variance in the stochastic errors in the real impedance and the variance in the stochastic errors in the imaginary impedance. The distributions of the errors in the components of the complex impedance are, in all cases, Gaussian, and can therefore be defined by variances, but the variances of the distributions may not be equal. For additive errors, it was shown that the variances of real and imaginary components were equal when the FRA technique was used, but not when the PSD technique was used.

5.1. Source of Bias Errors in Phase-Sensitive Detection

The distribution of errors in the real impedance calculated via the PSD technique on current and voltage signals with additive Gaussian noise introduced into the input and output signals with respect to the theoretical impedance spectrum is presented in Fig. 9. The PSD calculations introduced a bias error in that the mean value of the distribution is not equal to zero.

If error is present in the PSD measurement of the impedance magnitude, then the integral of the product of signals in eqn (11) can be rewritten to be

$$e_{out} = \frac{\omega}{2\pi}\int_0^{2\pi/\omega} AS dt = \frac{2A_0}{\pi}\cos\left(\phi_A - \phi_S + \varepsilon_\phi\right) + \varepsilon_X. \tag{22}$$

where ε_X is the error in the integration (an additive error) and ε_ϕ is the error in the phase angle measurement (due to error in the phase angle controller). If the difference in phase angles is equal to zero, then the observed value of the signal amplitude will be

$$A_{0,ob} = \frac{\pi}{2\cos\left(\varepsilon_\phi\right)}\left(e_{out} - \varepsilon_X\right). \tag{23}$$

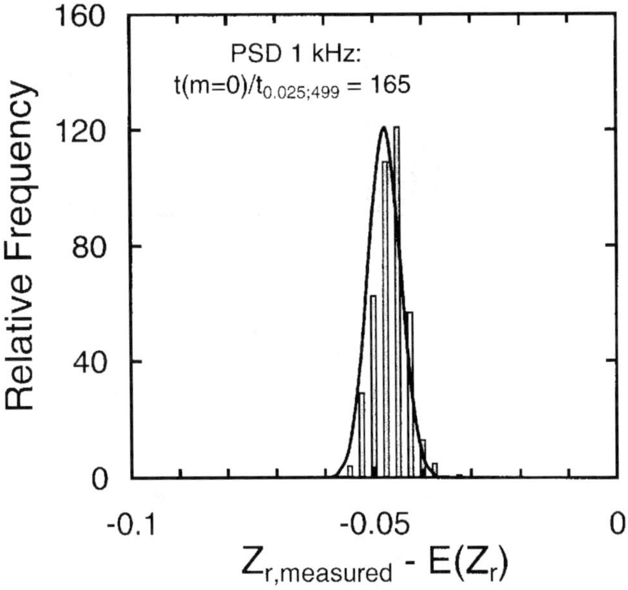

Fig. 9 *Distribution of stochastic errors in the real impedance of 1kHz calculated using the PSD technique. The errors were calculated as the difference between the measured impedance and the theoretical impedance spectrum (used as the system in the simulation).*

The error in magnitude estimate, ε_A, can be written as the difference between the observed value of the magnitude and the expectation value

$$\varepsilon_A = A_{0,ob} - E(A_0) = \frac{\pi}{2} e_{out} - \frac{\pi}{2\cos(\varepsilon_\phi)}(e_{out} - \varepsilon_X). \tag{24}$$

Therefore, the error in the magnitude is a function of the integral, the error in the phase angle controller and error in the integral

$$\varepsilon_A = \frac{\pi}{2} e_{out}\left(1 - \frac{1}{\cos(\varepsilon_\phi)}\right) + \frac{\pi}{2}\left(\frac{\varepsilon_X}{\cos(\varepsilon_\phi)}\right). \tag{25}$$

Note that the error in phase angle propagates into the magnitude through a cosine. Thus, while the error in phase angle may be distributed normally about zero, the cosine of those errors will not be normally distributed since the cosine of a negative number is the same as the cosine of the absolute value of that number (see Fig. 10). The skewness of the distribution of the cosine of the errors in phase angle that are normally distributed about zero is quite negative. As a result of this cosine, the error in amplitude will be biased away from zero. This is the cause for the inequality of variances in the real and imaginary impedances calculated via PSD. The bias in the frequency–domain errors is therefore not characteristic of the error distribution. The bias is an artifact caused by the PSD algorithm.

5.2. Condition on Spectroscopy Errors by the Kramers–Kronig Relations

Durbha *et al.* determined that the statistical expectation of the variance of the stochastic errors in the real impedance is equal to the expectation of the variance of the stochastic

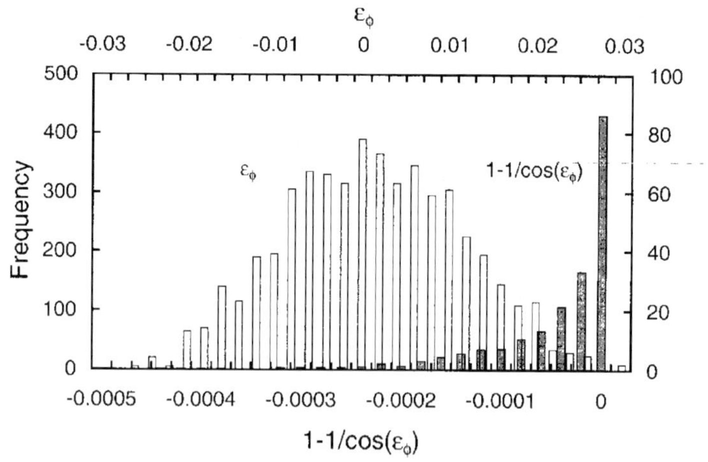

Fig. 10 *Distribution of errors in phase angle and distribution of the cosine of those errors. A Gaussian distribution is skewed and biased by the cosine operation.*

errors in the imaginary impedance [25]. The fundamental assumptions associated with their derivation were that the expectation of the frequency–domain errors was assumed to be equal to zero, that the errors were distributed normally, and that the real and imaginary parts were determined simultaneously. For example, the equality of the variances would not be expected for measurements of the complex refractive index because the real and imaginary parts are determined independently.

The results of the simulations on the systems for which Gaussian additive noise was introduced into the signals and the impedance was calculated using the FRA or LP techniques satisfied the assumptions made by Durbha *et al.* [25]. The spectral errors were distributed normally about zero. The real and imaginary parts of the spectrum were calculated simultaneously by these two methods. The equality of the variances was not observed for impedance calculated using the PSD technique on the same systems. As it was shown in the previous section that the expectation of the errors were not be equal to zero, the observation that under the PSD algorithm the variances were not equal is consistent with the derivation of Durbha *et al.* [25].

In separate work, the results of simulations on the systems with proportional noise was introduced into the signals were reported [31,34]. When the FRA or LP techniques were employed for impedance calculation, the equality of the variances was not observed. When the PSD technique was used for the calculation, the equality of the variances was observed. Proportional noise could be considered as resulting from stochastic errors in frequency. An unstated assumption in the derivation of Durbha *et al.* [25] was that the frequency of the measured impedance must be free of error.

6. Conclusions

The objective of this work was to investigate the manner in which frequency–domain errors arise from time–domain measurements. The results of the simulations and analytical derivations demonstrated that, in the absence of bias errors, the errors in the real and imaginary impedance are uncorrelated and the variances of the real and imaginary parts of the complex impedance are equal. The equality of variances was seen except for cases where the phase-sensitive detection measurement technology introduced bias errors. This work also showed that, independent of the distribution function for stochastic errors in the time–domain, stochastic errors in the frequency–domain have a normal distribution. The normal distribution of stochastic errors in the frequency–domain is a consequence of the Central Limit Theorem applied to the methodology used to measure the complex impedance.

The approach presented here was based on the modelling of the instruments actually used for the measurement of complex quantities and allows analysis of the propagation of errors from the original measurements in the time domain to the desired complex variables in the frequency domain. This work shows that, as long as the complex quantities are measured through similar physical principles, a general model for the propagation of error can be established. While the use of the model developed was demonstrated through applications to electrochemical impedance spectroscopy, it can be applied directly to measurement of other complex quantities.

7. Acknowledgements

The financial support of the Engineering Research Center for Particle Science and Technology at the University of Florida is gratefully acknowledged.

References

1. R. J. Sheppard, B. P. Jordan and E. H. Grant, *J. Phys. D: Appl. Phys.*, 1970, **3**, 1759–1764.
2. R. J. Sheppard, *J. Phys. D: Appl. Phys.*, 1973, **6**, 790–794.
3. J. R. Macdonald and J. A. Garber, *J. Electrochem. Soc.*, 1977, **124**, 1022–1030.
4. J. R. Macdonald and L. D. Potter, Jr., *Solid State Ionics*, 1987, **23**, 61–79.
5. P. Zoltowski, *J. Electroanal. Chem.*, 1984, **178**, 11.
6. B. A. Boukamp, *Solid State Ionics*, 1986, **20**, 31.
7. J. R. Macdonald and L. D. Potter, Jr., *Solid State Ionics*, 1987, **23**, 61.
8. P. Zoltowski, *J. Electroanal. Chem.*, 1989, **260**, 269.
9. P. Zoltowski, *J. Electroanal. Chem.*, 1989, **260**, 287.
10. J. R. Macdonald, *Electrochim. Acta*, 1990, **35**, 1483.
11. J. R. Macdonald and W. J. Thompson, *Commun. Statist. Simula.*, 1991, **20**, 843.
12. G. Spinolo, G. Chiodelli, A. Moghistris and U. A. Tamburini, *J. Electrochem. Soc.*, 1988, **135**, 1419
13. P. Agarwal, M. E. Orazem and L. H. García-Rubio, *J. Electrochem. Soc.*, 1992, **139**, 1917–1927.
14. P. Agarwal, M. E. Orazem and L. H. García-Rubio, in *Electrochemical Impedance: Analysis and Interpretation*, ASTM STP 1188 J. Scully, D. Silverman, M. Kendig, Eds. American Society for Testing and Materials, Philadelphia, 1993, 115–139.
15. P. Agarwal, O. C. Moghissi, M. E. Orazem and L. H. García-Rubio, *Corrosion*, 1993, **49**, 278–289.
16. P. Agarwal, Oscar D. Crisalle, M. E. Orazem and L. H. García-Rubio, *J. Electrochem. Soc.*, 1995, **142**, 4149–4158.
17. P. Agarwal, M. E. Orazem and L. H. García-Rubio, *J. Electrochem. Soc.*, 1995, **142**, 4159–4168.
18. R. W. Christy, *Am. J. Phys.*, 1972, **40**, 1403.
19. A. Jutan and L. H. Garcia-Rubio, *Process Control and Quality*, 1993, **4**, 235.
20. L. K. DeNoyer and J. G. Good, *Am. Labor.*, March, 1990. Software available from Spectrum Square Associates, Ithaca, NY, 114850.
21. M. Orazem, P. Agarwal, A. N. Jansen, P. T. Wojcik and L. H. García–Rubio, *Electrochim. Acta*, 1993, **38**, 1903.
22. M. E. Orazem, T. El Moustafid, C. Deslouis and B. Tribollet, *J. Electrochem. Soc.*, 1996, **143**, 3880.
23. J. R. Dygas, W. Bogusz, and F. Krok, presented at the *4th Int. Symp. on Electrochemical Impedance Spectroscopy*, Rio de Janeiro, Brazil, 2–7 August, 1998.
24. J. R. Macdonald, *Electrochim. Acta*, 1993, **38**, 1883–1890.
25. M. Durbha, M. E. Orazem and L. H. García-Rubio, *J. Electrochem. Soc.*, 1997, **144**, 48.
26. J. R. Macdonald and V. I. Piterbarg, *J. Electroanal. Chem.*, 1997, **428**, 1.
27. P. Eykhoff, *System Identification – Parameter and State Estimation*. John Wiley & Sons, New York, 1974.
28. R. H. Milocco and S. I. Biagiola, *Current Topics in Electrochemistry*, 1994, **3**, 553–567.
29. Gabrielli, Claude, *Identification of Electrochemical Processes by Frequency Response Analysis*, Technical Report Number 004/83, Solartron Instruments, Issue 2: August 1984.
30. W. H. Press, B. P. Flannery, S. A. Teukolsky and W. T. Vetterling, *Numerical Recipes: The Art of Scientific Computing FORTRAN Version*. Cambridge University Press, 1989.

31. S. L. Carson, Application of Complex Spectroscopic Techniques to the Characterization of particles in Suspension, Ph.D. dissertation, University of Florida, May, 1999.
32. W. H. Beyer, *CRC Standard Mathematical Tables and Formulae*, 29th Ed. CRC Press, Boca Raton, 1991.
33. S. L. Carson, M. E. Orazem, O. D. Crisalle and L. H. Garcia-Rubio, in preparation.
34. S. L. Carson, M. E. Orazem, O. D. Crisalle and L. H. Garcia-Rubio, in preparation.

List of Abbreviations

The following abbreviations occur in the text and in the Index of contents.

ANN	Artificial neural networks
BTA	Benzotriazole
CEC	Composite electrochemical coatings
CNLS	Complex non-linear least square
CPE	Constant phase element
CQI	Coating quality index
CV	Cyclic voltammetry
DPH	diphosphono heptane
DSC	Differential scanning calorimetry
e.c.	electrochemical
ECL	Electrogenerated chemiluminescence
EDAX	Energy dispersive X-ray analysis
EDTA	Ethylenediamine tetraacetic acid
EDXS	Energy dispersive X-ray spectroscopy
EEC	Equivalent electrical circuit
EIS	Electrochemical impedance spectroscopy
EM	Emeraldine
ENA	Electrochemical noise analysis
EPMA	Electron probe microanalysis
FFT	Fast Fourier transform
FRA	Frequency response analysis
ICP	Inductively coupled plasma spectroscopy
IEP	Ion exchange pigment
IEPS	isoelectric point
ippl	inner part of passive layer
KKR	Kramers-Kronig Relations
LE	Leucoemeraldine
LP	Lissajous paramatisation
MIC	Microbially influenced corrosion
NMP	N-methyl pyrrolidone
NLS	Non-linear least squares
OCP	Open circuit potential
OFLD	Open boundary finite length diffusion (element)
oppl	Outer (hydrated) part of passive layer
ORC	Oxidation reduction cycle
OTCX	One time-constant model
PANI	Polyaniline
PEG	Polyethylene glycol
PMT	Photomultiplier tube
PSD	Power spectral density
PVC	Pigment volume content
RH	Relative humidity
SAM	Self-assembling molecules
SEF	Surface enhancement factor
SEM	Scanning electron microscope
SERS	Surface enhanced Raman spectroscopy
SKP	Scanning Kelvin probe
S/N	Signal to noise ratio

TFR	Total film resistance
TL	Transmission line
TSRO	Topological short range order
TU	Thiourea
UV–VIS	Ultraviolet-visible (spectra)
XPS	X-ray photoelectron spectroscopy
ZRP	Zinc-rich paints

Index